T0255124

Lecture Notes in Computer Science 14231

Founding Editors

Gerhard Goos
Juris Hartmanis

The series Lecture Notes in Computer Science (LNCS), including its subseries Lecture Notes in Artificial Intelligence (LNAI) and Lecture Notes in Bioinformatics (LNBI), has established itself as a medium for the publication of new developments in computer science and information technology research, teaching, and education.

LNCS enjoys close cooperation with the computer science R & D community, the series counts many renowned academics among its volume editors and paper authors, and collaborates with prestigious societies. Its mission is to serve this international community by providing an invaluable service, mainly focused on the publication of conference and workshop proceedings and postproceedings. LNCS commenced publication in 1973.

Mauro Iacono · Marco Scarpa ·
Enrico Barbierato · Salvatore Serrano ·
Davide Cerotti · Francesco Longo
Editors

Computer Performance Engineering and Stochastic Modelling

19th European Workshop, EPEW 2023
and 27th International Conference, ASMTA 2023
Florence, Italy, June 20–23, 2023
Proceedings

Springer

Editors

Mauro Iacono (ID)
Università degli Studi della Campania Luigi
Vanvitelli
Caserta, Italy

Enrico Barbierato (ID)
Università Cattolica del Sacro Cuore
Brescia, Italy

Davide Cerotti (ID)
Università del Piemonte Orientale
Alessandria, Italy

Marco Scarpa (ID)
Università degli Studi di Messina
Messina, Italy

Salvatore Serrano (ID)
Università degli Studi di Messina
Messina, Italy

Francesco Longo (ID)
Universita degli Studi di Messina
Messina, Italy

ISSN 0302-9743   ISSN 1611-3349 (electronic)
Lecture Notes in Computer Science
ISBN 978-3-031-43184-5   ISBN 978-3-031-43185-2 (eBook)
https://doi.org/10.1007/978-3-031-43185-2

Preface

It is a pleasure for us to present another volume of our series on Performance Evaluation in LNCS and present to our audience the best contributions the performance engineering and formal methods communities proposed this year to the 27th International Conference on Analytical & Stochastic Modelling Techniques & Applications (ASMTA 2023) and the 19th European Performance Engineering Workshop (EPEW 2023), jointly held in Florence, Italy, in June 2023.

This book is the result of thorough discussions and revisions of the works of 33 researchers from 10 countries that submitted to ASMTA and 61 researchers from 12 countries that submitted to EPEW, and represents the current state of the research in performance evaluation of computer-based systems of a curious, attentive and lively community of computer scientists and engineers, applied mathematicians, telecommunication engineers and many others. Originally, the conferences received 35 submissions. Each submission was single-blind reviewed by at least 2, and on average 2.7, program committee members. The committee decided to accept 26 papers in total. The EPEW workshop received a total of 22 submissions. Eventually, the committee accepted fifteen papers for presentation. On the other hand, the ASMTA conference received 13 submissions. In this case, the committee decided to accept 11 papers for the conference.

This year our communities also had the chance for cross-fertilization and fruitful exchanges with the Italian computer performance engineering community (InfQ), which colocated the QualITA event of the CINI Working Group on Systems and Service Quality, and with the European Council for Modelling and Simulation, which hosted all the events together with its International ECMS Conference on Modelling and Simulation. Even more than in previous years, the dynamism and the friendly environment that characterize this research area emerged and gave birth to new interesting questions and new problems to deal with.

We were very happy to have Wil Van Der Aalst, from RWTH Aachen University, Germany, Christine Currie, from University of Southampton, UK, and Marco Gribaudo from Politecnico di Milano, Italy as keynote speakers. Their talks touched on different areas of performance engineering focusing on both simulation applications and theoretical aspects of performance engineering.

We would like to thank all the colleagues that contributed, in any sense, to the preparation of this volume and to all the work and events that allowed us to make it. We are especially grateful to the ECMS 2023 team, Romeo Bandinelli, Virginia Fani,

Enrico Vicario, Michele Mastroianni and Martina-Maria Seidel, and all the students that supported them.

June 2023

Mauro Iacono
Marco Scarpa
Enrico Barbierato
Salvatore Serrano
Davide Cerotti
Francesco Longo

Organization ASMTA 2023

General Chair

Mauro Iacono   Università degli Studi della Campania Luigi
Vanvitelli, Italy

Program Committee Chairs

Enrico Barbierato   Università Cattolica del Sacro Cuore, Italy
Salvatore Serrano   University of Messina, Italy

Program Committee

Nail Akar	Bilkent University, Turkey
Konstantin Avrachenkov	Inria, France
Simonetta Balsamo	Università di Venezia, Italy
Tejas Bodas	IIT Dharwad, India
Dieter Claeys	Ghent University, Belgium
Céline Comte	Eindhoven University of Technology, The Netherlands
Koen De Turck	Ghent University, Belgium
Antonis Economou	University of Athens, Greece
Dieter Fiems	Ghent University, Belgium
Marco Gribaudo	Politecnico di Milano, Italy
Irina Gudkova	People's Friendship University, Russia
Yezekael Hayel	University of Avignon, France
András Horváth	University of Turin, Italy
William Knottenbelt	Imperial College London, UK
Lasse Leskelä	Aalto University, Finland
Martin Lopez Garcia	University of Leeds, UK
Andrea Marin	Università Ca' Foscari Venezia, Italy
Jose Nino-Mora	Carlos III University of Madrid, Spain
Juan F. Perez	Universidad del Rosario, Colombia
Tuan Phung-Duc	University of Tsukuba, Japan
Liron Ravner	University of Haifa, Israel
Marie-Ange Remiche	University of Namur, Belgium

Jacques Resing	Eindhoven University of Technology, The Netherlands
Yutaka Sakuma	National Defence Academy of Japan, Japan
Konstantin Samouylov	People's Friendship University, Russia
Bruno Sericola	Inria, France
Janos Sztrik	University of Debrecen, Hungary
Miklos Telek	Budapest University of Technology and Economics, Hungary
Benny Van Houdt	University of Antwerp, Belgium
Joris Walraevens	Ghent University, Belgium
Jinting Wang	Central University of Finance and Economics, China

Organization EPEW 2023

General Chair

Marco Scarpa   University of Messina, Italy

Program Committee Chairs

Francesco Longo   University of Messina, Italy
Davide Cerotti   University of Piemonte Orientale, Italy

Program Committee

Salvador Alcaraz   Miguel Hernández University of Elche, Spain
Elvio Gilberto Amparore   Università degli studi di Torino, Italy
Paolo Ballarini   CentraleSupeléc, France
Enrico Barbierato   Università Cattolica del Sacro Cuore, Italy
Marco Bernardo   University of Urbino, Italy
Laura Carnevali   University of Florence, Italy
Dieter Fiems   Ghent University, Belgium
Di Pompeo Daniele   Università dell'Aquila, Italy
Matthew Forshaw   Newcastle University, UK
Jean-Michel Fourneau   University of Versailles, France
Pedro Pablo Garrido   Miguel Hernández University of Elche, Spain
Reinhard German   University of Erlangen–Nuremberg, Germany
Marco Gribaudo   Politecnico di Milano, Italy
Boudewijn Haverkort   Tilburg University, The Netherlands
András Horváth   University of Turin, Italy
Mauro Iacono   Università degli Studi della Campania Luigi
                                   Vanvitelli, Italy
Alain Jean-Marie   Inria, France
Carlos Juiz   University of the Balearic Islands, Spain
Lasse Leskelä   Aalto University, Finland
Andrea Marin   University of Venice, Italy
Nihal Pekergin   Université Paris-Est Créteil, France
Tuan Phung-Duc   University of Tsukuba, Japan
Agapios Platis   University of the Aegean, Greece

Anne Remke   WWU Münster, Germany
Markus Siegle   University of the Bundeswehr Munich, Germany
Miklos Telek   Budapest University of Technology and
                              Economics, Hungary
Joris Walraevens   Ghent University, Belgium
Katinka Wolter   Frei Universität Berlin, Germany

Contents

EPEW 2023

ASMTA 2023

Strategic Revenue Management for Discriminatory Processor Sharing Queues

Dieter Fiems[✉]

Department of Telecommunications and Information Processing,
Ghent University, Ghent, Belgium
Dieter.Fiems@UGent.be

Abstract. We investigate optimal revenue management for Markovian discriminatory processor sharing (DPS) queues. The server receives revenue per customer, as well as an additional fee if customers opt to receive premium service. We first study the parameter allocation of the DPS discipline which optimises the server's revenue, assuming that all customers rationally select between premium and non-premium service. We then extend revenue management to DPS queues with heterogenous customers that are also allowed to balk. It is shown that the optimal DPS discipline is a strict priority discipline when customers cannot balk, while a non-degenerate DPS discipline is optimal with balking.

Keywords: Discriminatory processor sharing · Queueing game · Revenue management

1 Introduction

Businesses can increase their revenue by offering both standard and premium tiers of service to customers. By providing a choice between standard and premium tiers, businesses can offer faster access to the service at an additional cost, thereby catering to customers with varying needs and preferences. However, in order to optimise revenue, it is important for service providers to carefully select the premium service fee, as well as the service differentiation mechanism. Effective revenue management involves considering how customers respond to the service differentiation mechanism. Customers must make a trade-off between the cost of the premium service and the benefits it brings, such as reduced waiting times. This trade-off determines how many customers are willing to pay for premium service, and therefore also the additional revenue service differentiation can generate. Setting the premium fee too high will discourage customers to opt for premium service, while potential revenue is lost by setting the fee too low. Setting the right price point for the compelling benefit of reduced waiting times is the key to successful revenue management.

Revenue management is well investigated for various priority queueing systems, including preemptive and non-preemptive priority queues, see e.g. [7,13] and the references therein. Additionally, various authors also focus on strategic customer behaviour in priority queues. Altmann et al. [3] investigate rational priority selection in a Markovian priority queue with multiple priority classes and multiple customer classes, each customer class having distinct utilities and therefore also distinct strategies in selecting its preferred priority class. Van den Berg et al. [5] considers revenue management with and without preemptive priority where pricing depends on the service time. Within each service class, a processor sharing discipline is assumed. Upon arrival the customers choose between balking, low-priority and high-priority service, accounting for the length of the service time. The optimal strategy is a threshold strategy: high priority if the service time is short, low priority for longer service times, and balking for even longer service times. Finally, rational balking of low-priority customers in a priority queue with two classes is investigated by Xu et al. [23], both when the low-priority customers can observe the system, as well as in a scenario where the queues are not observable.

Priority queues offer a rather crude type of service differentiation: customers either have priority or not, and the server does not start serving any non-priority customer as long as there are priority customers in the queue. The service differentiation mechanism has no parameters to tune how much faster premium customers are served. To assess revenue management when service differentiation can be tuned, we here investigate revenue management for a discriminatory processor sharing (DPS) queueing system. Processor sharing queues and their multi-class extensions like discriminatory and generalised processor sharing queues naturally arise as a convenient abstraction for many networking and computer systems. Processor sharing queues gained particular popularity in modelling flow-level resource sharing in the Internet, see for example [16,17,19]. Literature on processor-sharing queueing systems is extensive and we refer to the surveys on processor sharing [24] and discriminatory processor sharing [1] for more pointers to the theory and applications of processor sharing queues and limit the present discussion to processor sharing queueing games.

Altman and Shimkin [2] consider the processor sharing game, where customers decide on joining the processor sharing queue, when they can observe the number of customers already present. In a concert queueing game, customers choose their arrival time as to minimise some cost which also depends on the arrival times of the other customers. In [18], customers choose their arrival times in a processor sharing queue, where the cost depends on the sojourn time, and the deviation from the customer's preferred departure time. Games with processor sharing are also studied as models for distributed non-cooperative load balancing. In such load-balancing scenarios, either customers choose a server from a set of servers [9], or are sent to one of the servers by a limited number of dispatchers [4,6,9]. Finally, a time- and load-dependent fluid processor sharing queue is used for modelling the use of park-and-ride systems during rush hour [11,12].

The literature on discriminatory processor games is more limited. Hayel and Tuffin [14] consider revenue management for a DPS queue where customers belong to fixed classes. Each customer represents a Poisson stream of packets, and more customers join till their utility drops below a threshold. The server's revenue is then determined by the number of customers of both classes that join, as customers cannot choose their class. Class selection is studied by Wu et al.

[22] for a DPS queue with multiple classes. The authors impose a specific cost for joining the different classes and show that a Nash equilibrium exists in the M/M/1-DPS queue setting, before focusing on the system in heavy traffic. Similar costs are imposed in [20] where a DPS queueing game is proposed to study pricing for virtual operators in 5G networks.

In this paper, we first revisit the class selection problem of [22]. We fully characterise the Wardrop equilibrium for two classes in the Markovian setting and show that revenue management is optimal when DPS degenerates to a pure priority discipline. We identify multiple Wardrop equilibria and draw upon ideas from evolutionary game theory such that the most profitable equilibrium is attained. We then consider the class selection problem when customers are heterogenous. The heterogeneity is introduced by including a random weight factor for the customer waiting times in the customer utility. Also in this heterogeneous setting, priority queueing is optimal. Finally, we also allow for customers to balk. In contrast to the system without balking, the parameter setting which optimises revenue for the system with balking is a non-degenerate DPS policy.

The remainder of the paper is organised as follows. In the next section, we consider the single-server DPS game and its optimal revenue management. Sections 3 and 4 then extend the DPS game analysis to scenarios with heterogenous customers and customers that balk, respectively. Finally conclusions are drawn in Sect. 5.

2 Revenue Management Without Balking

2.1 Queueing Model

We consider a queueing system which uses discriminatory processor sharing to offer normal and premium service. Customers arrive in accordance with a Poisson process with rate λ, and the customer service times constitute a sequence of independent and identically exponentially distributed random variables with rate $\mu > \lambda$. When there are n_1 premium customers and n_2 non-premium customers in the system, the class 1 and class 2 customers service rates equal,

$$\mu_1 = \frac{\alpha\mu}{\alpha n_1 + n_2}, \quad \mu_2 = \frac{\mu}{\alpha n_1 + n_2},$$

respectively. Here $\alpha > 1$ denotes the class 1 weight of the discriminatory processor sharing discipline. Customers can opt to buy the premium service offer upon joining the queue. Let λ_1 and λ_2 denote the arrival rate for the premium and non-premium classes respectively. Following [10], the mean sojourn times of the premium and non-premium classes then equal,

$$\bar{T}_1 = \frac{1}{\mu - \lambda_1 - \lambda_2} \left(1 - \frac{\lambda_2(\alpha - 1)}{\alpha(\mu - \lambda_1) + (\mu - \lambda_2)} \right),$$

$$\bar{T}_2 = \frac{1}{\mu - \lambda_1 - \lambda_2} \left(1 + \frac{\lambda_1(\alpha - 1)}{\alpha(\mu - \lambda_1) + (\mu - \lambda_2)} \right), \quad (1)$$

for $\lambda_1 + \lambda_2 < \mu$.

The customers' decisions to buy the premium service depends on the expected sojourn time in the system, as well as on an additional admittance fee F for the premium customers and the reward of receiving service R. For convenience, we express the fee and reward in terms of sojourn times. The reward corresponds to the time one is willing to wait. Similarly, the admittance fee corresponds to the additional time one is willing to wait for the monetary value of the admittance fee. The utility of joining the premium and non-premium services then equal,

$$U_1 = R - F - \bar{T}_1, \quad U_2 = R - \bar{T}_2, \tag{2}$$

respectively.

Customers are rational, and therefore select the best available option. Hence, in Wardrop equilibrium, both options yield the same utility if they are chosen with positive probability (such that $\lambda_1 > 0$ and $\lambda_2 > 0$). If all customers choose the same option, the utility of that option cannot be smaller than the utility of the option that is not chosen.

Let f_ℓ and f_h denote,

$$f_\ell = \frac{(\alpha - 1)\lambda}{(\mu - \lambda)(\alpha\mu + \mu - \lambda)}, \quad f_h = \frac{(\alpha - 1)\lambda}{(\mu - \lambda)(\mu + \alpha(\mu - \lambda))}.$$

Comparing the utilities of the different options then shows that there are three equilibrium strategies. For $F \leq f_h$, all customers opting for premium service is an equilibrium strategy. For $F \geq f_\ell$, all customers opting for normal service is an equilibrium strategy. Finally, for $f_\ell \leq F \leq f_h$, the mixed strategy where normal service is chosen with probability ϕ is an equilibrium strategy,

$$\phi = \frac{1}{(\mu - \lambda)F} - \frac{\mu + \alpha(\mu - \lambda)}{(\alpha - 1)\lambda}. \tag{3}$$

Note that the choice only depends on differences in utilities, hence the reward R has no influence on the equilibrium solution. Moreover, for $f_\ell \leq F \leq f_h$ there are three distinct equilibria. The two pure equilibria, as well as the mixed equilibrium.

Clearly, the presence of multiple equilibria complicates revenue management. We however can reduce the complexity by excluding equilibria that cannot be attained in an evolutionary context where customers adopt their strategy based on previous experience. An equilibrium strategy is an evolutionary stable strategy (ESS) if it is resistant to small mutations. That is, if a small proportion of the customers follows a different strategy than the equilibrium strategy, their utility will be lower and they will therefore benefit from playing the equilibrium strategy. The expected utility of a customer playing strategy $\tilde{\boldsymbol{\lambda}} = (\tilde{\lambda}_1, \tilde{\lambda}_2)$, if every one else follows the strategy $\boldsymbol{\lambda} = (\lambda_1, \lambda_2)$ equals,

$$U(\tilde{\boldsymbol{\lambda}}, \boldsymbol{\lambda}) = \frac{\tilde{\lambda}_1}{\lambda} U_1(\boldsymbol{\lambda}) + \frac{\tilde{\lambda}_2}{\lambda} U_2(\boldsymbol{\lambda}).$$

where we make the $\boldsymbol{\lambda}$-dependence of U_1 and U_2 as defined in (2) explicit. Following [15,21], an equilibrium strategy $\boldsymbol{\lambda}$ is an ESS if for any mutant strategy

$\tilde{\lambda} \neq \lambda$ and some $0 < \epsilon < 1$, one is better off playing λ if a fraction ϵ follows the mutant strategy,

$$U(\tilde{\lambda}, (1 - \epsilon)\lambda + \epsilon\tilde{\lambda}) < U(\lambda, (1 - \epsilon)\lambda + \epsilon\tilde{\lambda}).$$

Note that in the current setting $U(\tilde{\lambda}, \lambda)$ is not linear in λ, so that we cannot simplify the criterion above in the first and second order conditions for evolutionary stable strategies, see [15, 21].

For the mixed equilibrium, set $\tilde{\lambda}_1 = (1 - \phi)\lambda - \delta$ and $\tilde{\lambda}_2 = \phi\lambda + \delta$. Then, for a fraction ϵ playing strategy $\tilde{\lambda}$, we find,

$$U(\tilde{\lambda}, (1 - \epsilon)\lambda + \epsilon\tilde{\lambda}) - U(\lambda, (1 - \epsilon)\lambda + \epsilon\tilde{\lambda}) = \frac{1}{\lambda} \frac{\delta^2 F^2 \epsilon(\mu - \lambda)}{\lambda + \delta F\epsilon(\mu - \lambda)} > 0,$$

which shows that the mixed strategy is not an ESS. In contrast, the pure strategies are both evolutionary stable: the premium strategy is an ESS for $F < f_h$ and the normal strategy is an ESS for $F > f_\ell$.

2.2   Revenue Management

Now assume that the service provider receives a monetary profit C per customer served, as well as the additional fee cF for every premium customer served. Here c is a constant, which converts waiting times in monetary values. Retaining only the evolutionary stable strategies, the server receives a payoff $P = cF + C$ per customer, if all customers opt for premium service, and a payoff $P = C$ per customer if all customers opt for normal service. From the vantage point of the server it is preferred that all customers opt for premium service, with the highest fee possible. Clearly, the server can choose any $F < f_\ell$ as premium service is then preferred by all customers. Indeed, for $F < f_\ell$, premium service is the only equilibrium strategy. As the premium strategy is evolutionary stable for $F < f_h$, we can also select a higher fee. For $F < f_h$ and all customers opting for premium strategy, a small fraction changing to the normal strategy will be worse off.

In fact, we can increase the fee F, even if not all customers have yet adopted the premium strategy. To evaluate how customers change their strategy over time, evolutionary game theory proposes various game dynamics, including replicator dynamics and imitation dynamics [15]. While the replicator dynamics mimics the effect of natural selection, the imitation dynamics models how successful strategies can spread through imitation. While the former is the best known dynamic, the latter is more appropriate in the present context. The imitation dynamic models how customers adopt their strategy by interchanging information on their experiences with normal and premium service. Moreover, replicator dynamics are a special type of imitation dynamics. Let $p_1 = \lambda_1/\lambda$ ($p_2 = \lambda_2/\lambda$) denote the fraction of customers opting for premium (normal) service. Following [15], the evolution of the rates p_1 and p_2 over time is described by the set of differential equations,

$$\dot{p}_1 = p_1[f_{12}(\lambda) - f_{21}(\lambda)]p_2$$
$$\dot{p}_2 = p_1[f_{21}(\lambda) - f_{12}(\lambda)]p_2$$

with $\boldsymbol{\lambda} = (p_1\lambda, p_2\lambda)$, with

$$f_{ij}(\boldsymbol{\lambda}) = \psi(U_i(\boldsymbol{\lambda}) - U_j(\boldsymbol{\lambda}))$$

and with $\psi(x) = \beta_1 |x|_2^\beta \operatorname{sgn}(x)$ for some $\beta_1 > 0$ and $\beta_2 \geq 0$. Parameter β_1 relates to the speed of the adaptation, while β_2 is a parameter of the imitation process. For $\beta_2 = 1$, the imitation dynamics correspond to the replicator dynamics, in the limiting case $\beta_2 = 0$, customers always imitate the better strategy. Note that the evolution of the strategies depends on the fee F, as the utilities U_1 and U_2 depend on the fee.

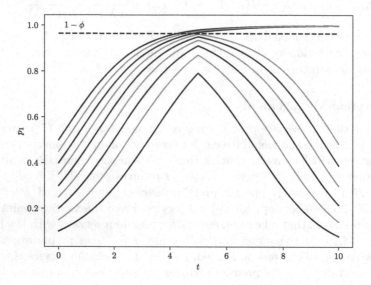

Fig. 1. Evolution of the fraction of customers p_1 opting for premium rate. The curves correspond to different values of the initial fraction of premium customers.

To illustrate that we can indeed increase the fee to values approaching f_h, Fig. 1 shows the evolution of the premium arrival rates, with the imitation process parameters set to $\beta_1 = 1$ and $\beta_2 = 0.1$. The arrival rate equals $\lambda = 0.8$ and the service rate equals $\mu = 1$, which corresponds to a server load of 80%. The DPS-weight of the premium service equals $\alpha = 2$. With these values, we have $f_\ell \approx 1.82$ and $f_h \approx 2.86$. We therefore consider a scenario where the fee is initially set to $F = 1.8$. At time $t = 5$, the fee is updated to $F = 2.8$. This situation corresponds to a strategy where the premium service is offered at a reduced cost for some time, followed by a price increase to optimise revenue. The curves clearly illustrate that if the fraction of premium customers is sufficiently high at the time of the change, the imitation dynamics converge to the premium ESS ($p_1 = 1$). However, if the fraction of customers opting for premium service is lower, the imitation dynamics converge to the normal ESS ($p_1 = 0$). The fraction of normal customers that separates convergence to the different

ESS equals ϕ, the mixed equilibrium of the preceding section (calculated with the fee $F = 2.8$). A dashed line shows the corresponding fraction of premium customers $1 - \phi = 0.96$ on the figure.

So far, we have investigated optimal revenue management for a given α. This weight is however also controlled by the server, so one can choose the α that yields the highest revenue. In view of the preceding discussion, the additional revenue per customer that can be attained equals f_h. We have $\partial f_h / \partial \alpha > 0$ for all α, so f_h is an increasing function of α. Hence, $\alpha = \infty$ is optimal: this is the system where the premium class has preemptive priority over the non-premium class. The corresponding optimal fee F_o equals,

$$F_o = \lim_{\alpha \to \infty} f_h = \frac{\lambda}{(\mu - \lambda)^2}.$$

As all customers opt for premium service, the revenue per customer equals $P = C + c F_o$.

3 Heterogenous Customers

The model in Sect. 2 assumes that all customers evaluate sojourn times in the same way. Some customers however may find the inconvenience of longer sojourn times less important than others. In this case, the influence of the fee on the choice is more outspoken. To evaluate the utilities of the customers, we now introduce a sojourn time weight W for each customer, the weights of the consecutive customers constituting a sequence of independent and identically distributed non-negative random variables. Accounting for the weights, the utility of joining the premium and normal queue now equal,

$$U_1 = R - F - W\bar{T}_1, \quad U_2 = R - W\bar{T}_2,$$

respectively. For ease of analysis, we assume that the weight distribution is continuous with support $\mathbb{R}^+$. In this case, we can be sure that there are always both normal and premium customers, whatever the unweighted utilities. As before, let ϕ denote the fraction of customers that opt for normal service, then an equilibrium requires,

$$\phi = \Pr[U_1 \leq U_2] = \Pr[R - F - W\bar{T}_1(\phi) \leq R - W\bar{T}_2(\phi)],$$

or equivalently,

$$\phi = \Pr\left[W \leq \frac{F}{\bar{T}_2(\phi) - \bar{T}_1(\phi)}\right]. \tag{4}$$

Here, we made the dependence of $\bar{T}_1$ and $\bar{T}_2$ on ϕ explicit. Recall that $\bar{T}_1$ and T_2 are defined in (1), with $\lambda_1 = (1 - \phi)\lambda$ and $\lambda_2 = \phi\lambda$. The expression above simply states that if a fraction ϕ opts for normal service, then there is a fraction ϕ whose utility for normal service exceeds their utility for premium service. Let

$W(x)$ denote the distribution function of the weight, plugging the definitions of $\bar{T}_i$ in the former expression then yields,

$$\phi = W\left(\frac{F(\mu - \lambda)\left((\alpha + 1)\mu + \lambda\phi(\alpha - 1) - \alpha\lambda\right)}{\lambda(\alpha - 1)}\right). \tag{5}$$

Clearly, the left hand side of the equality above linearly increases from 0 to 1 when ϕ increases from 0 to 1. The right hand side increases from $w_0 > 0$ to $w_1 < 1$ with,

$$w_0 = W\left(\frac{F(\mu - \lambda)\left((\alpha + 1)\mu - \alpha\lambda\right)}{\lambda(\alpha - 1)}\right),$$

$$w_1 = W\left(\frac{F(\mu - \lambda)\left((\alpha + 1)\mu - \lambda\right)}{\lambda(\alpha - 1)}\right).$$

Hence, we conclude there is a ϕ^* that solves (5). Note that it is in general possible that there are multiple equilibria. We then write $\phi^*(\alpha, F)$ for the smallest equilibrium, which makes the dependence on α and F explicit. In the examples in the remainder, we do not encouter scenarios with multiple equilibriums. The smallest equilibrium is then also the unique equilibrium.

Close inspection of (5) yields that ϕ^* strictly increases for increasing F. Indeed, the right hand side of (5) is an increasing function of F. Plotting the right hand side of (5) as a function of ϕ for different values of F then shows that the (smallest) intersect of this function with the identity function is larger for larger F. In other words, ϕ^* increases with F. This is not unexpected: fewer customers opt for premium service if the fee is higher. We can further express the derivative of ϕ^* with respect to α in terms of the derivative of ϕ^* with respect to F. Differentiating (4) with respect to α and F yields,

$$\frac{\partial\phi^*}{\partial\alpha} = -\frac{(2\mu - \lambda)F}{\lambda(\alpha - 1)^2\phi(\alpha, F) + (\alpha - 1)\left(\alpha(\mu - \lambda) + \mu\right)}\frac{\partial\phi^*}{\partial F}.$$

This shows that ϕ^* is a strictly decreasing function of α.

Borrowing the notation from the preceding section, the revenue per customer equals $P = C + c(1 - \phi^*)F$. Moreover, we have $\phi^* \to 1$ for $F \to \infty$. In view of (5), the revenue is bounded by,

$$C + cF(1 - w_1) \leq P \leq C + cF(1 - w_0).$$

In the remainder, we assume that the weight distribution adheres $\lim_{x\to\infty} x(1 - W(x)) = 0$. This assumption assures that $\lim_{F\to\infty} P = C$, as both bounds converge to C. In other words, the optimal revenue is obtained for finite F.

As ϕ^* is a decreasing function of α, the optimal revenue is obtained for $\alpha \to \infty$. In line with the queueing system with homogeneous customers of Sect. 2, the pure priority queueing system outperforms the DPS queue in terms of revenue. For $\alpha \to \infty$, the fraction of customers $\phi^\circ(F) \doteq \phi^*(\infty, F)$ that opt for normal service is the unique solution of

$$\phi = W\left(\frac{F(\mu - \lambda)(\mu + \lambda\phi - \lambda)}{\lambda}\right). \tag{6}$$

We now determine the fee that maximises the revenue per customer. Additionally assuming that $W(x)$ is differentiable, we have $\partial P/\partial F = 0$ for the optimal F, or

$$h\left(\frac{F\left(\mu - \lambda\right)\left(\mu + \lambda\phi - \lambda\right)}{\lambda}\right) = \frac{\lambda}{\mu(\mu - \lambda)F}, \tag{7}$$

where $h(x) = W'(x)/(1 - W(x))$ is the hazard rate function of the weight. The former expression clearly shows the possibility of multiple local maxima, as the hazard rate function is not necessarily monotone.

3.1 Exponentially Distributed Weights

In the special case that the weights are exponentially distributed with rate ν, we can explicitly determine the fee F that yields the largest revenue. Indeed, the hazard rate is then constant and equal to the rate ν. We therefore find the following optimal fee,

$$F = \frac{\lambda}{\mu(\mu - \lambda)\nu}.$$

The corresponding fraction of customers that opt for normal service is the solution of,

$$\phi = 1 - \exp\left(-\frac{\mu + \lambda\phi - \lambda}{\mu}\right). \tag{8}$$

We can solve for ϕ explicitly,

$$\phi = 1 + \frac{\mu}{\lambda}\,\Omega\left(-\frac{\lambda}{\mu}\exp(-1)\right).$$

Here, Ω is the principal branch of the Lambert W function [8]. This result is surprising: the fraction of customers that opt for normal service only depends on the system load λ/μ and does not depend on the rate parameter ν of the weight distribution.

3.2 Numerical Example

We now study how the weight distribution affects revenue and revenue management. Here and in the remaining numerical examples, we use fixed point iteration to find the fraction of customers that opt for premium and normal service. The optimal fee is then determined by means of the Broyden-Fletcher-Goldfarb-Shanno algorithm.

Figure 2 displays the optimal fee F and corresponding revenue P as a function of the arrival rate λ. We assume that the weights are gamma distributed with mean 1 and different variances are assumed as indicated. The case $\sigma^2 = 1$ corresponds to exponentially distributed weights. The service rate equals $\mu = 1$, and we set $C = 0$ and $c = 1$. Hence, the profit that is displayed corresponds to the additional profit generated by offering premium service, expressed in terms of waiting times.

The figure clearly shows that the optimal fee as well as the profit (per customer) increase with the arrival rate. This is explained by noting that the offer of a premium service only makes sense if there is congestion. In the absence of congestion, customers have no incentive to opt for premium service. The influence of customer heterogeneity on profit is negative: there is less profit if the variance of the weight distribution is higher. In contrast, the optimal fee is not a monotone function of the weight variance. Comparing the curves at $\lambda = 0.75$ shows that the optimal fee first decreases and then increases.

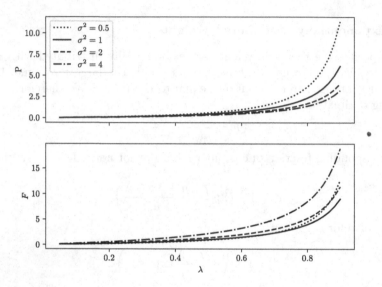

Fig. 2. Optimal revenue P and fee F vs. the arrival rate λ, for gamma distributed weights with mean 1 and variance σ^2 as indicated.

For the parameter values and weight distributions in the numerical example above, numerical experimentation reveals that there is a unique globally optimal fee F, and there are no other local maxima. In general, as already mentioned above, neither uniqueness nor the absence of local maxima can be excluded. Figure 3 illustrates the presence of multiple local maxima as well as an example where the global optimum is not unique. In view of Eq. (7), consider the following hazard rate function $h(t) = (t - a)^2$. The corresponding weight distribution equals

$$W(x) = 1 - \exp\left(-\int_0^x h(t)dt\right) = 1 - \exp\left(-\frac{a^3 + (x - a)^3}{3}\right).$$

We set $a = 2$ in Fig. 3. Moreover, we assume $C = 0$, $c = 1$ and $\mu = 1$, in line with the parameter values of the preceding numerical example. Different arrival rates are assumed as depicted. By a simple optimisation procedure, we found that there are two global optima for an arrival rate $\lambda_0 \approx 0.899$. The figure further illustrates that the profit approaches zero if F is to high: almost no customers then select premium service. Further, the profit per customer is higher if λ is higher. This is another illustration of the observation that congestion is required to get additional revenue by service differentiation.

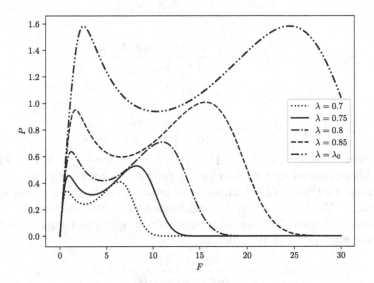

Fig. 3. Optimal revenue P vs. the fee F for different arrival rates λ as indicated.

4 Heterogeneous Customers and Balking

The models in the preceding sections assume a constant arrival rate, independent of the utility that is offered. Even if the customers' utility is negative, the customers have to choose between premium and normal service. In this section, we offer a third option: customers balk when the expected utility is negative. In particular, we adopt the assumptions of Sect. 2, regarding the heterogeneity of the customers, but now assume that customers do not enter if their expected utility is negative. Let ϕ_1 and ϕ_2 denote the fraction of customers that opt for premium and normal service, respectively. Then a fraction $1 - \phi_1 - \phi_2$ of the customers balk. We retain the assumption that the weight distribution has support $\mathbb{R}^+$. In the present setting it is possible that no customer opts premium or normal service.

Clearly, customers opt for premium (normal) service when the utility of premium (normal) service is non-negative and exceeds the utility of normal (premium) service. Therefore, we have,

$$\phi_1 = \Pr[U_1 \geq U_2, U_1 \geq 0], \quad \phi_2 = \Pr[U_2 \geq U_1, U_2 \geq 0].$$

We can again express these probabilities in terms of the weight distribution $W(x)$. By means of some simple calculations, we find the following set of equations,

$$\phi_1 = \max(W(w_2) - W(w_1), 0) \doteq \Phi_1(\phi_1, \phi_2),$$
$$\phi_2 = W(\min(w_1, w_3)) \doteq \Phi_2(\phi_1, \phi_2), \tag{9}$$

with,

$$w_1 = F(\mu - \phi_1\lambda - \phi_2\lambda)\frac{(\alpha+1)\mu - \alpha\phi_1\lambda - \phi_2\lambda}{(\alpha-1)(\phi_1+\phi_2)\lambda},$$

$$w_2 = (R - F)(\mu - \phi_1\lambda - \phi_2\lambda)\frac{(\alpha+1)\mu - \alpha\phi_1\lambda - \phi_2\lambda}{(\alpha+1)\mu - \alpha(\phi_1+\phi_2)\lambda},$$

$$w_3 = R(\mu - \phi_1\lambda - \phi_2\lambda)\frac{(\alpha+1)\mu - \alpha\phi_1\lambda - \phi_2\lambda}{(\alpha+1)\mu - (\phi_1+\phi_2)\lambda}.$$

By construction, for any ϕ_1 and ϕ_2 with $0 \le \phi_1 + \phi_2 \le 1$, the continuous mapping above retains that ϕ_1 and ϕ_2 are probabilities with $0 \le \phi_1 + \phi_2 \le 1$. Hence, Brouwer's theorem for compact sets guarantees the existence of at least one fixed point. We have however not established that there is a unique solution.

We further characterise the solution(s) of the system of equations above. First, consider solutions of (9), such that

$$\phi_1 + \phi_2 < \frac{F(\alpha+1)\mu}{\lambda(F + (\alpha-1)R)} \doteq L.$$

If this inequality holds, we have $w_2 < w_1$ and $w_3 < w_1$. Vice versa, if either $w_2 < w_1$ or $w_3 < w_1$, then inequality (9) holds. Therefore, if there is a fixed point that satisfies the inequality, we have $\phi_1 = 0$ and

$$\phi_2 = W(R(\mu - \lambda\phi_2)).$$

Note that the right-hand side decreases from $W(R\mu)$ for $\phi_2 = 0$ to $W(R(\mu - \lambda))$ for $\phi_2 = 1$. Hence, there exists a unique solution $\phi_2^* \in [0,1]$. We conclude that $(0, \phi_2^*)$ is a fixed point of (Φ_1, Φ_2) if

$$F > \frac{\phi_2^*\lambda(\alpha-1)R}{(\alpha+1)\mu - \phi_2^*\lambda}.$$

In other words, we have a fixed point which combines normal service with probability ϕ_2^* and balking with probability $1 - \phi_2^*$ if the fee is sufficiently large.

Now consider solution of (9), such that

$$\phi_1 + \phi_2 \ge L. \tag{10}$$

We then have $w_2 \ge w_1$ and $w_3 \ge w_1$. After some calculations, Eq. (9) simplifies to,

$$\phi_1 = W\left((R - F)(\mu - (\phi_1 + \phi_2)\lambda)\frac{(\alpha+1)\mu - (\alpha-1)\phi_1\lambda - (\phi_1+\phi_2)\lambda}{(\alpha+1)\mu - \alpha(\phi_1+\phi_2)\lambda}\right) - \phi_2,$$

$$\phi_2 = W\left(F(\mu - \phi_1\lambda - \phi_2\lambda)\frac{(\alpha+1)\mu - \alpha\phi_1\lambda - \phi_2\lambda}{(\alpha-1)(\phi_1+\phi_2)\lambda}\right),$$

or, equivalently, with $\tilde{\phi} = \phi_1 + \phi_2$,

$$\tilde{\phi} = W\left((R-F)(\mu - \tilde{\phi}\lambda)\frac{(\alpha+1)\mu - \alpha\tilde{\phi}\lambda + (\alpha-1)\phi_2\lambda}{(\alpha+1)\mu - \alpha\tilde{\phi}\lambda}\right),$$

$$\phi_2 = W\left(F(\mu - \tilde{\phi}\lambda)\frac{(\alpha+1)\mu - \alpha\tilde{\phi}\lambda + (\alpha-1)\phi_2\lambda}{(\alpha-1)\tilde{\phi}\lambda}\right). \tag{11}$$

The right-hand side of the second equation is increasing in ϕ_2 for fixed $\tilde{\phi}$, from a positive value for $\phi_2 = 0$ to a value below 1 for $\phi_2 = 1$. Hence, for each $\tilde{\phi}$ there exists a $\phi_2 = \zeta(\tilde{\phi})$ that solves the second equation. In fact, after some calculations, we can explicitly express ζ as follows,

$$\zeta(\tilde{\phi}) = W\left(W^{-1}(\tilde{\phi})\frac{F}{R-F}\frac{(\alpha+1)\mu - \alpha\tilde{\phi}\lambda}{(\alpha-1)\tilde{\phi}\lambda}\right).$$

We further have $\zeta(\tilde{\phi}) \leq \tilde{\phi}$ if and only if inequality (10) holds. By introducing ζ into (11), we find that $\tilde{\phi}$ solves the following equation,

$$\tilde{\phi} = W\left((R-F)(\mu - \tilde{\phi}\lambda)\frac{(\alpha+1)\mu - \alpha\tilde{\phi}\lambda + (\alpha-1)\zeta(\tilde{\phi})\lambda}{(\alpha+1)\mu - \alpha\tilde{\phi}\lambda}\right).$$

For any solution $\tilde{\phi} \in [L, 1]$ of the equation above, $(\tilde{\phi} - \zeta(\tilde{\phi}), \zeta(\tilde{\phi}))$ is an equilibrium of the DPS queueing game with balking.

4.1 Numerical Example

We now investigate revenue management by means of a numerical example. In contrast to the models without balking, we obtain a non-degenerate DPS discipline. Figure 4 is a contour plot of the profit P per arriving customer. The contour lines represent (α, F) pairs that yield the same profit. The arrival and service rates are equal to $\lambda = 0.95$ and $\mu = 1$, respectively. Hence, there is 95% system load. The weights are exponentially distributed with rate $\nu = 0.5$ and the customer reward is equal to $R = 5$. The revenue per normal customer is $C = 15$ and we set $c = 1$. With these parameters, we find that the optimal DPS weight equals $\alpha \approx 3.67$, with optimal fee $F \approx 0.81$. The optimal (α, F)-pair is marked on the contour plot. Further experimentation reveals that depending on the parameter values, either there is an optimal $\alpha < \infty$, or the most profit is obtained for $\alpha \to \infty$, like in the preceding sections.

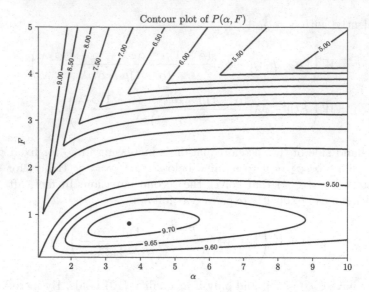

Fig. 4. Contour plot of the profit as a function of the DPS weight α and the premium fee F.

5   Conclusion

We considered revenue management for a DPS queue. In contrast to priority queueing disciples, discriminatory processor sharing allows for fine tuning the service differentiation. Three different queueing models were discussed. In the first two models, customers were not allowed to balk, and it was found that priority scheduling was optimal. Priority scheduling is obtained if the weight of the DPS discipline is sent to infinity. This observation corresponds to the intuitive notion that premium service should differ as much as possible from normal service, in order to optimise revenue. When customers can balk, the tuning possibilities of DPS can increase revenue, by giving enough incentive to opt for premium service and at the same time mitigate the effects of balking customers on profits.

References

1. Altman, E., Avrachenkov, K., Ayesta, U.: A survey on discriminatory processor sharing. Queueing Syst. **53**(1–2), 53–63 (2006). https://doi.org/10.1007/s11134-006-7586-8
2. Altman, E., Shimkin, N.: Individual equilibrium and learning in processor sharing systems. Oper. Res. **46**(6), 776–784 (1998)
3. Altmann, J., Daanen, H., Oliver, H., Suárez, A.S.B.: How to market-manage a QoS network. In: IEEE INFOCOM 2002. Twenty-First Annual Joint Conference of the IEEE Computer and Communications Societies. Proceedings, vol. 1, pp. 284–293 (2002)

4. Ayesta, U., Brun, O., Prabhu, B.J.: Price of anarchy in non-cooperative load balancing games. Perform. Eval. **68**(12), 1312–1332 (2011)
5. Van den Berg, H., Mandjes, M., Nunez-Queija, R.: Pricing and distributed QoS control for elastic network traffic. Oper. Res. Lett. **35**, 297–307 (2007)
6. Brun, O., Prabhu, B.: Worst-case analysis of non-cooperative load balancing. Ann. Oper. Res. **239**, 471–495 (2016)
7. Chamberlain, J., Starobinski, D.: Strategic revenue management of preemptive versus non-preemptive queues. Oper. Res. Lett. **49**(2), 184–187 (2021). https://doi.org/10.1016/j.orl.2020.12.011
8. Corless, R.M., Gonnet, G.H., Hare, D.E.G., Jeffrey, D.J., Knuth, D.E.: On the Lambert W function. Adv. Comput. Math. **5**, 329–359 (1996)
9. Altman, E., Ayesta, U., Prabhu, B.J.: Load balancing in processor sharing systems. Telecommun. Syst. **47**(1–2), 35–48 (2011). https://doi.org/10.1007/s11235-010-9300-8
10. Fayolle, G., Mitrani, I., Iasnogorodski, R.: Sharing a processor among many job classes. J. ACM **27**(3), 519–532 (1980). https://doi.org/10.1145/322203.322212
11. Fiems, D., Prabhu, B., De Turck, K.: Travel times, rational queueing and the macroscopic fundamental diagram of traffic flow. Physica A-Stat. Mech. Appl. **524**, 412–421 (2019). https://doi.org/10.1016/j.physa.2019.04.127
12. Fiems, D., Prabhu, B.J.: Macroscopic modelling and analysis of flows during rush-hour congestion. Perform. Eval. **149–150** (2021). https://doi.org/10.1016/j.peva.2021.102218
13. Gurvich, I., Lariviere, M.A., Ozkan, C.: Coverage, coarseness, and classification: determinants of social efficiency in priority queues. Manage. Sci. **65**(3), 1061–1075 (2019)
14. Hayel, Y., Tuffin, B.: Pricing for heterogeneous services at a discriminatory processor sharing queue. In: Proceedings of Networking (2005)
15. Hofbauer, J.: Evolutionary Games and Population Dynamics. Cambridge University Press (1998)
16. Massoulié, L., Roberts, J.: Bandwidth sharing and admission control for elastic traffic. Telecommun. Syst. **15**(1–2), 185–201 (2000). https://doi.org/10.1023/A:1019138827659
17. Massoulié, L., Roberts, J.: Bandwidth sharing: objectives and algorithms. IEEE/ACM Trans. Networking **10**(3), 320–328 (2002). https://doi.org/10.1109/TNET.2002.1012364
18. Ravner, L., Haviv, M., Vu, H.L.: A strategic timing of arrivals to a linear slowdown processor sharing system. Eur. J. Oper. Res. **255**, 496–504 (2015)
19. Roberts, J.W.: A survey on statistical bandwidth sharing. Comput. Netw. **45**(3), 319–332 (2004). https://doi.org/10.1016/j.comnet.2004.03.010
20. Sacoto-Cabrera, E.J., Guijarro, L., Vidal, J.R., Pla, V.: Economic feasibility of virtual operators in 5G via network slicing. Futur. Gener. Comput. Syst. **109**, 172–187 (2020)
21. Weibull, J.: Evolutionary Game Theory. MIT Press (1995)
22. Wu, Y., Bui, L., Johari, R.: Heavy traffic approximation of equilibria in resource sharing games. IEEE J. Sel. Areas Commun. **30**(11), 2200–2209 (2012)
23. Xu, B., Xu, X., Zhong, Y.: Equilibrium and optimal balking strategies for low-priority customers in the M/G/1 queue with two classes of customers and preemptive priority. J. Ind. Manage. Optim. **15**(4), 1599–1615 (2019)
24. Yashkov, S.F., Yashkova, A.S.: Processor sharing: a survey of the mathematical theory. Autom. Remote Control **68**(9), 1662–1731 (2007)

Performance and Energy Evaluation for Solving a Schrödinger-Poisson System on Multicore Processors

Thomas Rauber[1]([⊠]) [iD] and Gudula Rünger[2] [iD]

[1] University Bayreuth, Bayreuth, Germany
rauber@uni-bayreuth.de
[2] Chemnitz University of Technology, Chemnitz, Germany
ruenger@informatik.tu-chemnitz.de

Abstract. In this article, a Fourier-Galerkin approach for solving a Schrödinger-Poisson system is considered. The Fourier-Galerkin approach leads to an approximation method with two steps consisting of a truncation of the Fourier-Galerkin series and the solution of the resulting ordinary differential equation with a Runge-Kutta solver. Both steps influence the numerical accuracy of the final solution as well as the performance and energy behavior. The numerical approximation software is implemented as a multi-threaded program. The exploitation of frequency scaling and the degree of parallelism provided by the number of cores may also result in different values of execution time and energy consumption. The goal of this article is to evaluate the performance behavior of the numerical approximation software based on measurements on three multicore platforms. The mutual interaction of the number of threads, the operational frequency, and the computational requirement for achieving a specific numerical accuracy of the solution are investigated. Experiments show a good scalability behavior for computing approximation solutions of different numerical accuracy.

1 Introduction

The Schrödinger-Poisson system is a time-dependent partial differential equation (PDE) that is used to model a large variety of physical phenomena, such as plasma physics or molecular dynamics [12]. In most cases, it is not possible to determine the exact solution of the PDE in closed form and, therefore, the solution is approximated by a numerical technique. Approximations usually produce an error due to the calculation technique, and it is a concern how good the approximation fits to the real solution. The quality of an approximation is usually studied by investigating the numerical accuracy, which captures the global error of the approximation in comparison with the exact solution.

The numerical solution of a PDE requires a considerable amount of computation time and energy and, thus, it is desirable to have a low execution time of the approximation computation while retaining a certain quality of the

M. Iacono et al. (Eds.): EPEW/ASMTA 2023, LNCS 14231, pp. 18–33, 2023.
https://doi.org/10.1007/978-3-031-43185-2_2

accuracy. Today, a low energy consumption of the computation is an additional important aspect. On recent multicore architectures, the execution time as well as the energy consumption can be influenced by the operational frequency. Frequency scaling often has opposite effects, such that a faster computation might require more energy. For numerical solution methods, such as PDE solvers, also the accuracy requested can influence the computation performance and this is the main concern of this article.

In this article, we consider spectral methods for the solution of PDEs. Spectral methods provide a global approximation of the solution of the PDE in form of a finite sum over an appropriate set of base functions with unknown coefficients which are to be determined. The advantage of spectral methods is the high accuracy, however, they might be more costly than other methods, such as finite differences. The accuracy of the approximation can be influenced by the summation length, which is then directly connected to the computational effort. For time-dependent PDEs, the spectral method leads to a system of ordinary differential equations (ODEs), which can be solved by some standard ODE solver, such as a Runge-Kutta method [9,14]. The accuracy of the ODE solution can be guided by a given tolerance value and a step-size control mechanism, if needed. In this article, the spectral method is applied to a Schrödinger-Poisson system with periodic boundary conditions, which is solved with a Fourier-Galerkin spectral method and an embedded Runge-Kutta method for computing the time-dependent coefficients. The numerical software is implemented as a multi-threaded program suitable for current multicore processors.

The focus of the article is the investigation of the multi-threaded numerical solution process with respect to execution time and energy needed to compute an approximation solution. Execution time and energy consumption may depend on both, architectural aspects, such as the number of cores employed or the operational frequency chosen, but also on the desired accuracy of the numerical approximation solution to be computed. The accuracy depends on the truncation error as well as on the tolerance value required for the ODE solver. The dependence of the performance of the numerical computation on these factors is studied on three multicore processors: an Intel Broadwell server processor with 10 cores, an Intel Cascade Lake server processor with 40 cores and an Intel Coffee Lake desktop processors with 8 cores.

The contributions of this article include: (a) The investigation of the performance behavior of the approximation method with respect to execution time, power and energy consumption, and the influence of parallelism and frequency scaling on these metrics. (b) The investigation of the time and energy that needs to be invested to obtain a certain accuracy of the numerical approximation solution, which depends on specific tolerance value used for the ODE solver and the summation length of the spectral approximation. (c) The exploration of the dependence of the execution time and the energy consumption on the desired accuracy of the numerical solution to be computed. The main novelty of the paper lies in the exploration of the interaction between the numerical accuracy of the solution and the corresponding performance and energy consumption required to reach this accuracy on different multicore platforms.

The rest of the article is structured as follows. Section 2 presents related work. Section 3 describes the computational structure of the spectral method and the numerical solution process of the Schrödinger-Poisson system with a Fourier-Galerkin method. Section 4 presents and evaluates the performance behavior and the energy consumption of the parallel implementation with respect to the accuracy of the resulting numerical solution. Section 5 concludes the paper.

2   Related Work

The Schrödinger-Poisson equation has been used to describe a large variety of phenomena in physics, e.g., in astrophysics. The article [12] gives an intensive overview of these applications and concentrates itself on the use of the Schrödinger-Poisson to approximate the Vlasov-Poisson equation. Several numerical methods to approximate the solution of the Schrödinger-Poisson equation have been proposed in the last decades, including [10] and [5].

Runge-Kutta (RK) methods are popular solution methods for ODEs and implementations of RK methods are contained in several numerical libraries. Especially the implementation of DOPRI5 provided by Hairer and Wanner [9] is often used in practice. Sequential implementations of RK methods are provided by RKSUITE, Matlab, and IMSL. The PETSc library [1] provides parallel implementations. The parallelization approach of PETSc uses the MPI library, and a parallelization based on multi-threading is not supported. Parallelism in solution methods for ODEs has been considered in [3,7,22], focussing on the assignment of stage vectors to different execution units. In contrast, this article exploits data parallelism in classical RK methods. Thus, the degree of parallelism is limited only by the size of the ODE system to be solved, which can be quite large, e.g., if ODEs resulting from discretized PDEs such as the Schrödinger-Poisson equation are considered. An analysis of different versions of RK methods has been given in [16] considering a reaction-diffusion equation as example.

The energy consumption of processors has gained increasing interest during the last years [19]. To improve the energy consumption, several power-management techniques have been developed and used, including power capping [8], clock gating [2], and DVFS [20]. Energy metrics for parallel application codes are described in [17]. An experimental evaluation of the memory hierarchy of AMD EPYC Rome and Intel Xeon Cascade Lake SP server processors is given in [23]. The Cascade Lake processor is one of the processors used for the experimental evaluation in this article.

3   Numerical Approximation Method

The numerical solution method is a spectral method which starts with a series expansion of the unknown solution of a PDE using an appropriate set of base functions. The specific expansion is to be chosen according to the PDE problem.

3.1   Fourier-Galerkin Problem

For the solution of the Schrödinger-Poisson equation, the Fourier-Galerkin method has been chosen. Fourier-Galerkin spectral methods with base functions $h_l(x) = \frac{1}{\sqrt{L}} \exp(i\lambda_l x)$ and $\lambda_l = (\frac{2\pi l}{L})^2$ are well-suited to represent a global approximation of a time-dependent PDE with initial condition and periodic boundary condition with period L in space. The global representation of the solution is an infinite summation which is truncated to provide an approximate solution. To obtain the approximate solution, the weak solution is considered which results in a system of ordinary differential equations (ODEs). The following coupled Schrödinger-Poisson equation [13], also known as Schrödinger-Newton equation, is considered:

$$i\hbar \frac{\partial \Psi}{\partial t} = -\frac{\hbar}{2m} \frac{\partial^2 \Psi}{\partial x^2} + e\Phi\Psi \tag{1}$$

$$\frac{\partial^2 \Phi}{\partial x^2} = \frac{e\, n_0}{\varepsilon_0}(1 - \Psi\Psi^*) \tag{2}$$

where $\hbar$ is the reduced Planck constant, m is the electron mass, e is the electron charge, ε_0 is the permittivity of vacuum, and $i = \sqrt{-1}$. The value n_0 is the density of a uniform positively charged background of ions. An L–periodic boundary condition $\Psi(0,t) = \Psi(L,t)$ in the space dimension is used. The normalization of the Schrödinger wave function Ψ is taken as $\int_0^L \Psi\Psi^* dx = L$ where Ψ^* is the conjugate complex of Ψ. The potential is $\Phi(x,t)$. The electron density is evaluated from the Schrödinger wave function as $n(x,t) = \Psi(x,t)\Psi^*(x,t)$. The global approximation solution with truncation number $N \in \mathbb{N}$ is represented by two summations

$$\Psi_N(t,x) = \sum_{|l| \leq N} \alpha_l(t)h_l(x), \quad \Phi_N(t,x) = \sum_{|l| \leq N} \beta_l(t)h_l(x), \tag{3}$$

and the weak solution Ansatz brings us to the following ODE system for the unknown coefficients $\alpha(t) = (\alpha_N(t),..\alpha_0(t),..\alpha_{-N}(t))$ and $\beta(t) = (\beta_N(t),..\beta_0(t),..\beta_{-N}(t))$:

$$i\frac{d\alpha_l}{dt} = -\lambda_l \alpha_l + \sum_{|j| \leq N} \alpha_l \cdot (\Phi_N h_j, h_l), \quad l = 0, \pm 1, .., \pm N \,,$$

$$\beta_l = \frac{1}{\lambda_l}(|\Psi_N|^2, h_l), \quad l = \pm 1, \ldots, \pm N \,, \tag{4}$$

where $(h,g) = \left(\int_0^L h \cdot g \cdot dx\right)^{1/2}$ is the scalar product. The ODE System (4), referred to as SP ODE in the following, is of size $2N + 1$, and since the Ansatz functions are complex, this leads to a system of real-valued ODEs of size $4N + 2$. Thus, a higher truncation number N leads to a larger ODE system so that a higher accuracy of the approximate solution might lead to higher computation time.

The ODE system can be solved by a suitable numerical method. In this article, Runge-Kutta solvers are used due to their stable numerical behavior, see Sect. 3.2 and [9]. The required accuracy of the coefficients α and β are influenced by a tolerance value TOL and the step size control of the Runge-Kutta solver. The effect of both influencing factors on the accuracy are considered in the next subsections.

3.2   Runge-Kutta Solution Method

The computation effort of the Fourier-Galerkin method is caused by the solution of the system of ODEs resulting for the weak solution Ansatz. A larger truncation number N leads to a larger ODE system to be solved and, thus, to a higher computational effort.

Explicit Runge-Kutta (RK) methods are suitable for the solution of the resulting ODE system. RK methods compute a series of approximation vectors $\mathbf{y}_0, \mathbf{y}_1, \mathbf{y}_2 \ldots$ which approximate the exact solution at discrete x-values $x_0, x_1, x_2, \ldots$. One approximation step computing $\mathbf{y}_{\kappa+1}$ from $\mathbf{y}_\kappa$, $\kappa = 0, 1, 2, \ldots$, has the form:

$$\mathbf{k}_l = \mathbf{f}(x_\kappa + c_l h_\kappa, \mathbf{y}_\kappa + h_\kappa \sum_{i=1}^{l-1} a_{li} \mathbf{k}_i) \text{ for } l = 1, \ldots, s. \tag{5}$$

In formula (5), the vectors $\mathbf{k}_1, \ldots, \mathbf{k}_s$ denote the stage vectors. The value h_κ is the step-size used in the specific approximation step κ, i.e., $x_{\kappa+1} = x_\kappa + h_\kappa$, and $\mathbf{y}_\kappa$ is the previous approximation vector. The stage vectors $\mathbf{k}_1, \ldots, \mathbf{k}_s$, computed according to Eq. (5), are needed to compute the next approximation vector $\mathbf{y}_{\kappa+1}$ and an additional approximation vector $\hat{\mathbf{y}}_{\kappa+1}$ used for error control and step-size adaption:

$$\mathbf{y}_{\kappa+1} = \mathbf{y}_\kappa + h_\kappa \sum_{l=1}^{s} b_l \mathbf{k}_l, \quad \hat{\mathbf{y}}_{\kappa+1} = \mathbf{y}_\kappa + h_\kappa \sum_{l=1}^{s} \hat{b}_l \mathbf{k}_l. \tag{6}$$

The computation scheme (5) and (6) uses the following coefficients: s-dimensional vectors $b = (b_1, \ldots, b_s)$, $\hat{b} = (\hat{b}_1, \ldots, \hat{b}_s)$, and $c = (c_1, \ldots, c_s)$, as well as an $s \times s$ matrix $A = (a_{il})$ which are specific for the particular RK method chosen and are usually depicted in the Butcher tableau [9]. For explicit RK methods, the matrix A is a strictly lower triangular matrix. The order r of the approximation vector $\mathbf{y}_{\kappa+1}$ and the order $\hat{r}$ of the second approximation vector $\hat{\mathbf{y}}_{\kappa+1}$ typically differ by 1 so that $r = \hat{r} + 1$ holds. An asymptotic estimate of the local error in the lower order approximation is computed by the difference between the two approximations $\mathbf{y}_{\kappa+1}$ and $\hat{\mathbf{y}}_{\kappa+1}$. The local error is used for a stepsize control [6]. The approximation vector of the current step κ is accepted, if a suitable weighted norm of the local error estimate lies within the predefined tolerance level. Although the estimate of the local error is in the lower order approximation, the more accurate approximation is usually taken to advance the integration (local extrapolation).

Several embedded RK methods have been proposed in the literature, including the methods of Dormand and Prince (DOPRI), e.g. DOPRI5 of order 5(4) or DOPRI8 of order 8(7), and Verner's methods DVERK of order 6(5) [4,9]. The investigations in this article are performed for the popular DOPRI5 method, which uses 7 stages. An error control mechanism and a stepsize selection method are integrated into the RK methods according to [15], so that an approximation of a step for which the resulting error would be too large is rejected and the step is repeated with a smaller stepsize.

3.3 Accuracy of the Numerical Solution

For the Fourier-Galerkin method, there are two possibilities to influence the accuracy of the numerical solution. First, the truncation of the series in formula (3) leads to global approximate solutions of different accuracy, typically with increasing accuracy for a higher truncation number N. Second, the approximation error for computing the coefficients can be guided by the TOL value for the step-size control of the ODE method solving the resulting ODE system (4). The accuracy achieved for the numerical solution of the Schrödinger-Poisson system is given in Fig. 1. The diagram depicts the global error for different system sizes $n = 4N + 2$, shown on the x-axis, and different TOL values between 10^{-2} and 10^{-6}, resulting in a family of curves. As Runge-Kutta solver, the popular DOPRI5 method has been used [9]. The global error is computed by using an exact solution known for a special case. The diagram shows that the global error steadily decreases with the size N and also with decreasing TOL values. For large values of N, the accuracy obtained is significantly larger than the TOL value used and a high accuracy can be already reached for a TOL value of 10^{-2}. For example, using $N = 5000$ with TOL value of 10^{-2} leads to a global error of $0.95 \cdot 10^{-14}$. Thus, the diagram confirms the expected behavior of the accuracy

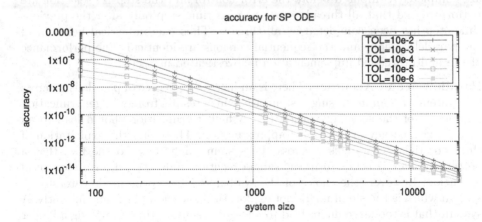

Fig. 1. Accuracy for the Fourier-Galerkin spectral method for different system sizes and different tolerance values TOL.

when solving the non-linear PDE (1) with the spectral Ansatz in (3). In the following sections, the computational effort and the energy needed to achieve a certain accuracy are studied. Moreover it is investigated how a parallel execution and frequency scaling can be exploited to reduce the execution time and the energy consumption.

3.4 Multi-threaded Implementation

The RK methods are sequential in the simulation time steps $\kappa = 0, 1, 2, \ldots$, since each new time step $\kappa+1$ needs the result of the previous time step κ, $\kappa \geq 0$. However, within each time step there is a potential of parallelism across the system. This kind of parallelism assigns single equations (or set of equations) to different processing units. Since ODE solvers are step-size methods, this assignment can be repeated in every time step. We have developed a multi-threaded implementation of the RK methods using the Pthreads library. The implementation exploits parallelism across the system as described in the following.

Multi-threaded Program: In the parallel implementations of the RK method, the computations of the components of the argument vectors, the stage vectors, and the approximation vectors are distributed in a block-wise way over different threads. The error control and stepsize selection for the next time step is performed by a single thread at the end of the current time step. If the error control observes that the error is too large, the previous time step is repeated with a newly computed smaller step size. Synchronization operations are included to ensure numerical correctness. A first barrier synchronization is used after the computation of each stage vector of the RK method so that the computation of the next stage vector uses the most recent values of the preceding stage vectors. A barrier synchronization is also used before and after the error control and the stepsize selection, which ensures that the approximation vectors $\mathbf{y}_{\kappa+1}$ and $\hat{\mathbf{y}}_{\kappa+1}$ are completely computed before the error control and the stepsize selection are performed and that all threads start the next time step only after the previous time step has been completed by all threads. Thus, the numerical behavior of the parallel versions and the sequential versions are identical. The performance of the multi-threaded implementation is investigated in Sect. 4.

Parallelism and Accuracy: There is an increasing degree of parallelism across the system with an increasing system size, which results from a larger truncation number N of the series expansion (3). A higher truncation should lead to an increasing accuracy of the global approximation. Therefore, the exploitation of the potential of parallelism across the system might lead to the fact that a better accuracy is achieved in the same execution time needed for a sequential execution. However, the choice of the system size has to be made carefully. A system which is too small might not exploit the execution platform efficiently. A system that is too large might lead to a larger execution time providing a higher accuracy which might not be required.

4   Performance and Energy Behavior

The multi-threaded numerical software described in the previous section is now evaluated with respect to its performance behavior in terms of execution time and energy consumption. The execution time as well as the energy consumption may depend on several influencing factors, which are the number of threads employed, the number of base functions, and frequency scaling. The dependencies of the performance on these parameters are studied.

4.1   Metrics and Parameters

For multicore processor providing DVFS (dynamic voltage and frequency scaling), the operational frequency f influences the execution time t_{end}, the power P and the energy consumption E of the computation. Also the number of cores p has an impact on the execution time t_{end}, usually with decreasing execution time for greater p, and on the power P, often with a higher power consumption when more cores are employed. Thus, for DVFS systems with p_{max} cores and operational frequencies f ranging between a minimum frequency f_{min} and a maximum frequency f_{max}, the energy consumption $E(p, f)$ depends on the parameters p and f and can expressed as

$$E(p, f) = \int_{t=0}^{t_{end}(p,f)} P(p, f)(t)dt, \qquad (7)$$

where $t_{end}(p, f)$ denotes the execution time of the program when using p processors and operational frequency f, and $P(p, f)(t)$ denotes the power consumption at time t. A linear dependence between the energy consumption $E(p, f)$ and the execution time t_{end} cannot be assumed since the power consumption $P(p, f)$ may vary during the execution time.

In this section, the influence of the parameters p and f on the performance metrics are investigated. Furthermore, the influence of application specific parameters on the metrics are studied. These parameters include the size $4N + 2$ of the ODE system to be solved. Another influencing parameter is the tolerance value TOL, which is used for a specific run of the ODE solver, and measurements for different TOL-values $TOL = 10^{-2}, 10^{-3}, 10^{-4}, 10^{-5}, 10^{-6}$ are conducted in the performance experiments, leading to a set of curves depicted in the diagrams.

4.2   Experimental Setup

For the experimental evaluation, three different processors have been used: (i) an Intel Broadwell i7-6950X server CPU (Broadwell architecture) with 10 cores, (ii) an Intel Core i7-9700 desktop CPU with 8 cores (Coffee Lake architecture), and (iii) an Intel Intel Xeon Gold 6248 server CPU with 40 cores (Cascade Lake architecture) on two sockets, see Table 1 for more details. The compilation has been performed with gcc (Version 7.5.0) using optimization level -O3.

Table 1. Characteristics of the processors used for the experimental evaluation.

	Xeon i7 6950X	Core i7 9700	Xeon Gold 6248
architecture	Broadwell	Coffee Lake	Cascade Lake
year of release	2016	2019	2019
minimum frequency	1.2 GHz	0.8 GHz	1.0 GHz
maximum frequency	3.0 GHz	3.0 GHz	2.5 GHz
TDP	140 W	65 W	150 W
physical cores	10	8	40
L1 data cache	32 KB	32 KB	32 KB
L2 cache	256 KB	256 KB	1 MB
L3 shared cache	25 MB	12 MB	28 MB
RAM size	32 GB	16 GB	376 GB

The time and energy measurements have been performed using the Running Average Power Limit (RAPL) interface and sensors of the Intel architecture [11,18]. RAPL sensors can be accessed by control registers, known as Model Specific Registers (MSRs) [11]. In particular, we have used the likwid tool-set, especially the likwid-perfctr tool in Version 4.3.2 [21], which provides access to the MSRs. The execution threads are pinned to specific cores for the time and energy measurements. To keep disturbance effects as small as possible, the runtime and energy measurements have been performed with no other user on the system and no other process except the operating system running.

4.3   Dependence on the Number of Threads

Behavior of Execution Time on Broadwell: Figure 2 shows the execution time (left) and speedup (right) for solving the SP ODE system with $N = 5000$ using different numbers of threads, shown on the x-asis, and different tolerance values between 10^{-2} and 10^{-6} on the Broadwell processor. The speedup is computed by using a purely sequential program version, which has no threading overhead. For all TOL values, the same integration interval has been used. The diagrams show the good scalability of the method. The speedup values increase with decreasing TOL values, which can be explained by the larger number of time steps needed for a smaller TOL value. The number of time steps increases with decreasing TOL values, however with a different rate: the number of time steps needed for $TOL = 10^{-2}$ to $TOL = 10^{-6}$ are 25,40,59,88, and 137, respectively A higher number of time steps leads to a larger portion of computations that can be parallelized and a better compensation of the sequential overhead.

Behavior of Energy and Power on Broadwell: Figure 3 shows the energy and power consumption corresponding to the situation in Fig. 2. The power depicted in the right diagram shows an almost linear increase with the number of

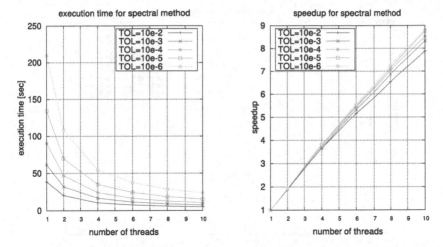

Fig. 2. Execution time (left) and speedup (right) of SP ODE for different numbers of cores on Broadwell and different tolerance values between 10^{-2} and 10^{-6}.

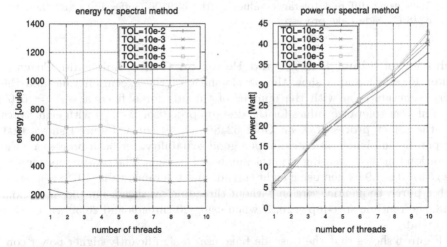

Fig. 3. Energy consumption (left) and power consumption (right) of SP ODE for different numbers of cores on Broadwell and different tolerance values between 10^{-2} and 10^{-6}.

threads used, which indicates that each additional core is active with work. The good speedup values shown in Fig. 2 indicate that this work is efficiently invested. The energy consumption in the left diagram of Fig. 3 is nearly constant and varies only slightly with the number of threads, which means that the higher power consumption is compensated by a corresponding reduction of the execution time.

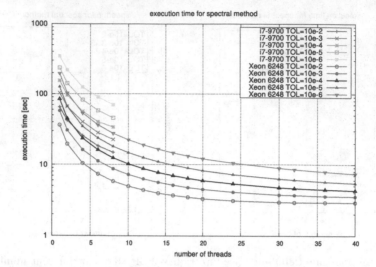

Fig. 4. Development of the execution time with the number of threads for a fixed $N = 5000$ using different tolerance values on the i7-9700 Coffee Lake and the Xeon Gold 6248 Cascade Lake processor.

Behavior of Time, Energy, and Power on Coffee Lake and Cascade Lake: Figures 4 and 5 show the development of the execution time and the energy consumption with the number of threads for a fixed size $N = 5000$ and different tolerance values for the desktop processor i7-9700 with eight cores and the server processor Xeon Gold 6248 with 40 cores. Figure 4 shows that the parallel implementation exhibits a good scalability: for both processors, the execution time is decreasing with the number of threads. The runtime speedup on the i7-9700 is 4.95 when using eight threads and tolerance value 10^{-6}, compared with a purely sequential version without threading overhead. On the Xeon Gold 6248, the runtime speedup is 16.27 when using 20 threads and 26.39 when using 40 threads.

Figure 6 shows that the Cascade Lake uses a significantly higher power consumption than the Coffee Lake processor. For both processors, the power consumption increases with the number of threads. For the Cascade Lake system, there is a steep increase of the power consumption if more than 20 threads are employed due to the fact that the second socket is addressed. Figures 4 shows that the execution on the Cascade Lake processor is much faster than on the Coffee Lake processor. In Fig. 5, it can be seen that the Coffee Lake system has a much smaller energy consumption than the Cascade Lake system despite the larger execution time, which is caused by the difference in the power consumption.

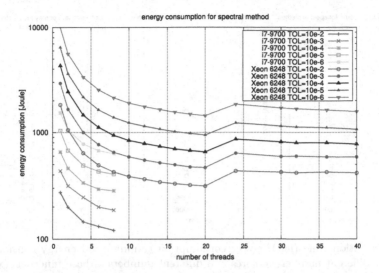

Fig. 5. Development of the energy consumption with the number of threads for a fixed $N = 5000$ using different tolerance values on the i7-9700 Coffee Lake and the Xeon Gold 6248 Cascade Lake processor.

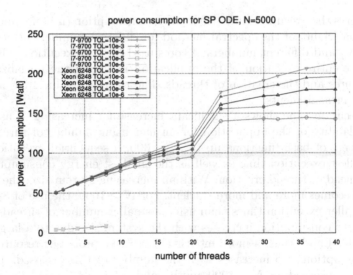

Fig. 6. Development of the power consumption with the number of threads for a fixed $N = 5000$ using different tolerance values on the i7-9700 Coffee Lake and the Xeon Gold 6248 Cascade Lake processor.

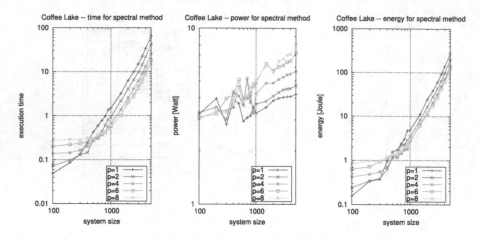

Fig. 7. Execution time (left), power consumption (middle) and energy consumption (right) for different numbers of cores and different numbers of base functions (x-axis) on Coffee Lake processor.

4.4 Increasing Number of Base Functions

Figure 7 shows the execution time (left), power consumption (middle) and energy consumption (right) of the spectral method for different numbers of base functions (x-axis) and different numbers of cores used for the execution. The figure captures the entire execution of the application including the initialization of data structures and the creation of threads. The tolerance value 10^{-2} has been used for all measurements.

The execution time shows a significant increase for increasing N as well as a good scalability of the application for an increasing number of threads. For small numbers of base functions up to $N = 300$, a sequential execution leads to the smallest execution time as well as the smallest energy consumption due to the overhead of thread creation. With an increasing system size, the parallel execution becomes more and more efficient. For $N \geq 1000$, the use of 8 threads leads to smaller execution times than using a smaller number of threads.

The power consumption increases with the system size due to a larger computational load and, thus, a more intensive use of the processor resources. The power consumption also increases with the number of threads used. This can especially be observed for $N = 1000$ and beyond.

For the energy consumption, different numbers of threads lead to the smallest energy consumption for different system sizes: up to $N = 300$: one thread; for $N = 500$: two threads; for $600 \leq N \leq 1000$: four threads; for $1500 \leq N \leq 2000$: six threads; for $2500 \leq N \leq 5000$: eight threads. A similar behavior can be observed for other tolerance values and the other processors considered.

4.5   Pareto Analysis of Frequency Scaling

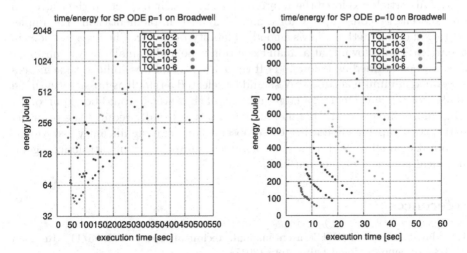

Fig. 8. Phase diagram for execution time and energy consumption using different tolerance values and frequencies for a sequential execution (left) and a parallel execution with 10 threads (right).

Figure 8 (left) shows the sequential execution time (x-axis) and energy consumption (y-axis) of the spectral method on the Broadwell processor using $N = 5000$ for different tolerance values between 10^{-2} and 10^{-6} (shown in different colors) and all available operational frequencies between 1.2 GHz and 2.9 GHz. Figure 8 (right) shows the same information for a parallel execution with 10 threads. Each dot in the diagram corresponds to a pair of values (time, energy) for an execution with a fixed frequency and a predefined tolerance value. The diagram shows five convex curves for executions that are performed with the same tolerance value $TOL \in \{10^{-2}, 10^{-3}, 10^{-4}, 10^{-5}, 10^{-6}\}$. For each curve, the frequencies decrease from left to right.

In Fig. 8, it can be seen that the smallest energy consumption of 42.6 J results for an execution with one thread using operational frequency 1.6 GHz and tolerance value $TOL = 10^{-2}$. The smallest energy consumption for an execution with 10 threads is 57.2 J obtained for operational frequency 1.2 GHz and tolerance value $TOL = 10^{-2}$. The smallest execution time is 5.0 s, reached with 10 threads using frequency 2.9 GHz and tolerance value $TOL = 10^{-2}$. However, the energy consumption is significantly larger in this case (191.1 J).

5   Conclusions

Spectral methods can influence the accuracy of the resulting solution of a PDE via two mechanisms: the truncation number N and the tolerance value for controlling the stepsize and error correction mechanism for solving the resulting

ODE system. There are mutual interactions between these influencing factors and the best constellation to achieve a desired numerical accuracy is not a priori clear. This article explores these interactions and their impact on the execution time and energy consumption for a multi-threaded implementation. The implementation of the method shows a good scalability behavior. The investigations in this article have shown that a small execution time and a small energy consumption can be obtained as follows: (1) If there are several possibilities to achieve a predefined minimum accuracy, N should be selected as small as possible, even if a smaller tolerance value, i.e., a larger number of time steps, would be required. (2) Using more threads in the parallel implementation usually reduces the resulting energy consumption. Following these observations can significantly reduce the computational effort and, thus, the energy consumption, while attaining a good numerical accuracy of the solution.

References

1. Balay, S.A., et al.: PETSc users manual. Technical report ANL-95/11 - Revision 3.8, Argonne National Laboratory (2017)
2. Bezati, E., Casale-Brunet, S., Mattavelli, M., Janneck, J.W.: Clock-gating of streaming applications for energy efficient implementations on FPGAs. IEEE Trans. Comput. Aided Des. Integr. Circuits Syst. **36**(4), 699–703 (2017)
3. Burrage, K.: Parallel methods for initial value problems. Appl. Numer. Math. **11**, 5–25 (1993)
4. Dormand, J.R., Prince, P.J.: A family of embedded Runge-Kutta formulae. J. Comput. Appl. Math. **6**(1), 19–26 (1980)
5. Ehrhardt, M., Zisowsky, A.: Fast calculation of energy and mass preserving solutions of schrödinger-poisson systems on unbounded domains. J. Comput. Appl. Math. **187**(1), 1–28 (2006)
6. Enright, W.H., Higham, D.J., Owren, B., Sharp, P.W.: A Survey of the Explicit Runge-Kutta Method. Technical report 94–291, University of Toronto, Department of Computer Science (1995)
7. Gear, C.W.: Massive parallelism across space in ODEs. Appl. Numer. Math. **11**, 27–43 (1993)
8. Haidar, A., Jagode, H., Vaccaro, P., YarKhan, A., Tomov, S., Dongarra, J.: Investigating power capping toward energy-efficient scientific applications. Concurr. Comput. Pract. Exp. **31**(6), e4485 (2019). e4485 cpe.4485
9. Hairer, E., Nørsett, S.P., Wanner, G.: Solving Ordinary Differential Equations I: Nonstiff Problems. Springer-Verlag, Berlin, Heidelberg (1993). https://doi.org/10.1007/978-3-540-78862-1
10. Harrison, R., Moroz, I., Tod, K.P.: A numerical study of the Schrödinger-Newton equations. Nonlinearity **16**(1), 101–122 (2002)
11. Intel. Intel 64 and IA-32 Architecture Software Developer's Manual, System Programming Guide, 2011
12. Mocz, P., Lancaster, L., Fialkov, A., Becerra, F., Chavanis, P.: Schrödinger-poisson-vlasov-poisson correspondence. Phys. Rev. D **97**, 083519 (2018)
13. Nguyen, V.T., Bertrand, P., Izrar, B., Fijalkow, E., Feix, M.R.: Integration of vlasov equation by quantum mechanical formalism. Comput. Phys. Commun. **34**, 295–301 (1985)

14. Rauber, T., Rünger, G.: Parallel execution of embedded and iterated Runge-Kutta methods. Concurr. Pract. Exp. **11**(7), 367–385 (1999)
15. Rauber, T., Rünger, G.: On the energy consumption and accuracy of multithreaded embedded Runge-Kutta methods. In: Proceedings of the The International Conference on High Performance Computing & Simulation (HPCS 2019), vol. 15, pp. 382–389. IEEE, July 2019
16. Rauber, T., Rünger, G.: Modeling the effect of application-specific program transformations on energy and performance improvements of parallel ODE solvers. J. Comput. Sci. **51** (2021)
17. Rauber, T., Rünger, G., Stachowski, M.: Performance and energy metrics for multithreaded applications on DVFS processors. Sustain. Comput. Inform. Syst. **17**, 55–68 (2017)
18. Rotem, E., Naveh, A., Ananthakrishnan, A., Rajwan, D., Weissmann, E.: Power-management architecture of the intel microarchitecture code-named sandy bridge. IEEE Micro **32**(2), 20–27 (2012)
19. Saxe, E.: Power-efficient software. Commun. ACM **53**(2), 44–48 (2010)
20. Tarplee, K.M., Friese, R., Maciejewski, A.A., Siegel, H.J., Chong, E.K.P.: Energy and makespan tradeoffs in heterogeneous computing systems using efficient linear programming techniques. IEEE Trans. Parallel Distrib. Syst. **27**(6), 1633–1646 (2016)
21. Treibig, J., Hager, G., Wellein, G.: LIKWID: a lightweight performance-oriented tool suite for x86 multicore environments. In: 39th International Conference on Parallel Processing Workshops, ICPP '10, pp. 207–216. IEEE Computer Society (2010)
22. van der Houwen, P., Sommeijer, B.: Parallel Iteration of high-order Runge-Kutta Methods with stepsize control. J. Comput. Appl. Math. **29**, 111–127 (1990)
23. Velten, M., Schöne, R., Ilsche, T., Hackenberg, D.: Memory performance of AMD EPYC Rome and intel cascade lake SP server processors. In: Feng, D., Becker, S., Herbst, N., Leitner, P. (eds.), ICPE '22: ACM/SPEC International Conference on Performance Engineering, Bejing, China, 9–13 April 2022, pp. 165–175. ACM (2022)

Stochastic Analysis of Rumor Spreading with Multiple Pull Operations in Presence of Non-cooperative Nodes

Sébastien Kilian[1], Emmanuelle Anceaume[2]([⊠]), and Bruno Sericola[1]

[1] Centre Inria de l'Univ. de Rennes, Irisa, Rennes, France
[2] CNRS, Univ. Rennes, Irisa, Rennes, France
emmanuelle.anceaume@irisa.fr

Abstract. The recent rise of interest in distributed applications has highlighted the importance of effective information dissemination. The challenge lies in the fact that nodes in a distributed system are not necessarily synchronized, and may fail at any time. This has led to the emergence of randomized rumor spreading protocols, such as push and pull protocols, which have been studied extensively. The k-pull operation, which allows an uninformed node to ask for the rumor from a fixed number of other nodes in parallel, has been proposed to improve the pull algorithm's effectiveness. This paper presents and studies the performance of the k-pull operation in the presence of a certain fraction f of non-cooperative nodes. Our goal is to understand the impact of k on the propagation of the rumor despite the presence of a fraction f of non-collaborative nodes.

1 Introduction

The dissemination of information in distributed systems has been an active area of research in recent years. With the rise of distributed applications, efficient and robust methods for information dissemination have become increasingly important. In a distributed system, the nodes are not necessarily synchronized, and can fail at any time. This makes the dissemination of information a challenging problem. This problem, often called *rumor spreading* or *gossip spreading*, is the process of sending a message to all the nodes in a network [5, 7]. The nodes in the network are anonymous, meaning that they can not be designated in advance, and any two nodes cannot tell whether they already interacted together or not. Different variants of randomized rumor spreading protocols have been studied. The push protocol provides a single operation, called the push operation, that allows an informed node to contact some randomly chosen node and sends it the rumor. The pull protocol, on the other hand allows, through the pull operation, an uninformed node to contact some random node to ask for the rumor. The same node can perform both operations according to whether it knows or not the rumor, which corresponds to the push-pull protocol.

One of the important questions raised by these randomized rumor spreading protocols is the spreading time, that is the time needed for all the nodes of the network to become informed.

Several models have been considered to answer this question. The synchronous model assumes that all the nodes of the network act in synchrony, which allows the algorithms designed in this model to divide time in synchronized rounds. During each synchronized round, each node i of the network selects at random one of its neighbors j and either sends to j the rumor if i knows it (push operation) or gets the rumor from j if j knows the rumor (pull operation). In this model, the spreading time of a rumor is defined as the number of synchronous rounds necessary for all the nodes to become informed. Analyses have been conducted when the underlying communication graph is complete (e.g., [10,11], and in different topologies (e.g., [2,4,9,15]), in the presence of link or nodes failures as in [8], in dynamic graphs as in [3]. Another alternative consists for the nodes to make more than one call during the push-pull operations [16]. In large scale and open networks, assuming that all nodes act in synchrony is a strong assumption since it requires that all the nodes have access to some global synchronization mechanism and that message delays are upper bounded. Several authors, including [1,6,12,14,17], suppose that nodes asynchronously trigger operations with randomly chosen nodes. In this model, the spreading time of a rumor is defined as the number of operations necessary for all the nodes to know the rumor. In [17], the authors model a multiple call by tuning the clock rate of each node with a given probability distribution.

Regarding the type of interaction, the pull algorithm has attracted very little attention because this algorithm was long considered inefficient to spread a rumor within a large scale network [19]. However, it is very useful in systems fighting against message saturation (see for instance [22]). The ineffectiveness of the pull protocol stems from the fact that it takes some time before the rumor reaches a phase of exponential growth. In the line of Panagiotou et al.'s work [16], Robin et al [18] have extended the pull operation with the k-pull operation, which allows an uninformed node to ask for the rumor to a fixed number $k - 1$ of other nodes in parallel.

The objective of this paper is to push further this line of inquiry by presenting and studying the performance of the k-pull operation in presence of a certain fraction f of non collaborative nodes. A non-cooperative node is a node that refuses to learn and thus to propagate the rumor. Our aim is to understand the impact of k on the propagation of the rumor despite the presence of a proportion f of non-collaborative nodes.

The remaining of the paper is organized as follows. Section 2 presents the k-pull protocol in presence of non-cooperative nodes. Section 3 analyses the rumor spreading time when $k = 2$. Section 4 presents a numerical study of the influence of larger values of k on the expected spreading time and on the distribution of the k-pull operation. Finally Sect. 5 presents future works.

2 The k-Pull Protocol in Presence of Non-cooperative Nodes

We consider a system with interacting subpopulations of informed, uninformed, and non-cooperative nodes. The total number of nodes is equal to n. The communication graph among the n nodes is complete. Nodes are anonymous meaning that that they do not use identifiers and thus cannot determine whether any two interactions have occurred with the same nodes or not. However, for ease of presentation the nodes are numbered $1, 2, \ldots, n$. We assume a thoroughly mixed population, so that nodes encounter each other at random, with uniform probability. Initially, a single node knows the rumor and wishes to propagate it (this is not a non-cooperative node) to uninformed nodes. However, among uninformed nodes, a constant fraction f of them are non-cooperative, i.e., they do not want to learn the rumor and thus to propagate it further.

The k-pull protocol is defined as follows. At each discrete time t, a single uninformed node s contacts $k - 1$ distinct nodes, chosen at random uniformly among the $n - 1$ other nodes, and applies the following rule:

– If at least one of the $k - 1$ contacted nodes knows the rumor and s is not non-cooperative then node s becomes informed.

Note that despite the fact that non-cooperative nodes do not want to learn the rumor, they trigger the k-pull operation. Their motivation is to increase the spreading time of the rumor. At any time we suppose that $nf \geq k$, otherwise we come back to Robin et al.'s analysis [18]. Note that in practice, nf is an integer, but this is not necessary for the analysis. The protocol halts by itself once all the $n(1 - f)$ nodes are informed.

2.1 The k-Pull Protocol in Absence of Non-cooperative Nodes

To analyze the k-pull protocol, the authors in [18] have introduced the discrete-time stochastic process $Y = \{Y_t, \ t \geq 0\}$ where Y_t represents the number of informed nodes at time t. Stochastic process Y is a discrete-time homogeneous Markov chain with n states where states $1, \ldots, n-1$ are transient and state n is absorbing. When the Markov chain Y is in state i at time t, then at time $t + 1$, either it remains in state i if none of the $k - 1$ chosen nodes know the rumor or it transits to state $i + 1$ if at least one of the $k - 1$ chosen nodes know the rumor. Let P be the transition probability matrix of Markov chain Y. The non zero entries of matrix P are thus $P_{i,i}$ and $P_{i,i+1}$, for any $i = 1, \ldots, n - 1$. We denote by $T_{k,n}$ the random variable defined by

$$T_{k,n} := \inf\{t \geq 0 \mid Y_t = n\},$$

which represents the spreading time, that is the total number of k-pull operations needed for all the nodes in the network to know the rumor. The spreading time distribution can thus be expressed as a sum of independent random variables $S_{k,n}(i)$, where $S_{k,n}(i)$ is the sojourn time of Markov chain Y in state i. For all

$i = 1, \ldots, n - k$, $S_{k,n}(i)$ follows a geometric distribution with parameter $P_{i,i+1}$ which we denote more simply by $p_{k,n}(i)$. It is shown in [18] that

$$p_{k,n}(i) = 1 - P_{i,i} = 1 - \prod_{h=1}^{k-1} \left(1 - \frac{i}{n-h} \right). \tag{1}$$

2.2 The k-Pull Protocol in Presence of nf Non-cooperative Nodes

We keep the same notation used in the previous subsection. We just suppose that when $f \neq 0$, we have $nf \geq k$. In presence of a proportion f of non-cooperative node, when $Y_t = i$, i.e. when i nodes are informed of the rumor at time t, we have $Y_{t+1} = i$ if and only if, at time $t+1$, either the selected node is a non-cooperative node (with probability $nf/(n-i)$) or the selected node is not a non-cooperative node (with probability $1 - nf/(n-i)$) and the set of $k - 1$ chosen nodes (i.e. $k - 1$ among $n - 1$) must be chosen among the $n - 1 - i$ non informed nodes which corresponds to the situation where $f = 0$. More formally, if M_t denotes the status of the selected node at time t (1 if it is non-cooperative and 0 otherwise), we have, for every $t \geq 0$, using (1)

$$
\begin{aligned}
P_{i,i} &= \mathbb{P}\{Y_{t+1} = i \mid Y_t = i\} \\
&= \mathbb{P}\{Y_{t+1} = i \mid M_t = 1, Y_t = i\}\mathbb{P}\{M_t = 1 \mid Y_t = i\} \\
&\quad + \mathbb{P}\{Y_{t+1} = i \mid M_t = 0, Y_t = i\}\mathbb{P}\{M_t = 0 \mid Y_t = i\} \\
&= \frac{nf}{n-i} + \left(1 - \frac{nf}{n-i} \right) \prod_{h=1}^{k-1} \left(1 - \frac{i}{n-h} \right)
\end{aligned}
$$

Observe that Y has now $n(1 - f)$ states where states $1, \ldots, n(1 - f) - 1$ are transient and state $n(1 - f)$ is absorbing. In the same way, we introduce the notation

$$p_{k,n}(i) = 1 - P_{i,i} = \left(1 - \frac{nf}{n-i} \right) \left(1 - \prod_{h=1}^{k-1} \left(1 - \frac{i}{n-h} \right) \right) \tag{2}$$

and we have

$$T_{k,n} = \sum_{i=1}^{n(1-f)-1} S_{k,n}(i), \tag{3}$$

where $S_{k,n}(i)$ has a geometric distribution with parameter $p_{k,n}(i)$.

It is well-known, see for instance [20], that the distribution of $T_{k,n}$ is given, for every integer $t \geq 0$, by

$$\mathbb{P}\{T_{k,n} > t\} = \alpha Q^t \mathbb{1}, \tag{4}$$

where α is the row vector containing the initial probabilities of states $1, \ldots, n(1 - f) - 1$, that is $\alpha_i = \mathbb{P}\{Y_0 = i\} = \mathbb{1}_{\{i=1\}}$, Q is the matrix obtained from the transition matrix P by only keeping the transition probabilities between transient states, i.e. by removing the last line and the last column of P and $\mathbb{1}$ is the column vector of dimension $n(1 - f) - 1$ with all its entries equal to 1.

3   Analysis of the Rumor Spreading Time When $k = 2$

In this section, we analyze the two first moments and the distribution of the rumor spreading time $T_{2,n}$ and their asymptotic behavior when n goes to infinity. We denote by H_n the harmonic series defined, for every $n \geq 1$, by $H_n = \sum_{i=1}^{n} 1/i$ and we recall that the Euler-Mascheroni constant γ is given by $\gamma = \lim_{n \to \infty} (H_n - \ln(n))$, which is approximately equal to 0.5772156649.

3.1   Asymptotic Mean and Variance of the Rumor Spreading Time

In the case where $k = 2$, we have from relation (2),

$$p_{2,n}(i) = \left(1 - \frac{nf}{n-i}\right) \frac{i}{n-1} = \frac{(n(1-f)-i)\,i}{(n-1)(n-i)}. \tag{5}$$

The asymptotic expected rumor spreading time is obtained in the following theorem.

Theorem 1.
$$\mathbb{E}(T_{2,n}) \underset{n \to \infty}{\sim} \frac{(1+f)n\ln(n)}{1-f}$$

and

$$\lim_{n \to \infty} \left(\frac{\mathbb{E}(T_{2,n})}{n} - \frac{(1+f)\ln(n)}{1-f}\right) = \frac{(1+f)\,(\gamma + \ln(1-f))}{1-f}.$$

Proof. The expected value of the spreading time $T_{2,n}$ is given, using Relation (3), by

$$\mathbb{E}(T_{2,n}) = \sum_{i=1}^{n(1-f)-1} \frac{1}{p_{2,n}(i)} = (n-1) \sum_{i=1}^{n(1-f)-1} \frac{n-i}{(n(1-f)-i)\,i}$$

$$= (n-1)\left[\sum_{i=1}^{n(1-f)-1} \frac{1}{i} + nf \sum_{i=1}^{n(1-f)-1} \frac{1}{(n(1-f)-i)\,i}\right].$$

Observing that

$$\frac{1}{(n(1-f)-i)\,i} = \frac{1}{n(1-f)} \left(\frac{1}{i} + \frac{1}{n(1-f)-i}\right), \tag{6}$$

we obtain

$$\mathbb{E}(T_{2,n}) = (n-1)\left[\sum_{i=1}^{n(1-f)-1} \frac{1}{i} + \frac{2f}{1-f} \sum_{i=1}^{n(1-f)-1} \frac{1}{i}\right]$$

$$= \frac{(1+f)(n-1)H_{n(1-f)-1}}{1-f}$$

$$\underset{n \to \infty}{\sim} \frac{(1+f)n\ln(n)}{1-f}. \tag{7}$$

Moreover, we have

$$
\begin{aligned}
\frac{\mathbb{E}(T_{2,n})}{n} - \frac{(1+f)\ln(n)}{1-f} &= \frac{1+f}{1-f}\left[\frac{(n-1)H_{n(1-f)-1}}{n} - \ln(n)\right] \\
&= \frac{1+f}{1-f}\left[H_{n(1-f)-1} - \ln(n) - \frac{H_{n(1-f)-1}}{n}\right] \\
&= \frac{1+f}{1-f}\left[H_{n(1-f)-1} - \ln(n(1-f)-1)\right] \\
&\quad + \frac{1+f}{1-f}\left[\ln(n(1-f)-1) - \ln(n) - \frac{H_{n(1-f)-1}}{n}\right] \\
&= \frac{1+f}{1-f}\left[H_{n(1-f)-1} - \ln(n(1-f)-1)\right] \\
&\quad + \frac{1+f}{1-f}\left[\ln(1-f-1/n) - \frac{H_{n(1-f)-1}}{n}\right].
\end{aligned}
$$

The second term in square brackets tends to $\ln(1-f)$ when n tends to infinity, so by definition of γ we have

$$
\lim_{n \longrightarrow \infty}\left(\frac{\mathbb{E}(T_{2,n})}{n} - \frac{(1+f)\ln(n)}{1-f}\right) = \frac{(1+f)\,(\gamma + \ln(1-f))}{1-f},
$$

which completes the proof.

We consider now the variance of $T_{2,n}$ and its limiting value when n goes to infinity.

Theorem 2.

$$
\mathbb{V}(T_{2,n}) \underset{n \longrightarrow \infty}{\sim} \frac{(1+f^2)\pi^2 n^2}{6(1-f)^2},
$$

Proof. Using Relation (3), the variance of $T_{2,n}$ is given by

$$
\begin{aligned}
\mathbb{V}(T_{2,n}) &= \sum_{i=1}^{n(1-f)-1}\frac{1 - p_{2,n}(i)}{(p_{2,n}(i))^2} \\
&= (n-1)^2 \sum_{i=1}^{n(1-f)-1}\left(\frac{n-i}{i\,(n(1-f)-i)}\right)^2 - \mathbb{E}(T_{2,n}). \quad (8)
\end{aligned}
$$

Using relation (6), we write

$$
\begin{aligned}
\frac{n-i}{i\,(n(1-f)-i)} &= \frac{1}{i} + \frac{nf}{i\,(n(1-f)-i)} = \frac{1}{i} + \frac{f}{1-f}\left(\frac{1}{i} + \frac{1}{n(1-f)-i}\right) \\
&= \left(\frac{1}{1-f}\right)\frac{1}{i} + \left(\frac{f}{1-f}\right)\frac{1}{n(1-f)-i}
\end{aligned}
$$

and thus

$$
\left(\frac{n-i}{i\,(n(1-f)-i)}\right)^2 = \frac{1}{(1-f)^2}\left(\frac{1}{i^2} + \frac{f^2}{(n(1-f)-i)^2} + \frac{2f}{i\,(n(1-f)-i)}\right).
$$

Using again relation (6) for the third term of this last expression, we get

$$\left(\frac{n-i}{i\,(n(1-f)-i)}\right)^2 = \frac{1}{(1-f)^2}\left(\frac{1}{i^2}+\frac{f^2}{(n(1-f)-i)^2}\right)$$
$$+\frac{2f}{(1-f)^3 n}\left(\frac{1}{i}+\frac{1}{n(1-f)-i}\right).$$

It follows that

$$\sum_{i=1}^{n(1-f)-1}\left(\frac{n-i}{i\,(n(1-f)-i)}\right)^2 = \frac{1+f^2}{(1-f)^2}\sum_{i=1}^{n(1-f)-1}\frac{1}{i^2}+\frac{4f}{(1-f)^3 n}\sum_{i=1}^{n(1-f)-1}\frac{1}{i}.$$

Inserting this result in (8), we obtain using (7)

$$\mathbb{V}(T_{2,n}) = \frac{(1+f^2)(n-1)^2}{(1-f)^2}\sum_{i=1}^{n(1-f)-1}\frac{1}{i^2}$$
$$+\left(\frac{4f(n-1)}{(1-f)^3 n}-\frac{(1+f)}{1-f}\right)(n-1)H_{n(1-f)-1}.$$

The second term of this sum is in $O(n\ln(n))$, thus

$$\mathbb{V}(T_{2,n}) \underset{n\to\infty}{\sim} \frac{(1+f^2)\pi^2 n^2}{6(1-f)^2},$$

which completes the proof.　　∎

3.2　Asymptotic Distribution of the Rumor Spreading Time

This section provides the explicit limiting distribution of $(T_{2,n}-\mathbb{E}(T_{2,n}))/n$ when n tends to infinity. To prove the main result of this section, we need the following lemma.

Lemma 1.

$$\lim_{m\to\infty}\limsup_{\ell\to\infty}\frac{(1-f)^2}{(2\ell+1)^2}\sum_{i=m}^{\ell}\frac{1}{p_{2,(2\ell+1)/(1-f)}^2(i)} = 0.$$

and

$$\lim_{m\to\infty}\limsup_{\ell\to\infty}\frac{(1-f)^2}{(2\ell+1)^2}\sum_{i=m}^{\ell}\frac{1}{p_{k,(2\ell+1)/(1-f)}^2(2\ell+1-i)} = 0.$$

Proof. From (5), we have, by taking $n=(2\ell+1)/(1-f)$,

$$p_{2,(2\ell+1)/(1-f)}(i) = \frac{(2\ell+1-i)\,i}{\left(\dfrac{2\ell+1}{1-f}-1\right)\left(\dfrac{2\ell+1}{1-f}-i\right)}.$$

Introducing the notation

$$\Delta_{\ell,m} = \frac{(1-f)^2}{(2\ell+1)^2} \sum_{i=m}^{\ell} \frac{1}{p_{2,(2\ell+1)/(1-f)}^2(i)}$$

we obtain, after some algebra, and since $f \leq 1$,

$$\Delta_{\ell,m} = \frac{(2\ell+f)^2}{(1-f)^2(2\ell+1)^2} \sum_{i=m}^{\ell} \left(\frac{1}{i} + \frac{f}{2\ell+1-i}\right)^2$$

$$\leq \frac{1}{(1-f)^2} \sum_{i=m}^{\ell} \left(\frac{1}{i} + \frac{1}{2\ell+1-i}\right)^2 \tag{9}$$

$$= \frac{1}{(1-f)^2} \left(\sum_{i=m}^{2\ell+1-m} \frac{1}{i^2} + \frac{2}{2\ell+1} \sum_{i=m}^{2\ell+1-m} \frac{1}{i}\right)$$

$$\leq \frac{1}{(1-f)^2} \left(\sum_{i=m}^{2\ell+1-m} \frac{1}{i^2} + \frac{2(1+\ln(2\ell+1-m))}{2\ell+1}\right).$$

The $\lim_{m \to \infty} \limsup_{\ell \to \infty}$ of both terms is 0 because $\sum_{i \geq 1} 1/i^2$ is a converging series. This proves the first relation.

Concerning the second one, from (5), we have, by taking $n = (2\ell+1)/(1-f)$,

$$p_{2,(2\ell+1)/(1-f)}(2\ell+1-i) = \frac{i(2\ell+1-i)}{\left(\frac{2\ell+1}{1-f}-1\right)\left(\frac{2\ell+1}{1-f}-(2\ell+1-i)\right)}.$$

Introducing the notation

$$\Lambda_{\ell,m} = \frac{(1-f)^2}{(2\ell+1)^2} \sum_{i=m}^{\ell} \frac{1}{p_{k,(2\ell+1)/(1-f)}^2(2\ell+1-i)}$$

we obtain as we did for $\Delta_{\ell,m}$,

$$\Lambda_{\ell,m} = \frac{(2\ell+f)^2}{(1-f)^2(2\ell+1)^2} \sum_{i=m}^{\ell} \left(\frac{f}{i} + \frac{1}{2\ell+1-i}\right)^2$$

$$\leq \frac{1}{(1-f)^2} \sum_{i=m}^{\ell} \left(\frac{1}{i} + \frac{1}{2\ell+1-i}\right)^2,$$

which is exactly (9). This completes the proof. ∎

We are now able to prove the following theorem, where $\xrightarrow{\mathcal{L}}$ means the convergence in law.

Theorem 3. *Let $(Z_i)_{i\geq 1}$ be a sequence of i.i.d. random variables exponentially distributed with rate 1 and let W be defined by*

$$W = \sum_{i=1}^{\infty} \frac{Z_i - 1}{i}.$$

We then have

$$\frac{T_{2,n} - \mathbb{E}(T_{2,n})}{n} \xrightarrow{\mathcal{L}} \frac{1}{1-f}W^{(1)} + \frac{f}{1-f}W^{(2)} \ as \ n \longrightarrow \infty$$

where $W^{(1)}$ and $W^{(2)}$ are i.i.d. with the same distribution as W.

Proof. For a fixed value of i, we have $\lim_{n\longrightarrow\infty} p_{2,n}(i) = 0$. It follows that for every $x \geq 0$, we have

$$\mathbb{P}\{p_{2,n}(i)S_{k,n} > x\} = \mathbb{P}\{S_{2,n} > x/p_{2,n}(i) > x\} = (1 - p_{2,n}(i))^{\lfloor x/p_{k,n}(i)\rfloor}$$

which tends to e^{-x} when n tends to infinity, since $p_{2,n}(i)$ tends to 0. If Z_i is a random variable exponentially distributed with rate 1, we have shown that $p_{2,n}(i)S_{2,n} \xrightarrow{\mathcal{L}} Z_i$ when $n \longrightarrow \infty$. Moreover since the $(S_{2,n}(i))_{i=1,\ldots,n(1-f)-1}$ are independent, the $(Z_i)_{i=1,\ldots,n(1-f)-1}$ are also independent. In the same way, we have $\lim_{n\longrightarrow\infty} np_{2,n}(i) = (1-f)i$. Defining $R_{2,n}(i) = S_{2,n}(i) - \mathbb{E}(S_{2,n}(i))$ we obtain, since $\mathbb{E}(S_{2,n}(i)) = 1/p_{2,n}(i)$,

$$\frac{R_{2,n}(i)}{n} = \frac{S_{2,n}(i) - \mathbb{E}(S_{2,n}(i))}{n} = \frac{p_{2,n}(i)S_{2,n}(i) - 1}{np_{2,n}(i)} \xrightarrow{\mathcal{L}} \frac{Z_i - 1}{(1-f)i}. \quad (10)$$

In the same way, replacing i by $n(1-f) - i$ in (2) leads to

$$p_{2,n}(n(1-f) - i) = \frac{i(n(1-f) - i)}{(n-1)(nf + i)}.$$

It follows that

$$\lim_{n\longrightarrow\infty} p_{2,n}(n(1-f) - i) = 0 \text{ and } \lim_{n\longrightarrow\infty} np_{2,n}(n(1-f) - i) = \frac{i(1-f)}{f}$$

and thus

$$\frac{R_{2,n}(n(1-f) - i)}{n} = \frac{p_{2,n}(n(1-f) - i)S_{2,n}(n(1-f) - i) - 1}{np_{2,n}(n(1-f) - i)} \xrightarrow{\mathcal{L}} \frac{(Z_i - 1)f}{(1-f)i}. \quad (11)$$

Suppose that $n(1-f)$ is odd, i.e. $n(1-f) = 2\ell + 1$. We then have from (3)

$$\frac{T_{2,n} - \mathbb{E}(T_{2,n})}{n} = \frac{1-f}{2\ell + 1} \left(\sum_{i=1}^{\ell} R_{2,n}(i) + \sum_{i=\ell+1}^{2\ell} R_{2,n}(i) \right)$$

$$= \frac{1-f}{2\ell + 1} \left(\sum_{i=1}^{\ell} R_{2,n}(i) + \sum_{i=1}^{\ell} R_{2,n}(2\ell + 1 - i) \right)$$

$$= V_\ell + \overline{V}_\ell, \quad (12)$$

where

$$V_\ell = \frac{1-f}{2\ell+1} \sum_{i=1}^{\ell} R_{2,n}(i) \text{ and } \overline{V}_\ell = \frac{1-f}{2\ell+1} \sum_{i=1}^{\ell} R_{2,n}(2\ell+1-i).$$

Observe that the random variables $V_{2,\ell}$ and $\overline{V}_{2,\ell}$ are independent. The rest of the proof consists in checking the hypothesis of the principle of accompanying laws of Theorem 3.1.14 of [21]. We introduce the notation

$$V_{\ell,m} = \frac{1-f}{2\ell+1} \sum_{i=1}^{m-1} R_{2,n}(i) \text{ and } \overline{V}_{\ell,m} = \frac{1-f}{2\ell+1} \sum_{i=1}^{m-1} R_{2,n}(2\ell+1-i).$$

Using the fact that $\mathbb{E}(R_{2,n}(i)) = 0$ and the $R_{2,n}(i)$ are independent, we have

$$\mathbb{E}\left((V_\ell - V_{\ell,m})^2\right) = \mathbb{V}\left(\frac{1-f}{2\ell+1} \sum_{i=m}^{\ell} R_{2,n}(i)\right)$$

$$= \frac{(1-f)^2}{(2\ell+1)^2} \sum_{i=m}^{\ell} \mathbb{V}(R_{2,n}(i)) = \frac{(1-f)^2}{(2\ell+1)^2} \sum_{i=m}^{\ell} \mathbb{V}(S_{2,n}(i))$$

$$= \frac{(1-f)^2}{(2\ell+1)^2} \sum_{i=m}^{\ell} \frac{1 - p_{2,(2\ell+1)/(1-f)}(i)}{p_{2,(2\ell+1)/(1-f)}^2(i)} \leq \frac{(1-f)^2}{(2\ell+1)^2} \sum_{i=m}^{\ell} \frac{1}{p_{2,(2\ell+1)/(1-f)}^2(i)}$$

and, in the same way,

$$\mathbb{E}\left((\overline{V}_\ell - \overline{V}_{\ell,m})^2\right) \leq \frac{(1-f)^2}{(2\ell+1)^2} \sum_{i=m}^{\ell} \frac{1}{p_{2,(2\ell+1)/(1-f)}^2(2\ell+1-i)}.$$

Using Lemma 1, we have

$$\lim_{m \to \infty} \limsup_{\ell \to \infty} \mathbb{E}((V_\ell - V_{\ell,m})^2) = \lim_{m \to \infty} \limsup_{\ell \to \infty} \mathbb{E}((\overline{V}_\ell - \overline{V}_{\ell,m})^2) = 0.$$

Using now the Markov inequality, we obtain, for all $\varepsilon > 0$,

$$\mathbb{P}\{|V_\ell - V_{\ell,m}| \geq \varepsilon\} = \mathbb{P}\{(V_\ell - V_{\ell,m})^2 \geq \varepsilon^2\} \leq \frac{\mathbb{E}((V_\ell - V_{\ell,m})^2)}{\varepsilon^2}$$

and

$$\mathbb{P}\{|\overline{V}_\ell - \overline{V}_{\ell,m}| \geq \varepsilon\} = \mathbb{P}\{(\overline{V}_\ell - \overline{V}_{\ell,m})^2 \geq \varepsilon^2\} \leq \frac{\mathbb{E}((\overline{V}_\ell - \overline{V}_{\ell,m})^2)}{\varepsilon^2}.$$

Putting together these results, we deduce that for all $\varepsilon > 0$, we have

$$\lim_{m \to \infty} \limsup_{\ell \to \infty} \mathbb{P}\{|V_\ell - V_{\ell,m}| \geq \varepsilon\} = 0 \tag{13}$$

$$\lim_{m \to \infty} \limsup_{\ell \to \infty} \mathbb{P}\{|\overline{V}_\ell - \overline{V}_{\ell,m}| \geq \varepsilon\} = 0. \tag{14}$$

Let us introduce the notation

$$W_m = \frac{1}{1-f}\sum_{i=1}^{m-1}\frac{Z_i-1}{i} \text{ and } \overline{W}_m = \frac{f}{1-f}\sum_{i=1}^{m-1}\frac{Z_i-1}{i}.$$

Using (10) and (11) and the fact that the $R_{k,n}(i)$ are independent, we have

$$V_{\ell,m} \xrightarrow{\mathcal{L}} W_m \text{ and } \overline{V}_{\ell,m} \xrightarrow{\mathcal{L}} \overline{W}_m \text{ as } \ell \longrightarrow \infty. \tag{15}$$

The hypothesis of the principle of accompanying laws of Theorem 3.1.14 of [21] are properties (13), (14) and (15). We can thus conclude that

$$V_\ell \xrightarrow{\mathcal{L}} \frac{1}{1-f}W \text{ and } \overline{V}_\ell \xrightarrow{\mathcal{L}} \frac{f}{1-f}W \text{ as } \ell \longrightarrow \infty.$$

This means, from relation (12), that

$$\frac{T_{2,n}-\mathbb{E}(T_{2,n})}{n} \xrightarrow{\mathcal{L}} \frac{1}{1-f}W^{(1)} + \frac{f}{1-f}W^{(2)} \text{ as } n \longrightarrow \infty,$$

where $W^{(1)}$ and $W^{(2)}$ are independent and identically distributed as W. The same reasoning applies in the case where $n(1-f)=2\ell$. ∎

Corollary 1. *For all $x \in \mathbb{R}$ we have*

$$\lim_{n\to\infty}\mathbb{P}\left\{\frac{T_{2,n}-\mathbb{E}(T_{2,n})}{n}\le x\right\} = \int_0^\infty \exp\left(-t-t^{-f}e^{-(1-f)x-\gamma(1+f)}\right)dt.$$

Proof. L. Gordon has proved in [13] that

$$-\gamma + \sum_{i=1}^{+\infty}\frac{1-Z_i}{i} \stackrel{\mathcal{L}}{=} \ln(Z_1),$$

where (Z_i) are i.i.d. exponential with rate 1. Thus, by definition of W in Theorem 3, we have

$$W \stackrel{\mathcal{L}}{=} -\gamma - \ln(Z_1).$$

Introducing $W^{(1)} \stackrel{\mathcal{L}}{=} -\gamma - \ln(Z_1)$ and $W^{(2)} \stackrel{\mathcal{L}}{=} -\gamma - \ln(Z_2)$, we obtain from Theorem 3, for all $x \in \mathbb{R}$,

$$\begin{aligned}\lim_{n\to\infty}\mathbb{P}\left\{\frac{T_{2,n}-\mathbb{E}(T_{2,n})}{n}\le x\right\} &= \mathbb{P}\left\{\frac{1}{1-f}W^{(1)}+\frac{f}{1-f}W^{(2)}\le x\right\}\\ &= \mathbb{P}\{-\ln(Z_1)-f\ln(Z_2)\le(1-f)x+\gamma(1+f)\}\\ &= \mathbb{P}\left\{Z_1Z_2^f \ge e^{-(1-f)x-\gamma(1+f)}\right\}\\ &= \int_0^\infty \mathbb{P}\left\{Z_1\ge t^{-f}e^{-(1-f)x-\gamma(1+f)}\right\}e^{-t}dt\\ &= \int_0^\infty \exp\left(-t-t^{-f}e^{-(1-f)x-\gamma(1+f)}\right)dt,\end{aligned}$$

which completes the proof. ∎

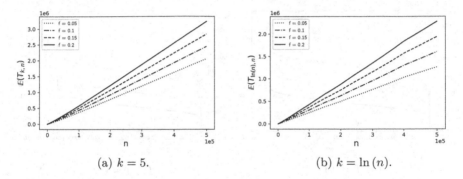

Fig. 1. Expected spreading time $T_{k,n}$ as a function of the number n of nodes in the system for different values of f, when $k = 5$ on the left, and $k = \ln(n)$ on the right.

4   General Case

Generalizing to higher values of k has shown to be more difficult. Indeed, as a node can interact with a parameterized number of other nodes, its behavior during an interaction becomes harder to predict. This can be observed in (1) where k has an influence on the number of terms in the product of the relation. Consequently, we numerically study the influence of larger values of k on the expected spreading time and on its distribution.

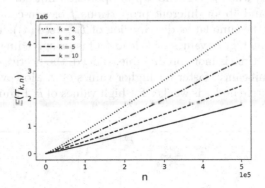

Fig. 2. Expected spreading time $T_{k,n}$ as a function of the number n of nodes in the system for different values of k when $f = 0.1$.

As expected, Fig. 1a shows that the mean spreading time increases with the proportion f. A realistic assumption would be to have k as a function of n. Figure 1b shows that having k as a function of n does not provide significant improvement for small values of n. More interestingly, k has a significant influence on the spreading time, independently from the size of the network, which is illustrated by Fig. 2. Using a k-pull operation helps to mitigate the influence of non-cooperative node compared to a regular asynchronous pull protocol ($k = 2$).

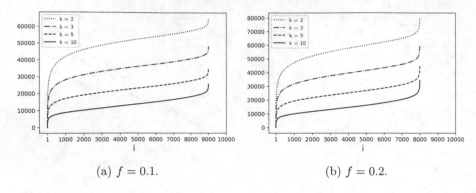

(a) $f = 0.1$.   (b) $f = 0.2$.

Fig. 3. $\sum_{j=1}^{i} \mathbb{E}(T_{k,j})$: Expected spreading time to reach i informed nodes, as a function of i, with $n = 10,000$ nodes.

One of the major caveats of the pull operation is that the diffusion of the rumor during the first interactions is very slow. This phenomenon is made worse with the presence of non-cooperative nodes. This is due to the fact that the few informed and cooperative nodes need to be selected to be able to propagate the rumor, as opposed to the push operation. This is highlighted in Fig. 3a for $k = 2$, where we can see that the propagation is quite slow during the first interactions. On the other hand, when $k > 2$, the k-pull operation mitigates this problem as shown in Figs. 3a and 3b for different proportions f of non-cooperative nodes.

Figure 4 shows the cumulative distribution of $T_{k,n}$ (see (4)). This distribution shows not only that higher values of k lead to faster spreading times, but also that this spreading time is more predictable. Indeed, the decrease of the function $\mathbb{P}\{T_{k,n} > t\}$ is significantly faster for higher values of k. Moreover, the influence of the non-cooperative nodes is weaker for high values of k. Figure 4a and Fig. 4b

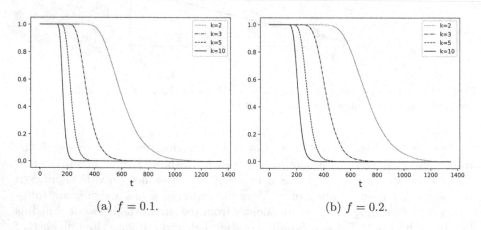

(a) $f = 0.1$.   (b) $f = 0.2$.

Fig. 4. $\mathbb{P}\{T_{k,n} > t\}$ with $n = 100$ nodes.

Table 1. Values of $t = \inf\{t \geq 0 \mid \mathbb{P}\{T_{k,n} > t\} < \varepsilon\}$ for different values of k, f and ε with $n = 100$ nodes.

k	$\varepsilon = 0.1$		$\varepsilon = 0.01$		$\varepsilon = 0.001$	
	$f = 0.1$	$f = 0.2$	$f = 0.1$	$f = 0.2$	$f = 0.1$	$f = 0.2$
$k = 2$	790	928	1020	1099	1131	1146
$k = 3$	448	536	582	689	699	829
$k = 5$	283	347	345	426	410	496
$k = 10$	200	258	232	313	264	354

provide this intuition, which is confirmed in Table 1. This table gives the smallest value of t such that $\mathbb{P}\{T_{k,n} > t\} < \varepsilon$.

5   Discussion

In this paper, we have considered the presence of non-cooperative nodes. Such nodes have an impact on the spreading time of a rumor, but do not endanger the content of the rumor. We are currently studying the impact of a proportion f of malicious nodes whose objective is to modify the rumor and propagate this modified rumor further. Concretely, when a malicious node is chosen during a k-pull operation, if this node knows the rumor, it will modify it and send it to the initiator of the k-pull operation. The objective is to analyze the necessary (and sufficient) conditions for the propagation of the initial rumor in presence of a fraction f of malicious nodes. The value of k should be predominant to enable the initiator of a k-pull operation to choose which rumor to learn during a k-pull operation.

References

1. Acan, H., Collevecchio, A., Mehrabian, A., Wormald, N.: On the push & pull protocol for rumour spreading. Trends Math. **6**, 3–10 (2017)
2. Chierichetti, F., Lattanzi, S., Panconesi, A.: Rumor spreading in social networks. Theor. Comput. Sci. **412**(24), 2602–2610 (2011)
3. Clementi, A., Crescenzi, P., Doerr, C., Fraigniaud, P., Pasquale, F., Silvestri, R.: Rumor spreading in random evolving graphs. Random Struct. Algorithms **48**(2), 290–312 (2015)
4. Daum, S., Kuhn, F., Maus, Y.: Rumor spreading with bounded indegree. In: Proceedings of the International Colloquium on Structural Information and Communication Complexity (SIROCCO) (2016)
5. Demers, A.J., et al.: Epidemic algorithms for replicated database maintenance. PODC **87**, 1–12 (1987)
6. Doerr, B., Fouz, M., Friedrich, T.: Experimental analysis of rumor spreading in social networks. MedAlg **2012**, 159–173 (2012)

48   S. Kilian et al.

7. Doerr, B., Kostrygin, A.: Randomized rumor spreading revisited. In: Chatzigian-nakis, I., Indyk, P., Kuhn, F., Muscholl, A. (eds.), ICALP 2017, vol. 80, pp. 138:1–138:14. Schloss Dagstuhl-Leibniz-Zentrum fuer Informatik (2017)
8. Feige, F., Peleg, D., Raghavan, P., Upfal, E.: Randomized broadcast in networks. Random Struct. Algorithms 1(4), 447–460 (1990)
9. Fountoulakis, N., Panagiotou, K.: Rumor spreading on random regular graphs and expanders. Random Struct. Algorithms 43(2), 201–220 (2013)
10. Frieze, A., Grimmet, G.: The shortest-path problem for graphs with random arc-lengths. Discret. Appl. Math. 10(1), 57–77 (1985)
11. Giakkoupis, G.: Tight bounds for rumor spreading in graphs of a given conduc-tance. In: Proceedings of the International Symposium on Theoretical Aspects of Computer Science (STACS) (2011)
12. Giakkoupis, G., Nazari, Y., Woelfel, P.: How asynchrony affects rumor spreading time. In: PODC '16, pp. 185–194 (2016)
13. Gordon, L.: Bounds for the distribution of the generalized variance. Ann. Stat. 17(4), 1684–1692 (1989)
14. Mocquard, Y., Robert, S., Sericola, B., Anceaume, E.: Analysis of the propagation time of a rumour in large-scale distributed systems. In: NCA 2016 (2016)
15. Panagiotou, K., Perez-Gimenez, X., Sauerwald, T., Sun, H.: Randomized rumor spreading: the effect of the network topology. Comb. Probab. Comput. 24(2), 457–479 (2015)
16. Panagiotou, K., Pourmiri, A., Sauerwald, T.: Faster rumor spreading with multiple calls. Electron. J. Comb. 22 (2015)
17. Pourmiri, A., Ramezani, F.: Brief announcement: ultra-fast asynchronous random-ized rumor spreading. SPAA 2019 (2019)
18. Robin, F., Sericola, B., Anceaume, E., Mocquard, Y.: Stochastic analysis of rumor spreading with multiple pull operations. Methodol. Comput. Appl. Probab. 24, 2195–2211 (2022)
19. Sanghavi, S., Hajek, B., Massoulié, L.: Gossiping with multiple messages. IEEE Trans. Inf. Theory 53(123) (2007)
20. Sericola, B.: Markov Chains. Theory, Algorithms and Applications. John Wiley & Sons, Hoboken (2013)
21. Stroock, D.W.: Probability Theory: An Analytic View, second edition. Cambridge University Press, Cambridge (2010)
22. Yao, G., Bi, J., Wang, S., Zhang, Y., Li, Y.: A pull model IPv6 duplicate address detection. In: LCN 2010 (2010)

Analysis of a Two-State Markov Fluid Model with 2 Buffers

Peter Buchholz[1] (ID), Meszaros Andras[2,3] (ID), and Miklos Telek[2,3](✉) (ID)

[1] Technische Universität Dortmund, Dortmund, Germany
peter.buchholz@cs.tu-dortmund.de
[2] Department of Networked Systems and Services,
Budapest University of Technology and Economics, Budapest, Hungary
[3] ELKH-BME Information Systems Research Group, Budapest, Hungary
{meszarosa,telek}@hit.bme.hu

Abstract. Single buffer Markov fluid models are well understood in the literature, but the extension of those results for multiple buffers is still an open research problem. In this paper we consider one of the simplest Markov fluid models (MFM) with 2 buffers of infinite capacity, where the fluid rates ensure that the fluid level of buffer 1 is never larger than fluid level of buffer 2. In spite of these restrictions, the stationary analysis is non straightforward with the available analysis tools. We provide an analysis approach based on the embedded time points at the busy-idle cycles of buffer 1.

Keywords: Markov fluid model with 2 buffers · embedded process · Laplace transform

1 Introduction

In the 1980's, the evolution of the telecommunication systems turned the attention of traffic engineers towards queueing models with "continuous" buffers, which are commonly referred to as fluid queues [2]. The associated fertile research effort resulted in many solution techniques for such queues from spectral decomposition based ones [7] to matrix analytic method based ones [8] by the 2010's, the computational methods for various performance measures of fluid models have also been enhanced [3,9].

During this evolution of single buffer fluid models and related solution techniques, many practical problems popped up where multiple fluid buffers are present, but the solution of such models is not available in general. In some special cases, e.g., when one of the buffer is allowed to be negative as well, promising analytical approaches are proposed [4,5], but results are still not available for one of the simplest fluid models, a system which has two infinite buffers whose levels

---

This work is partially supported by the Hungarian Scientific Research Fund OTKA K-138208 project.

M. Iacono et al. (Eds.): EPEW/ASMTA 2023, LNCS 14231, pp. 49–64, 2023.
https://doi.org/10.1007/978-3-031-43185-2_4

restricted to be non-negative and ita fluid rates are modulated by a background Markov chain.

In order to pave the road for the analysis of fluid models with 2 buffers, in this paper, we focus on a rather simple fluid model with two infinite buffers, whose levels are restricted to be non-negative, the fluid rates are such that the fluid level of buffer 1 is never larger than the level of buffer 2 and the background Markov chain has two states.

The rest of the paper is organized as follows. Section 2 introduces the considered fluid model and Sect. 3 presents its analysis at embedded time points. Based on the embedded measures, Sect. 4 provides the time stationary measures. Section 5 provides a numerical example whose results are compared with simulation results. Section 6 concludes the paper.

2   Model Description

We consider a system with 2 fluid buffers of infinite capacity whose fluid level is governed by a two-state background Markov chain (BMC). Let $X_1(t)$, $X_2(t)$ and $\phi(t)$ be the fluid level in buffer 1, buffer 2 and the state of the BMC, respectively. The state space of the BMC is composed of a state with positive fluid rates, $\mathcal{S}_+ = \{1\}$ and a state with negative fluid rates, $\mathcal{S}_- = \{2\}$. The generator matrix of the BMC is

$$\mathbf{Q} = \begin{bmatrix} -\lambda & \lambda \\ \mu & -\mu \end{bmatrix}. \tag{1}$$

The fluid accumulation is such that

$$\frac{d}{dt}X_1(t) = \frac{d}{dt}X_2(t) = 1 \text{ when } \phi(t) \in \mathcal{S}_+$$

and for $i = \{1, 2\}$

$$\frac{d}{dt}X_i(t) = -r_i < 0 \text{ when } \phi(t) \in \mathcal{S}_- \ \& \ X_i(t) > 0 \quad \text{and}$$

$$\frac{d}{dt}X_i(t) = 0 \text{ when } \phi(t) \in \mathcal{S}_- \ \& \ X_i(t) = 0$$

where $r_1 > r_2 > 0$. As a consequence $X_1(t) \leq X_2(t)$ holds for $\forall t > 0$ if $X_1(0) \leq X_2(0)$, which we assume in the paper. A trajectory of the system evolution is depicted in Fig. 1.

Our final goal is to compute the following time stationary measures

$$\tilde{W}_+(x, y) = \lim_{t \to \infty} Pr(\phi(t) \in \mathcal{S}_+, X_1(t) < x, X_2(t) < y)$$

$$\tilde{W}_-(x, y) = \lim_{t \to \infty} Pr(\phi(t) \in \mathcal{S}_-, X_1(t) < x, X_2(t) < y),$$

$$\tilde{V}(x) = \lim_{t \to \infty} Pr(\phi(t) \in \mathcal{S}_+, X_1(t) = X_2(t) < x),$$

$$\tilde{U}(y) = \lim_{t \to \infty} Pr(\phi(t) \in \mathcal{S}_-, X_1(t) = 0, X_2(t) < y),$$

$$P = \lim_{t \to \infty} Pr(\phi(t) \in \mathcal{S}_-, X_1(t) = X_2(t) = 0),$$

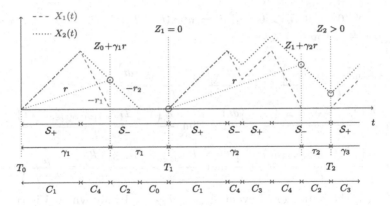

Fig. 1. Evolution of the buffer contents during the busy-idle cycles of buffer 1

and $W_+(x,y) = \frac{d}{dx}\frac{d}{dy}\tilde{W}_+(x,y)$, $W_-(x,y) = \frac{d}{dx}\frac{d}{dy}\tilde{W}_-(x,y)$, $V(x,y) = \frac{d}{dx}\tilde{V}(x)$, $U(x,y) = \frac{d}{dy}\tilde{U}(y)$, for $x,y > 0$. Due to the sign of the fluid rates the other measures are zero, e.g., $\lim_{t\to\infty} Pr(\phi(t) \in \mathcal{S}_+, X_1(t) = 0, X_2(t) = 0) = 0$.

3   Performance Analysis of the Fluid Model

3.1   Utilization of the Buffers and Stability Condition

The stationary probabilities of the BMC are $\lim_{t\to\infty} Pr(\phi(t) \in \mathcal{S}_+) = \frac{\mu}{\lambda+\mu}$ and $\lim_{t\to\infty} Pr(\phi(t) \in \mathcal{S}_-) = \frac{\lambda}{\lambda+\mu}$.

Assuming $X_1(0) = 0$, $\phi(0) \in \mathcal{S}_+$, we define the length of the busy period of buffer 1 (while the buffer is non-empty) as $\gamma = \min(t > 0 : X_1(t) = 0)$.

Theorem 1. *During the γ long busy period of buffer 1 the time spent in $\mathcal{S}_+$ and $\mathcal{S}_-$ are $\frac{r_1\gamma}{1+r_1}$ and $\frac{\gamma}{1+r_1}$.*

Proof. Since $X_1(0) = X_1(\gamma) = 0$, the fluid increase during the sojourn in $\mathcal{S}_+$ equals to the fluid decrease during the sojourn in $\mathcal{S}_-$ and $\frac{r_1\gamma}{1+r_1}(1) = \frac{\gamma}{1+r_1}(r_1)$.

The utilization of the buffer 1 is the stationary probability that the buffer is non-empty, that is $\rho_1 = \lim_{t\to\infty} Pr(X_1(t) > 0)$.

Theorem 2. *The utilization of the buffer 1 is $\rho_1 = \frac{\mu(1+\frac{1}{r_1})}{\lambda+\mu}$.*

Proof. Let $X_1(0) = 0$, $T > 0$ and $\check{T} = \max(t < T : X_1(t) = 0)$. For a stable queue, $\lim_{T \to \infty} \frac{\check{T}}{T} = 1$ and this way

$$
\lim_{T \to \infty} \frac{E\left(\int_{t=0}^{T} \mathcal{I}_{\{X_1(t)>0\}} dt\right)}{T} = \lim_{T \to \infty} \frac{E\left(\int_{t=0}^{\check{T}} \mathcal{I}_{\{X_1(t)>0\}} dt\right)}{\check{T}}
$$

$$
= \lim_{T \to \infty} \frac{E\left(\int_{t=0}^{\check{T}} \mathcal{I}_{\{X_1(t)>0, \phi(t) \in S_+\}} dt\right)}{\check{T}} + \lim_{T \to \infty} \frac{E\left(\int_{t=0}^{\check{T}} \mathcal{I}_{\{X_1(t)>0, \phi(t) \in S_-\}} dt\right)}{\check{T}}
$$

$$
= \lim_{T \to \infty} \frac{E\left(\int_{t=0}^{\check{T}} \mathcal{I}_{\{\phi(t) \in S_+\}} dt\right)}{\check{T}} + \frac{1}{r_1} \lim_{T \to \infty} \frac{E\left(\int_{t=0}^{\check{T}} \mathcal{I}_{\{\phi(t) \in S_+\}} dt\right)}{\check{T}} = \frac{\mu}{\lambda + \mu}\left(1 + \frac{1}{r_1}\right)
$$

where $\mathcal{I}_{\{A\}}$ is the indicator of event A (i.e., $\mathcal{I}_{\{A\}} = 1$ only when A is true) and $\int_{t=0}^{T} \mathcal{I}_{\{X_1(t)>0\}} dt$ denotes the time while $X_1(t) > 0$ in the $(0, T)$ interval. In the first term of the second step, we utilized that the fluid buffer is always non-empty when $\phi(t) \in S_+$. In the second term of the second step, similar to Theorem 1, we utilized the fact that the fluid increase equals to the fluid decrease in $(0, \check{T})$. In the last step, we utilized that the stationary probability of state 1 is $\frac{\mu}{\lambda + \mu}$.

Similarly, the utilization of the buffer 2 is $\rho_2 = \lim_{t \to \infty} Pr(X_2(t) > 0) = \frac{\mu(1 + \frac{1}{r_2})}{\lambda + \mu}$. Since $r_1 > r_2$, $\rho_1 < \rho_2$. The stability of both buffers are ensured when $\rho_2 < 1$, i.e., $\lambda r_2 > \mu$. We assume stable buffers along this paper.

Due to the $X_1(t) \leq X_2(t)$ inequality, for the joint stationary probabilities we have

$$
P = \lim_{t \to \infty} Pr(X_1(t) = 0, X_2(t) = 0) = \lim_{t \to \infty} Pr(X_2(t) = 0) = 1 - \rho_2, \quad (2)
$$

$$
\lim_{t \to \infty} Pr(X_1(t) > 0, X_2(t) = 0) = 0, \quad (3)
$$

$$
\int_0^{\infty} U(y) dy = \lim_{t \to \infty} Pr(X_1(t) = 0, X_2(t) > 0) = \rho_2 - \rho_1, \quad (4)
$$

$$
\int_0^{\infty} \int_0^{\infty} W_+(x, y) + W_-(x, y) dy dx = \lim_{t \to \infty} Pr(X_1(t) > 0, X_2(t) > 0)
$$

$$
= \lim_{t \to \infty} Pr(X_1(t) > 0) = \rho_1. \quad (5)
$$

That is, one of the performance measures of interest, P, is given by $P = 1 - \rho_2$. Further more, from Theorem 1 we have

$$
\int_0^{\infty} \int_0^{\infty} W_+(x, y) dy dx = \lim_{t \to \infty} Pr(X_1(t) > 0, X_2(t) > 0, \phi(t) \in S_+)
$$

$$
= \rho_1 \frac{r_1}{1 + r_1} = \frac{\mu}{\lambda + \mu}, \quad (6)
$$

$$
\int_0^{\infty} \int_0^{\infty} W_-(x, y) dy dx = \lim_{t \to \infty} Pr(X_1(t) > 0, X_2(t) > 0, \phi(t) \in S_-)
$$

$$
= \rho_1 \frac{1}{1 + r_1} = \frac{\mu}{r_1(\lambda + \mu)}. \quad (7)
$$

For the analysis of the remaining measures of interest we apply an embedded process based approach.

3.2 Analysis of Embedded Time Points

Let $X_1(0) = 0$, $X_2(0) = x$, $\phi(0) \in \mathcal{S}_+$, $\gamma = \min(t > 0 : X_1(t) = 0)$.

In $(0, \gamma)$, the time spent in $\mathcal{S}_+$ and $\mathcal{S}_-$ are $\frac{r_1\gamma}{1+r_1}$ and $\frac{\gamma}{1+r_1}$ (because the fluid inflow equals the fluid out flow during $(0, \gamma)$ and consequently, the ratio of time spent in $\mathcal{S}_+$ and $\mathcal{S}_-$ is $r_1 : 1$). This way

$$X_2(\gamma) = x + (1)\frac{r_1\gamma}{1+r_1} + (-r_2)\frac{\gamma}{1+r_1} = x + \frac{\gamma(r_1 - r_2)}{1+r_1} = x + \gamma r, \qquad (8)$$

where $r = \frac{r_1 - r_2}{1+r_1}$. Intuitively, r is the increase rate of buffer 2 during the busy period of buffer 1, as it is depicted in Fig. 1.

After time γ, buffer 1 remains idle for a random amount of time denoted by τ.

During the idle period of buffer 1, $X_2(t)$ decreases with rate r_2. At the end of the idle period of buffer 1, $X_2(\gamma + \tau) = \max\left(0, x + \frac{\gamma(r_1 - r_2)}{1+r_1} - r_2\tau\right)$.

Let $T_0 = 0, T_1, \ldots$ be the beginning of the busy-idle cycles of buffer 1 and $Z_n = X_2(T_n)$ the fluid level of buffer 2 in those instances. In Fig. 1, the first busy-idle cycle of buffer 1 is such that $Z_1 = 0$ and the second one is such that $Z_2 > 0$.

For Z_n we have

$$Z_{n+1} = \max\left(0, Z_n + \frac{\gamma_n(r_1 - r_2)}{1+r_1} - r_2\tau_n\right)$$

$$= \max\left(0, Z_n + r\gamma_n - r_2\tau_n\right) \qquad (9)$$

We are interested in the stationary behaviour of $Z = \lim_{n\to\infty} Z_n$. Let $p = Pr(Z - 0)$, $\tilde{Z}(x) = Pr(Z < x)$ and $Z(x) = \frac{d}{dx}\tilde{Z}(x)$ for $x > 0$, and we also introduce the Laplace transform of $Z(x)$, $Z^*(s) = \int_{x=0}^{\infty} e^{-sx} Z(x)dx$. For stable fluid models these quantities satisfy the normalizing equation

$$p + \int_{u=0}^{\infty} Z(u)du = p + Z^*(0) = 1. \qquad (10)$$

At this point, we would like to emphasize the difference between the time stationary and embedded stationary measures, that is

$$\lim_{t\to\infty} Pr(X_1(t) = 0, X_2(t) = 0) = 1 - \rho_2$$

$$\neq \lim_{n\to\infty} Pr(X_1(T_n) = 0, X_2(T_n) = 0) = \lim_{n\to\infty} Pr(Z_n = 0) = p.$$

According to the behaviour of the BMC in (1), τ is exponentially distributed with rate μ. The return measure of buffer 1,

$$\Psi(t) = \frac{d}{dt}Pr(\gamma < t | X_1(0) = 0),$$

is available in LT domain. With our BMC (commonly the related expressions are given in matrix from in the literature, e.g. in [1], which we specify here according to (1)) and fluid rates, $\Psi^*(s) \triangleq E(e^{-s\gamma}) = \int_0^\infty e^{-st}\Psi(t)dt$ is

$$\Psi^*(s) = \frac{s(r_1+1) + r_1\lambda + \mu - \sqrt{(s(r_1+1) + r_1\lambda + \mu)^2 - 4r_1\lambda\mu}}{2\mu}, \tag{11}$$

and $E(\gamma) = -\Psi^{*'}(0) = \frac{1+r_1}{\lambda r_1 - \mu}$.

Theorem 3. *At the beginning of stationary busy period of buffer 1 the buffer content in buffer 2 is characterized by*

$$Z^*(s) = \frac{p\mu\left(1 - \Psi^*(rs)\right)}{r_2 s - \mu + \mu\Psi^*(rs)}. \tag{12}$$

Proof. Considering that the density of γ is $\Psi(x)$ and the density and the CDF of τ are $f(x)$ and $F(x)$, from (9) we have

$$Z(x) = p\int_{y=x/r}^\infty \Psi(y)\frac{1}{r_2}f\left(\frac{ry-x}{r_2}\right)dy$$
$$+ \int_{u=0}^\infty Z(u)\int_{y=(x-u)^+/r}^\infty \Psi(y)\frac{1}{r_2}f\left(\frac{u+ry-x}{r_2}\right)dydu \tag{13}$$

$$p = p\int_{y=0}^\infty \Psi(y)\left(1 - F\left(\frac{ry}{r_2}\right)\right)dy$$
$$+ \int_{u=0}^\infty Z(u)\int_{y=0}^\infty \Psi(y)\left(1 - F\left(\frac{u+ry}{r_2}\right)\right)dydu. \tag{14}$$

Considering $f(x) = \mu e^{-\mu x}$ and $F(x) = e^{-\mu x}$, from (16) we have

$$p = p\int_{y=0}^\infty \Psi(y)\exp\left(-\mu\frac{ry}{r_2}\right)dy + \int_{u=0}^\infty Z(u)\int_{y=0}^\infty \Psi(y)\exp\left(-\mu\frac{u+ry}{r_2}\right)dydu$$
$$= p\int_{y=0}^\infty \Psi(y)\exp\left(-\mu\frac{ry}{r_2}\right)dy$$
$$+ \int_{u=0}^\infty Z(u)\exp\left(-\mu\frac{u}{r_2}\right)\int_{y=0}^\infty \Psi(y)\exp\left(-\mu\frac{ry}{r_2}\right)dydu$$
$$= p\Psi^*\left(\frac{r\mu}{r_2}\right) + Z^*\left(\frac{\mu}{r_2}\right)\Psi^*\left(\frac{r\mu}{r_2}\right) = \left(p + Z^*\left(\frac{\mu}{r_2}\right)\right)\Psi^*\left(\frac{r\mu}{r_2}\right)$$

That is

$$p = \left(p + Z^*\left(\frac{\mu}{r_2}\right)\right)\Psi^*\left(\frac{r\mu}{r_2}\right)$$

and

$$Z^*\left(\frac{\mu}{r_2}\right) = \frac{p\left(1 - \Psi^*\left(\frac{r\mu}{r_2}\right)\right)}{\Psi^*\left(\frac{r\mu}{r_2}\right)} \tag{15}$$

Similarly, from (13) we have

$$Z(x) = p\frac{\mu}{r_2}\int_{y=x/r}^{\infty} \Psi(y)\exp\left(-\mu\frac{ry-x}{r_2}\right)dy$$

$$+ \frac{\mu}{r_2}\int_{u=0}^{\infty} Z(u)\int_{y=(x-u)^+/r}^{\infty} \Psi(y)\exp\left(-\mu\frac{u+ry-x}{r_2}\right)dydu$$

$$= p\frac{\mu}{r_2}\int_{y=x/r}^{\infty} \Psi(y)\exp\left(-\mu\frac{ry}{r_2}\right)\exp\left(\mu\frac{x}{r_2}\right)dy$$

$$+ \frac{\mu}{r_2}\int_{u=0}^{\infty} Z(u)\int_{y=(x-u)^+/r}^{\infty} \Psi(y)\exp\left(-\mu\frac{u}{r_2}\right)\exp\left(-\mu\frac{ry}{r_2}\right)\exp\left(\mu\frac{x}{r_2}\right)dydu$$

A multiplying both sides with e^{-sx} and integrating from 0 to ∞ gives

$$Z^*(s) = p\frac{\mu}{r_2}\int_{x=0}^{\infty} e^{-sx}\int_{y=x/r}^{\infty} \Psi(y)\exp\left(-\mu\frac{ry}{r_2}\right)\exp\left(\mu\frac{x}{r_2}\right)dydx$$

$$+ \frac{\mu}{r_2}\int_{x=0}^{\infty} e^{-sx}\int_{u=0}^{\infty} Z(u)\int_{y=(x-u)^+/r}^{\infty} \Psi(y)\exp\left(-\mu\frac{u+ry-x}{r_2}\right)dydudx$$

$$= V_1^*(s) + V_2^*(s).$$

Using $\int_{x=0}^{\infty}\int_{y=x/r}^{\infty} \bullet dydx = \int_{y=0}^{\infty}\int_{x=0}^{ry} \bullet dxdy$, for the first term we have

$$V_1^*(s) = p\frac{\mu}{r_2}\int_{y=0}^{\infty} \Psi(y)\exp\left(-\mu\frac{ry}{r_2}\right)\underbrace{\int_{x=0}^{ry} \exp\left(\mu\frac{x}{r_2} - sx\right)dx}\, dy$$

$$= p\frac{\mu}{r_2}\int_{y=0}^{\infty} \Psi(y)\exp\left(-\mu\frac{ry}{r_2}\right)\underbrace{\frac{r_2}{r_2s - \mu}\left(1 - \exp\left(-y\frac{r(r_2s - \mu)}{r_2}\right)\right)}\, dy$$

$$= \frac{p\mu}{r_2s - \mu}\left(\Psi^*\left(\frac{r\mu}{r_2}\right) - \Psi^*\left(\frac{r\mu}{r_2} + \frac{r(r_2s - \mu)}{r_2}\right)\right)$$

$$= \frac{p\mu}{r_2s - \mu}\left(\Psi^*\left(\frac{r\mu}{r_2}\right) - \Psi^*(rs)\right).$$

For the second term we first refine the integrals

$$\int_{x=0}^{\infty}\int_{u=0}^{\infty}\int_{y=(x-u)^+/r}^{\infty}\bullet\,dydudx = \int_{u=0}^{\infty}\int_{x=0}^{\infty}\int_{y=(x-u)^+/r}^{\infty}\bullet\,dydxdu$$

$$= \int_{u=0}^{\infty}\int_{x=0}^{u}\int_{y=0}^{\infty}\bullet\,dydxdu + \int_{u=0}^{\infty}\int_{x=u}^{\infty}\int_{y=(x-u)/r}^{\infty}\bullet\,dydxdu$$

$$= \int_{y=0}^{\infty}\int_{u=0}^{\infty}\int_{x=0}^{u}\bullet\,dxdudy + \int_{u=0}^{\infty}\int_{y=0}^{\infty}\int_{x=u}^{u+yr}\bullet\,dxdydu$$

$$= \int_{y=0}^{\infty}\int_{u=0}^{\infty}\int_{x=0}^{u}\bullet\,dxdudy + \int_{y=0}^{\infty}\int_{u=0}^{\infty}\int_{x=u}^{u+yr}\bullet\,dxdudy$$

Based on this, the second term is

$$V_2^*(s) = black\frac{\mu}{r_2}\int_{y=0}^{\infty}\Psi(y)\exp\left(-y\frac{r\mu}{r_2}\right)\int_{u=0}^{\infty}Z(u)\exp\left(-u\frac{\mu}{r_2}\right)\int_{x=0}^{u}\exp\left(-x\left(s-\frac{\mu}{r_2}\right)\right)dx\,dudy$$

$$+ black\frac{\mu}{r_2}\int_{y=0}^{\infty}\Psi(y)\exp\left(-y\frac{r\mu}{r_2}\right)\int_{u=0}^{\infty}Z(u)\exp\left(-u\frac{\mu}{r_2}\right)\underbrace{\int_{x=u}^{u+yr}\exp\left(-x\left(s-\frac{\mu}{r_2}\right)\right)dx}\,dudy$$

$$= black\frac{\mu}{r_2}\int_{y=0}^{\infty}\Psi(y)\exp\left(-y\frac{r\mu}{r_2}\right)\int_{u=0}^{\infty}Z(u)\exp\left(-u\frac{\mu}{r_2}\right)\underbrace{\frac{r_2}{r_2s-\mu}\left(1-\exp\left(-u\left(s-\frac{\mu}{r_2}\right)\right)\right)}\,dudy$$

$$+ black\frac{\mu}{r_2}\int_{y=0}^{\infty}\Psi(y)\exp\left(-y\frac{r\mu}{r_2}\right)\int_{u=0}^{\infty}Z(u)\exp\left(-u\frac{\mu}{r_2}\right)$$

$$\cdot\underbrace{\frac{r_2}{r_2s-\mu}\left(\exp\left(-u\left(s-\frac{\mu}{r_2}\right)\right)-\exp\left(-u\left(s-\frac{\mu}{r_2}\right)-yr\left(s-\frac{\mu}{r_2}\right)\right)\right)}_{\frac{r_2}{r_2s-\mu}\exp(-us)\exp(u\frac{\mu}{r_2})\left(1-\exp(-yrs)\exp\left(yr\frac{\mu}{r_2}\right)\right)}\,dudy$$

$$= \frac{\mu}{r_2s-\mu}\Psi^*\left(\frac{r\mu}{r_2}\right)\left(Z^*\left(\frac{\mu}{r_2}\right)-Z^*(s)\right)$$

$$+ \frac{\mu}{r_2s-\mu}\int_{y=0}^{\infty}\Psi(y)\exp\left(-y\frac{r\mu}{r_2}\right)\int_{u=0}^{\infty}Z(u)\exp\left(-us\right)dudy$$

$$- \frac{\mu}{r_2s-\mu}\int_{y=0}^{\infty}\Psi(y)\exp\left(-yrs\right)\int_{u=0}^{\infty}Z(u)\exp\left(-us\right)dudy$$

$$= \frac{\mu}{r_2s-\mu}\left(\Psi^*\left(\frac{r\mu}{r_2}\right)\left(Z^*\left(\frac{\mu}{r_2}\right)-Z^*(s)\right)+\Psi^*\left(\frac{r\mu}{r_2}\right)Z^*(s)-\Psi^*(sr)Z^*(s)\right)$$

$$= \frac{\mu}{r_2s-\mu}\left(\Psi^*\left(\frac{r\mu}{r_2}\right)Z^*\left(\frac{\mu}{r_2}\right)-\Psi^*(sr)Z^*(s)\right).$$

Finally,

$$Z^*(s) = \frac{p\mu}{r_2s-\mu}\left(\Psi^*\left(\frac{r\mu}{r_2}\right)-\Psi^*(rs)\right)$$

$$+ \frac{\mu}{r_2s-\mu}\left(\Psi^*\left(\frac{r\mu}{r_2}\right)Z^*\left(\frac{\mu}{r_2}\right)-\Psi^*(rs)Z^*(s)\right),$$

Using (15)

$$Z^*(s) = \frac{p\mu}{r_2 s - \mu}\left(\Psi^*\left(\frac{r\mu}{r_2}\right) - \Psi^*(rs)\right) + \frac{\mu}{r_2 s - \mu}\left(p\left(1 - \Psi^*\left(\frac{r\mu}{r_2}\right)\right) - \Psi^*(rs)Z^*(s)\right)$$

$$= \frac{p\mu}{r_2 s - \mu}\left(-\Psi^*(rs)\right) + \frac{\mu}{r_2 s - \mu}\left(p - \Psi^*(rs)Z^*(s)\right)$$

$$= \frac{p\mu}{r_2 s - \mu}\left(1 - \Psi^*(rs)\right) - \frac{\mu}{r_2 s - \mu}Z^*(s)\Psi^*(rs)$$

$$= \frac{\mu}{r_2 s - \mu}\left(p - \bar{Z}^*(s)\Psi^*(rs)\right)$$

where $\bar{Z}^*(s) = p + Z^*(s)$.

$$\bar{Z}^*(s) = p + \frac{\mu}{\mu - r_2 s}\left(-p + \bar{Z}^*(s)\Psi^*(rs)\right)$$

$$= \left(1 - \frac{\mu}{\mu - r_2 s}\right)p + \frac{\mu}{\mu - r_2 s}\bar{Z}^*(s)\Psi^*(rs)$$

which results in (12).

Corollary 1. *The only unknown in (12) can be obtained as*

$$p = 1 - \frac{\mu(r_1 - r_2)}{r_2(\lambda r_1 - \mu)} \tag{16}$$

Proof. Since both, the numerator and the denominator of (12) are zero at $s = 0$, we use the L'Hopital's rule to obtain $Z^*(0)$

$$Z^*(0) \triangleq \lim_{s \to 0} Z^*(s) = \frac{\frac{d}{ds}p\mu\left(1 - \Psi^*(rs)\right)\Big|_{s \to 0}}{\frac{d}{ds}r_2 s - \mu + \mu\Psi^*(rs)\Big|_{s \to 0}} = \frac{-p\mu r\Psi^{*'}(0)}{r_2 + \mu r\Psi^{*'}(0)}. \tag{17}$$

Using $1 - p = Z^*(0)$ from (10), $r = \frac{r_1 - r_2}{1 + r_1}$ and $E(\gamma) = -\Psi^{*'}(0) = \frac{1 + r_1}{\lambda r_1 - \mu}$, we further have

$$1 - p = \frac{-p\mu r\Psi^{*'}(0)}{r_2 + \mu r\Psi^{*'}(0)} = \frac{p\mu(r_1 - r_2)}{r_2(\lambda r_1 - \mu) - \mu(r_1 - r_2)},$$

from which the corollary comes.

4 Time Stationary Behaviour

We aim to obtain the remaining time stationary measures based on the embedded stationary measure $Z(x)$. For the joint distribution of the buffers we have the following cases:

C0) $X_1(t) = X_2(t) = 0, \phi(t) \in \mathcal{S}_-$: this case can be obtained from the stationary analysis of buffer 2 in isolation. The associated stationary measure is P.

C1) $X_1(t) = X_2(t) > 0, \phi(t) \in \mathcal{S}_+$: this case can arise after an idle period of buffer 2. The associated stationary measure is $V(x)$.

C2) $X_1(t) = 0, X_2(t) > 0, \phi(t) \in \mathcal{S}_-$: this case can arise during the idle period of buffer 1. The associated stationary measure is $U(y)$.

C3) $X_2(t) \geq X_1(t) > 0, \phi(t) \in \mathcal{S}_+$: this case can arise during the busy period of buffer 1. The associated stationary measure is $W_+(x, y)$.

C4) $X_2(t) \geq X_1(t) > 0, \phi(t) \in \mathcal{S}_-$: this case can arise during the busy period of buffer 1. The associated stationary measure is $W_-(x, y)$.

4.1   Analysis of C1)

As it is demonstrated in Fig. 1, case C1) can occur only when $\phi(t) \in \mathcal{S}_+$, in those busy-idle cycles of buffer 1 where buffer 2 is idle at the beginning of the cycle. According to the ergodicity of the model

$$\tilde{V}(x) = \frac{E\left(\int_{t=0}^{\gamma} \mathcal{I}_{\{X_2(t)=X_1(t)<x\}} dt\right)}{E(\gamma + \tau)}, \tag{18}$$

where $E(\gamma + \tau) = E(\gamma) + E(\tau) = \frac{1+r_1}{\lambda r_1 - \mu} + \frac{1}{\mu}$.

We compute the numerator of the right hand side from the following conditional relation. Let Φ be the sojourn time during the first visit in $\mathcal{S}_+$ in a busy-idle cycle of buffer 1. If $\Phi = y$ and $X_2(0) = 0$, we have

$$\int_{t=0}^{\gamma} \mathcal{I}_{\{X_2(t)=X_1(t)<x\}} dt = \begin{cases} y & \text{if } y < x \\ x & \text{if } y > x \end{cases}$$

Let $F_\Phi(y) = \Pr(\Phi < y) = 1 - e^{-\lambda y}$ be the CDF of Φ and $\Pr(X_2(0) = 0) = p$. Then

$$E\left(\int_{t=0}^{\gamma} \mathcal{I}_{\{X_2(t)=X_1(t)<x\}} dt\right) = p \int_{y=0}^{x} y \, dF_\Phi(y) + p \int_{y=x}^{\infty} x \, dF_\Phi(y) \tag{19}$$

$$= p \int_{y=0}^{x} y \lambda e^{-\lambda y} dy + px \int_{y=x}^{\infty} \lambda e^{-\lambda y} dy = \frac{p(1 - e^{-\lambda x})}{\lambda}, \tag{20}$$

from which

$$\tilde{V}(x) = \frac{E\left(\int_{t=0}^{\gamma} \mathcal{I}_{\{X_2(t)=X_1(t)<x\}} dt\right)}{E(\gamma + \tau)} = \frac{p(1 - e^{-\lambda x})}{\lambda E(\gamma + \tau)}, \tag{21}$$

and $V(x) = \frac{d}{dx} \tilde{V}(x) = \frac{pe^{-\lambda x}}{E(\gamma + \tau)}$.

4.2   Analysis of C2)

According to Fig. 1, case C2) can occur only when $\phi(t) \in \mathcal{S}_-$. We recall from (9) that $Z = \max(0, Z + r\gamma - r_2\tau)$, where $r = \frac{r_1 - r_2}{1 + r_1}$, γ is the busy time of buffer 1 and τ is the idle time of buffer 1 in $\mathcal{S}_-$.

Similar to case C1), we compute $\tilde{U}(x)$ based on the analysis of the stationary busy-idle cycle of buffer 1 using the ergodicity property.

$$1 - \tilde{U}(x) = \frac{E\left(\int_{t=\gamma}^{\gamma+\tau} \mathcal{I}_{\{X_2(t)>x,X_1(t)=0\}}dt\right)}{E(\gamma+\tau)}. \tag{22}$$

The numerator of the right hand side is obtained from the following conditional relation. If $\tau = h, Z + r\gamma = y$ then

$$\int_{t=\gamma}^{\gamma+\tau} \mathcal{I}_{\{X_2(t)>x,X_1(t)=0\}}dt = \int_{t=0}^{\tau} \mathcal{I}_{\{X_2(t)>x,X_1(t)=0|X_2(0)=y,X_1(0)=0\}}dt$$

$$= \begin{cases} 0 & \text{if } y < x, \\ \frac{y-x}{r_2} & \text{if } x < y < x + hr_2, \\ h & \text{if } y > x + hr_2. \end{cases} \tag{23}$$

Let $G(y) = \Pr(Z + r\gamma < y)$ and $T(h) = \Pr(\tau < h) = 1 - e^{-\mu h}$ be the CDF of $Z + r\gamma$ and τ, respectively. Than

$$E\left(\int_{t=\gamma}^{\gamma+\tau} \mathcal{I}_{\{X_2(t)>x,X_1(t)=0\}}dt\right) \tag{24}$$

$$= \int_{h=0}^{\infty} \int_{y=0}^{\infty} E\left(\int_{t=\gamma}^{\gamma+\tau} \mathcal{I}_{\{X_2(t)>x,X_1(t)=0\}}dt | \tau = h, Z + r\gamma = y\right) dG(y)dT(h) \tag{25}$$

$$= \int_{h=0}^{\infty} \int_{y=x}^{x+r_2h} \frac{y-x}{r_2} dG(y)dT(h) + \int_{h=0}^{\infty} h \underbrace{\int_{y=x+hr_2}^{\infty} dG(y)\, dT(h)}_{1-G(x+hr_2)} \tag{26}$$

$$= \int_{h=0}^{\infty} \int_{y=0}^{r_2h} \frac{y}{r_2} dG(y+x)\mu e^{-\mu h}dh + \int_{h=0}^{\infty} (1 - G(x+hr_2))h\mu e^{-\mu h}dh \tag{27}$$

$$= \int_{y=0}^{\infty} \frac{y}{r_2} \underbrace{\int_{h=y/r_2}^{\infty} \mu e^{-\mu h}dh}_{e^{-\mu y/r_2}} dG(y+x) + \int_{h=0}^{\infty} (1 - G(x+hr_2))h\mu e^{-\mu h}dh \tag{28}$$

$$= \int_{y=0}^{\infty} \frac{y}{r_2} e^{-\mu y/r_2} G'(y+x)dy + \frac{1}{\mu} - \int_{h=0}^{\infty} G(x+hr_2)h\mu e^{-\mu h}dh \tag{29}$$

Finally,

$$U(x) = -\frac{d}{dx} \frac{\int_{y=0}^{\infty} \frac{y}{r_2} e^{-\mu y/r_2} G'(y+x)dy + \frac{1}{\mu} - \int_{h=0}^{\infty} G(x+hr_2)h\mu e^{-\mu h}dh}{E(\gamma+\tau)}$$

$$= \frac{\int_{h=0}^{\infty} G'(x+hr_2)h\mu e^{-\mu h}dh - \int_{y=0}^{\infty} \frac{y}{r_2} e^{-\mu y/r_2} G'''(y+x)dy}{E(\gamma+\tau)}. \tag{30}$$

For $G(x)$, we have

$$G^*(s) \triangleq \int_{x=0}^{\infty} e^{-sx} dG(x) = \int_{x=0}^{\infty} e^{-sx} G'(x) dx$$
$$= E(e^{-s(Z+r\gamma)}) = E(e^{-sZ})E(e^{-sr\gamma}) = (p+Z^*(s))\Psi^*(sr). \quad (31)$$

and $\int_{x=0}^{\infty} e^{-sx} G''(x) dx = sG^*(s) - G'(0)$, where $G'(0) = \lim_{s \to \infty} sG^*(s) = \frac{\lambda pr_1}{r_1 - r_2}$.
Since $G'(x)$ and $G''(x)$ are given only in Laplace transform domain, a numerical inverse Laplace transformation (NILT) is required (e.g., using [6]) to compute $U(x)$.

4.3   Analysis of Case C3) and C4)

Theorem 4. *If $X_1(0) = 0$, $\phi(0) = S_+$ and $t < \gamma$ then*

$$X_2(t) = X_2(0) + tr + X_1(t)(1-r),$$

where $r = \frac{r_1 - r_2}{1 + r_1}$.

Proof. If $X_1(0) = 0$, $\phi(0) = S_+$ and $X_1(t) = x$ for $t < \gamma$ then the process spent $\frac{r_1 t + x}{r_1 + 1}$ time in S_+ and $\frac{t-x}{r_1+1}$ time in S_- in the $(0,t)$ time interval, since $\frac{r_1 t + x}{r_1+1} + \frac{t-x}{r_1+1} = t$ and

$$X_1(t) = \underbrace{0}_{\text{initial fluid level}} + \underbrace{\frac{r_1 t + x}{r_1+1}(1)}_{\text{fluid increase}} + \underbrace{\frac{t-x}{r_1+1}(-r_1)}_{\text{fluid decrease}} = x.$$

In the mean time, buffer 2 increases with $\frac{r_1 t + x}{r_1+1} \times 1$ and decreases with $\frac{t-x}{r_1+1} \times r_2$, that is

$$X_2(t) = X_2(0) + \frac{r_1 t + x}{r_1+1} - \frac{t-x}{r_1+1} r_2 = X_2(0) + tr + x(1-r)$$

For $\circ \in \{+,-\}$, let the transient probabilities of buffer 1 during the first busy period be defined as

$$\Theta_\circ(t,x) = Pr(\phi(t) = S_\circ, X_1(t) < x, t < \gamma | X_1(0) = 0, \phi(0) = S_+),$$

and $\theta_\circ(t,x) = \frac{d}{dx}\Theta_\circ(t,x)$. Their Laplace transforms, $\theta_\circ^*(s,x) = \int_0^\infty e^{-st}\theta_\circ(t,x)dt$ are available in LT domain as

$$\theta_+^*(s,x) = e^{xK(s)} \quad \text{and} \quad \theta_-^*(s,x) = e^{xK(s)}\Psi(s), \quad (32)$$

where $K(s) = -\lambda - s + \mu\Psi(s)$. Similar to (11), $\theta_+^*(s,x)$, $\theta_-^*(s,x)$, and $K(s)$ are obtained from the matrix expressions of those fluid measures, e.g. in [1], which are specified here according to (1).

Applying the ergodic property again we have

$$\tilde{W}_+(x,y) = \frac{E\left(\int_{t=0}^{\gamma} \mathcal{I}_{\{X_1(t)<x,X_2(t)<y,\phi(t)\in\mathcal{S}_+\}}dt\right)}{E\left(\gamma+\tau\right)}. \tag{33}$$

The numerator of the right hand side is obtained from the following conditional relation. If $X_1(0) = 0$, $\phi(0) = \mathcal{S}_+$ and $t < \gamma$ then, according to Theorem 4, $X_2(t) = z + tr + X_1(t)(1-r)$. This way,

$$\mathcal{I}_{\{X_1(t)<x,X_2(t)<y\}} = \begin{cases} 1 & \text{if } X_1(t)<\min(x,\frac{y-z-tr}{1-r}), \\ 0 & \text{otherwise.} \end{cases}$$

Let $t^*(z) = \max\left(0, \frac{y-z-(1-r)x}{r}\right)$ then, for $t > 0$

$$\min\left(x, \frac{y-z-tr}{1-r}\right) = \begin{cases} x & \text{if } t \le t^*(z), \\ \frac{y-z-tr}{1-r} & \text{if } t^*(z) < t. \end{cases}$$

Using this and $\tilde{Z}(x) = \Pr(Z < x)$, we write

$$E\left(\int_{t=0}^{\gamma} \mathcal{I}_{\{X_1(t)<x,X_2(t)<y,\phi(t)\in\mathcal{S}_+\}}dt\right)$$

$$= \int_{t=0}^{\infty} \Pr(X_1(t) < x, X_2(t) < y, \phi(t) \in \mathcal{S}_+, t < \gamma)dt$$

$$= \int_{z=0}^{\infty} \int_{t=0}^{t^*(z)} \Pr(X_1(t) < x, \phi(t) \in \mathcal{S}_+, t < \gamma)dt d\tilde{Z}(z)$$

$$+ \int_{z=0}^{\infty} \int_{t=t^*(z)}^{\infty} \Pr\left(X_1(t) < \frac{y-z-tr}{1-r}, \phi(t) \in \mathcal{S}_+, t < \gamma\right)dt d\tilde{Z}(z)$$

$$= \int_{z=0}^{\infty} \int_{t=0}^{t^*(z)} \Theta_+(t,x)dt d\tilde{Z}(z) + \int_{z=0}^{\infty} \int_{t=t^*(z)}^{\infty} \Theta_+\left(t, \frac{y-z-tr}{1-r}\right)dt d\tilde{Z}(z) \tag{34}$$

which can be obtained from the transient measures of buffer 1.

4.4   Steps of the Numerical Procedure

Based on the above detailed analysis approach the stationary distribution of the model is obtained from the model parameters $(\lambda, \mu, r_1, r_2$, which satisfy $r_1 > r_2$ and the stability condition, $\lambda r_2 > \mu$) in the following steps:

1. compute P from (2),
2. compute p and $Z^*(s)$ from (16) and (12),
3. compute $V(x)$ from (21) using $E\left(\gamma+\tau\right) = \frac{1+r_1}{\lambda r_1 - \mu} + \frac{1}{\mu}$,
4. compute $G'(x)$ and $G''(x)$ from (31) via NILT,
5. compute $U(x)$ from (30),
6. compute $\Theta_+(t,x)$ and $\Theta_-(t,x)$ from (32) via NILT,
7. compute $W(x,y)$ using (34).

5   Numerical Example

To demonstrate the application of the procedure in Sect. 4.4, we present a numerical example which we also compare with simulation results. We consider a system with parameters $\lambda = 2$, $\mu = 1$, $r_1 = 3$, and $r_2 = 1$. The parameters in Steps 1, 2, and 3 of Sect. 4.4 can be calculated analytically. To compute $U(x)$ in Step 5, we need to calculate the integrals in (30) numerically, since $G'(x)$ and $G''(x)$ are only available as a result of NILT in a finite number of points. The required integrals could be calculated using, e.g., the Simpson integral formula. Instead, to accelerate the computation, we calculate $G'(x)$ and $G''(x)$ in 20 points for $G'(x)$ and 50 points for $G''(x)$ and fit functions $\hat{G}_1(x)$ and $\hat{G}_2(x)$ to them, respectively. Our numerical investigations show that a polynomial of form

$$\hat{G}_1(x) = \sum_{i=0}^{11} a_i x^i$$

provides a close fit for $G'(x)$. On the other hand, while $\hat{G}_2(x) = \frac{d\hat{G}_1(x)}{dx}$ gives acceptable results, a much better fit can be achieved using the form

$$\hat{G}_2(x) = b_0 + b_1 e^{-c_1 x} + b_2 e^{-c_2 x}.$$

Figure 2 shows the result of the fitting.

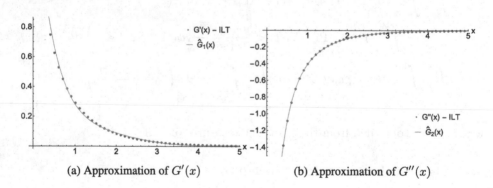

(a) Approximation of $G'(x)$ (b) Approximation of $G''(x)$

Fig. 2. Approximation of $G'(x)$ and $G''(x)$

Using the obtained $\hat{G}_1(x)$ and $\hat{G}_2(x)$ functions we can approximate $U(x)$ according to (30). To verify the results, we implemented a model specific simulation tool which computes $V(x)$ and $U(y)$ by discrete event simulation. We repeated the simulation 100000 times from simulation time 0 to 500 (where the time unit is defined by $\lambda = 2$ and $\mu = 1$).

Figure 3 compares the results of the procedure in Sect. 4.4 with the ones obtained from discrete event simulation. We note that in the figure the solid line corresponding to $V(x)$ is the result of fully analytical calculation, while $U(y)$ is approximated by simulation and the combination of NILT and numerical function fitting. Thus, Fig. 3a shows the precision of the simulation, while Fig. 3b shows that the simulation and the proposed numerical procedure gives quite similar results, thus verifying the obtained formulas and the validity of the proposed approach.

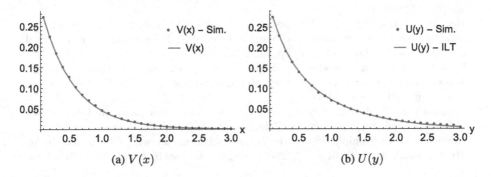

(a) $V(x)$ (b) $U(y)$

Fig. 3. Results for $U(y)$ and $V(x)$

6   Conclusion

The paper presents an embedded time points based analysis of a fluid model with two buffers, whose BMC is composed by two states and the fluid rates ensure that the content of one buffer is never less than the content of the other. This restriction significantly simplifies the analysis of fluid models with two buffers.

The evaluated numerical example verifies that the results of the proposed computational method closely fit with simulation results.

In the future, we intend to extend the analysis of this model with general BMC. The generalization of the embedded time points based approach does not seem to be straight forward because we miss an equation for obtaining the state probabilities of the BMC at embedded time points. Instead, we look for alternative analysis approaches which might be applicable for this fluid model also with general BMC.

References

1. Almousa, S.A.D., Horváth, G., Telek, M.: Transient analysis of piecewise homogeneous Markov fluid models. Ann. Oper. Res. **310**, 333–353 (2022). https://doi.org/10.1007/s10479-020-03831-1
2. Anick, D., Mitra, D., Sondhi, M.M.: Stochastic theory of a data-handling system. Bell Sys. Tech. J. **61**(8), 1871–1894 (1982)

3. Bean, N., O'Reilly, M., Taylor, P.: Hitting probabilities and hitting times for stochastic fluid flows: the bounded model. Probab. Eng. Inf. Sci. **23**, 121–147 (2009). https://doi.org/10.1017/S0269964809000102
4. Bean, N.G., O'Reilly, M.M.: A stochastic two-dimensional fluid model. Stoch. Model. **29**(1), 31–63 (2013). https://doi.org/10.1080/15326349.2013.750532
5. Bean, N.G., O'Reilly, M.M.: The stochastic fluid-fluid model: a stochastic fluid model driven by an uncountable-state process, which is a stochastic fluid model itself. Stoch. Process. Appl. **124**(5), 1741–1772 (2014). https://doi.org/10.1016/j.spa.2013.12.006
6. Horváth, I., Horváth, G., Almousa, S.A.D., Telek, M.: Numerical inverse Laplace transformation using concentrated matrix exponential distributions. Perform. Eval. (2019). https://doi.org/10.1016/j.peva.2019.102067
7. Kulkarni, V.G.: Fluid models for single buffer systems. In: Dshalalow, J.H. (ed.) Models and Applications in Science and Engineering, pp. 321–338. Frontiers in Queueing, CRC Press (1997)
8. Ramaswami, V.: Matrix analytic methods for stochastic fluid flows. In: International Teletraffic Congress, pp. 1019–1030. Edinburg (1999)
9. Remiche, M.A.: Compliance of the token-bucket model with Markovian traffic. Stoch. Model. **21**(2–3), 615–630 (2005). https://doi.org/10.1081/STM-200057884

A Variable-Cycle Traffic Light Queue

Freek Verdonck$^{(\boxtimes)}$ ⓘ, Herwig Bruneel ⓘ, and Sabine Wittevrongel$^{(\boxtimes)}$ ⓘ

Department of Telecommunications and Information Processing (TELIN),
SMACS Research Group, Ghent University (UGent),
Sint-Pietersnieuwstraat 41, 9000 Gent, Belgium
{freek.verdonck,herwig.bruneel,sabine.wittevrongel}@ugent.be

Abstract. In this paper we look at queueing phenomena occurring at
a traffic light. As opposed to the well-known and well-researched Fixed-
cycle Traffic Light, we consider a traffic light with variable lengths for
the red and green periods. More specifically we apply a method that
allows the red and green periods to be according to a stochastic distribution. We obtain the probability generating functions of the system
content at various observation epochs and a tail approximation of the
delay distribution.

Keywords: Queueing theory · Discrete-time · Traffic light · Variable
cycle length · System content · Delay

1 Introduction

Already in the 1950s queueing models were used to model traffic situations [1,
5,10,13,18]. A vast amount of traffic problems have been tackled with queueing
theory, the majority considering isolated intersections [5,13], but also networks
are investigated [11,14]. The so-called Fixed-Cycle Traffic Light is probably one
of the best researched problems in the traffic subdomain of queueing theory and
it still gathers considerable attention, see e.g. [5,6,10,12]. This problem consists
of analyzing the queue arising at a traffic light intersection, where the green and
red periods have a fixed length.

In this paper we will use and adapt our earlier developed methods [15] for
the system content and [16] for a tail approximation of the delay distribution
in multiserver queueing systems to investigate a Variable-Cycle Traffic Light,
where the lengths of the green and red periods are random variables.

2 Mathematical Model

We look at an intersection controlled by a traffic light and consider a single lane
in a specific direction and model it as a discrete-time queueing system. When
observing a queue at a traffic light, we assume that the delayed vehicles cross the
intersection within fixed intervals. We take such an interval as the slot length.

© The Author(s), under exclusive license to Springer Nature Switzerland AG 2023
M. Iacono et al. (Eds.): EPEW/ASMTA 2023, LNCS 14231, pp. 65–80, 2023.
https://doi.org/10.1007/978-3-031-43185-2_5

We furthermore assume that the lengths of the green and red phases are integer numbers of slots. The discrete-time assumption as applied here is common in the literature of the Fixed-Cycle Traffic Light, see e.g. [5, 6, 10].

Furthermore we assume that a vehicle arriving at the intersection during a green period does not experience any delay if the intersection is empty at the moment of arrival; its service time is then modelled as 0 slots. This assumption is sometimes referred to as the Traffic Light Assumption [6]. It is based on the observation that cars can pass a traffic light much faster when they do not need to slow down because of an existing queue and implies that if at some point during a green period the system becomes empty, the queue will remain empty at least until the beginning of the next red period.

During every slot, a number of cars can arrive at the traffic light. The number of arrivals during a slot is considered to be independent and identically distributed (i.i.d.) with mean λ and with the following distribution:

$$c(n) \triangleq \text{Prob}[n \text{ cars arrive during a slot}] ; \quad C(z) \triangleq \sum_{n=0}^{\infty} c(n)z^n . \qquad (1)$$

For modelling purposes, only two system states of a traffic light are relevant: the state when cars can cross the intersection (green state), and the state when cars cannot cross the intersection (red state). In many traffic situations, the length of red and green periods is not fixed. An example can be a railroad crossing. The duration of the green phase (for vehicles crossing the railroad) depends on the interarrival times of the trains, and the duration of the red phase depends on the speed and length of the train. In this model, the lengths of the green periods and red periods are stochastic, with mean $\bar{r}_{\text{green}}$ and $\bar{r}_{\text{red}}$ respectively and distributions characterized by

$$r_{\text{red}}(n) \triangleq \text{Prob}[\text{red period has } n \text{ slots}] ; \quad R_{\text{red}}(z) \triangleq \sum_{n=1}^{\infty} r_{\text{red}}(n)z^n ; \quad (2)$$

$$r_{\text{green}}(n) \triangleq \text{Prob}[\text{green period has } n \text{ slots}] ; \quad R_{\text{green}}(z) \triangleq \sum_{n=1}^{\infty} r_{\text{green}}(n)z^n .$$

$$(3)$$

The lengths of green and red periods during consecutive cycles are assumed i.i.d. and the lengths of a green period and a red period in the same cycle are also assumed independent. We will limit $R_{\text{green}}(z)$ to be a rational function of its argument so that it can be written as

$$R_{\text{green}}(z) = \frac{A_{\text{green}}(z)}{B_{\text{green}}(z)} , \qquad (4)$$

with $A_{\text{green}}(z)$ and $B_{\text{green}}(z)$ mutually prime polynomials of degree m_A and m_B respectively and with $A_{\text{green}}(1) = B_{\text{green}}(1) = 1$. Furthermore we introduce m_g as the maximum of m_A and m_B: $m_g \triangleq \max(m_A, m_B)$.

During red periods no servers (lanes) are available and during green periods 1 server (lane) is available. The probability that an arbitrary slot belongs to a green period is given by σ_{green} (and to a red period by σ_{red}):

$$\sigma_{\text{green}} = \frac{\bar{r}_{\text{green}}}{\bar{r}_{\text{red}} + \bar{r}_{\text{green}}} \; ; \quad \sigma_{\text{red}} = 1 - \sigma_{\text{green}} \,. \tag{5}$$

We are interested in the steady-state characteristics of this type of queueing system. The system is stable if the load ρ which is the ratio of the arrival intensity and the mean service capacity is smaller than 1: $\rho \triangleq \frac{\lambda}{\sigma_{\text{green}}} < 1$.

3   System Content Analysis

We first introduce some stochastic variables for the system content at specific observation epochs when the system has reached a steady state. We denote with g_0^{green} the system content at the beginning of a green period and with g_k^{green} (with $k \geq 1$) the system content at the end of the kth slot of a green period (i.e. at the beginning of the $(k+1)$st slot of a green period) with corresponding probability generating functions (pgfs) $G_0^{\text{green}}(z)$ and $G_k^{\text{green}}(z)$. Denoting with c_k^{green} the number of arrivals during the kth slot of a green period, we obtain the following recursive equation for $k \geq 1$:

$$g_k^{\text{green}} = \begin{cases} 0, & \text{if } g_{k-1}^{\text{green}} = 0; \\ g_{k-1}^{\text{green}} - 1 + c_k^{\text{green}}, & \text{if } g_{k-1}^{\text{green}} > 0. \end{cases} \tag{6}$$

Let us further introduce the stochastic variables g_0^{red} as the system content at the beginning of a red period and g_k^{red} (with $k \geq 1$) as the system content at the end of the kth slot of a red period (i.e. at the beginning of the $(k+1)$st slot of a red period) with corresponding pgfs $G_0^{\text{red}}(z)$ and $G_k^{\text{red}}(z)$. During a red slot no cars leave the system and we denote with c_k^{red} the number of vehicles arriving during the kth slot of a red period. We get the following recursive relation:

$$g_k^{\text{red}} = g_{k-1}^{\text{red}} + c_k^{\text{red}} \,, k \geq 1. \tag{7}$$

We can transform the above system equations into the z-domain. For the green periods we get from (6)

$$
\begin{aligned}
G_k^{\text{green}}(z) &= \frac{C(z)}{z} G_{k-1}^{\text{green}}(z) + \left[1 - \frac{C(z)}{z}\right] G_{k-1}^{\text{green}}(0) \\
&= \left(\frac{C(z)}{z}\right)^k G_0^{\text{green}}(z) + \left[1 - \frac{C(z)}{z}\right] \sum_{j=1}^{k} \left(\frac{C(z)}{z}\right)^{j-1} G_{k-j}^{\text{green}}(0) \,, \tag{8}
\end{aligned}
$$

which is valid for $k \geq 0$. For the red periods we easily get from (7)

$$
\begin{aligned}
G_k^{\text{red}}(z) &= C(z)\, G_{k-1}^{\text{red}}(z) \\
&= [C(z)]^k\, G_0^{\text{red}}(z) \,, k \geq 0. \tag{9}
\end{aligned}
$$

We can express that the system content at the end of a red period equals the system content at the beginning of a green period and vice versa. This leads to

$$G_0^{\text{green}}(z) = \sum_{k=1}^{\infty} r_{\text{red}}(k) G_k^{\text{red}}(z) = R_{\text{red}}(C(z)) G_0^{\text{red}}(z) \; ; \qquad (10)$$

$$G_0^{\text{red}}(z) = \sum_{k=1}^{\infty} r_{\text{green}}(k) G_k^{\text{green}}(z)$$

$$= R_{\text{green}}\left(\frac{C(z)}{z}\right) G_0^{\text{green}}(z) + \left[\frac{z}{C(z)} - 1\right] Q\left(\frac{C(z)}{z}\right), \qquad (11)$$

with $Q(z)$ unknown and defined by

$$Q(z) \triangleq \sum_{j=1}^{\infty} q(j) z^j \; ; \quad q(j) \triangleq \sum_{k=0}^{\infty} r_{\text{green}}(k+j) G_k^{\text{green}}(0) \; . \qquad (12)$$

Combination of (10) and (11) leads to

$$G_0^{\text{red}}(z) = \frac{\left[\frac{z}{C(z)} - 1\right] Q\left(\frac{C(z)}{z}\right)}{1 - R_{\text{red}}(C(z)) R_{\text{green}}\left(\frac{C(z)}{z}\right)} \; ; \quad G_0^{\text{green}}(z) = R_{\text{red}}(C(z)) G_0^{\text{red}}(z) \; . \qquad (13)$$

We introduce g^{green} as the steady-state system content at the beginning of an arbitrary green slot, and g^{red} as the steady-state system content at the beginning of an arbitrary red slot. The corresponding pgfs are denoted as $G^{\text{green}}(z)$ and $G^{\text{red}}(z)$. Given an arbitrary green (red) slot, the probability that it is the kth slot of a green (red) period is given by $\frac{\sum_{n=k}^{\infty} r_{\text{green}}(n)}{\bar{r}_{\text{green}}}$ $\left(\frac{\sum_{n=k}^{\infty} r_{\text{red}}(n)}{\bar{r}_{\text{red}}}\right)$. Using these probabilities, we can obtain expressions for $G^{\text{green}}(z)$ and $G^{\text{red}}(z)$ as follows:

$$G^{\text{green}}(z) = \sum_{k=1}^{\infty} \frac{\sum_{n=k}^{\infty} r_{\text{green}}(n)}{\bar{r}_{\text{green}}} \left\{ \left(\frac{C(z)}{z}\right)^{k-1} G_0^{\text{green}}(z) \right.$$

$$\left. + \left[1 - \frac{C(z)}{z}\right] \sum_{j=1}^{k-1} \left(\frac{C(z)}{z}\right)^{j-1} G_{k-j-1}^{\text{green}}(0) \right\}$$

$$= \frac{z\left[R_{\text{green}}\left(\frac{C(z)}{z}\right) - 1\right]}{[C(z) - z]\bar{r}_{\text{green}}} G_0^{\text{green}}(z) + \frac{Q(1) - \frac{z}{C(z)} Q\left(\frac{C(z)}{z}\right)}{\bar{r}_{\text{green}}} , \qquad (14)$$

and

$$G^{\text{red}}(z) = \sum_{k=1}^{\infty} \frac{\sum_{n=k}^{\infty} r_{\text{red}}(n)}{\bar{r}_{\text{red}}} [C(z)]^{k-1} G_0^{\text{red}}(z) = \frac{R_{\text{red}}(C(z)) - 1}{[C(z) - 1]\bar{r}_{\text{red}}} G_0^{\text{red}}(z) \; . \qquad (15)$$

The pgf $G(z)$ of the system content g at the beginning of an arbitrary slot is given by

$$G(z) = \sigma_{\text{green}} G^{\text{green}}(z) + \sigma_{\text{red}} G^{\text{red}}(z) . \tag{16}$$

The obtained pgfs still contain the unknowns that are present in the function $Q(z)$. In a similar way as in [4,15], it can be proven that the number of unknowns is finite. When $R_{\text{green}}(z)$ is a rational function of its argument, also $Q(z)$ is rational and more specifically it has the following shape:

$$Q(z) = \frac{\sum_{i=1}^{m_g} q_i z^i}{B_{\text{green}}(z)} . \tag{17}$$

The (common) denominator of $G_0^{\text{green}}(z)$ and $G_0^{\text{red}}(z)$ in (13) has exactly m_g zeros z_j within the complex unit disk, with $z_1 = 1$. As $G_0^{\text{green}}(z)$ and $G_0^{\text{red}}(z)$ are pgfs, they must be normalized and cannot have singularities within the complex unit disk. Expressing these conditions for either of the two pgfs leads to the following set of m_g equations to determine the unknown coefficients in $Q(z)$:

$$\begin{cases} \dfrac{Q(1)\,(1-\lambda)}{-\overline{r}_{\text{red}}\lambda + \overline{r}_{\text{green}}(1-\lambda)} = 1; \\ R_{\text{red}}(C(z_j)) \left[\dfrac{z_j}{C(z_j)} - 1\right] Q\left(\dfrac{C(z_j)}{z_j}\right) = 0, \quad j = 2,\ldots,m_g . \end{cases} \tag{18}$$

We can calculate several performance measures based on the obtained pgfs, such as the mean system content and the mean overflow queue length (which is defined as the queue length at the beginning of a red period). Also higher order moments can be obtained as well as the probability that the system is empty at the beginning of an arbitrary slot or at the beginning of a green or red period. The formulae for some common performance measures are listed below. The mean system content at the beginning of a green period is given by

$$E[g_0^{\text{green}}] = \frac{\psi_1 \lambda^2 + \psi_2 \lambda + \psi_3}{2\left[\overline{r}_{\text{green}}(1-\lambda) - \overline{r}_{\text{red}}\lambda\right]} , \tag{19}$$

with

$$\psi_1 = 2\overline{r}_{\text{green}}\overline{r}_{\text{red}} + 2Q(1)\,(1-\overline{r}_{\text{red}}) - 2Q'(1) + R_{\text{green}}''(1) + R_{\text{red}}''(1) ; \tag{20}$$

$$\psi_2 = -2\overline{r}_{\text{green}}\overline{r}_{\text{red}} - 2\overline{r}_{\text{green}} + 2Q(1)\,(\overline{r}_{\text{red}} - 1) + 4Q'(1) - R_{\text{green}}''(1) ; \tag{21}$$

$$\psi_3 = [\overline{r}_{\text{green}} + \overline{r}_{\text{red}} - Q(1)]\,C''(1) + \overline{r}_{\text{green}} + R_{\text{green}}''(1) - Q'(1) . \tag{22}$$

The mean system contents at the beginning of a red period, at the beginning of an arbitrary green slot, at the beginning of an arbitrary red slot and at the beginning of an arbitrary slot are given by

$$E\left[g_0^{\text{red}}\right] = E\left[g_0^{\text{green}}\right] - \bar{r}_{\text{red}}\lambda \, ; \tag{23}$$

$$E\left[g^{\text{green}}\right] = E\left[g_0^{\text{green}}\right] + \frac{(\lambda - 1)R''_{\text{green}}(1)}{2\bar{r}_{\text{green}}} + \frac{(\lambda - 1)\left[Q(1) - Q'(1)\right]}{\bar{r}_{\text{green}}} \, ; \tag{24}$$

$$E\left[g^{\text{red}}\right] = E\left[g_0^{\text{red}}\right] + \frac{R''_{\text{red}}(1)\lambda}{2\bar{r}_{\text{red}}} \, ; \tag{25}$$

$$E[g] = \sigma_{\text{green}}E\left[g^{\text{green}}\right] + \sigma_{\text{red}}E\left[g^{\text{red}}\right] \, . \tag{26}$$

An interesting metric for this queueing system is the probability that a customer does not experience any delay, which is the case when it arrives during a green slot and the system was empty at the beginning of that slot. The probability that an arbitrary customer arrives during the kth slot of a green period is given by $\pi_k^{\text{green}} = \sigma_{\text{green}}\frac{1-\sum_{j=1}^{k-1} r_{\text{green}}(j)}{\bar{r}_{\text{green}}}$. Therefore, the probability $w(0)$ that a vehicle experiences no delay can be calculated as

$$w(0) = \sum_{k=1}^{\infty} \sigma_{\text{green}}\frac{1 - \sum_{j=1}^{k-1} r_{\text{green}}(j)}{\bar{r}_{\text{green}}}G_{k-1}^{\text{green}}(0) = \frac{Q(1)}{\bar{r}_{\text{green}} + \bar{r}_{\text{red}}} \, . \tag{27}$$

4   Delay Analysis

In this section we will consider the delay analysis of a Variable-Cycle Traffic Light. The delay of a vehicle is counted as the number of slots between the last slot boundary of its arrival slot and the last slot boundary of its departure slot and is therefore always an integer number of slots.

We focus first on the conditional delay of a vehicle, given the circumstances of its arrival. We introduce the stochastic variable d_k^n as the delay of a delayed vehicle arriving during a green slot with n more slots until the next red slot (excluding the current slot) and with k vehicles in the queue in front of the tagged vehicle. The corresponding pgf is $D_k^n(z)$. We currently do not consider the vehicles that arrive at an empty system during a green slot and cross the intersection without delay. We get the following relation:

$$D_k^n(z) = \begin{cases} z^{k+1} \, , & \text{if } k < n \, ; \\ z^n R_{\text{red}}(z)\sum_{l=1}^{\infty} r_{\text{green}}(l)D_{k-n}^l(z) \, , & \text{if } k \geq n \, . \end{cases} \tag{28}$$

Let us now introduce the following auxiliary functions:

$$D_k(z) \triangleq \sum_{n=1}^{\infty} r_{\text{green}}(n)D_k^n(z) \, , \quad k \geq 0 \, ; \quad D(x, z) \triangleq \sum_{k=0}^{\infty} D_k(z)\, x^k \, . \tag{29}$$

Working out the above definition of $D(x, z)$ making use of (28) leads to

$$D(x, z) = R_{\text{green}}(xz)\, R_{\text{red}}(z)\, D(x, z) + z\frac{1 - R_{\text{green}}(xz)}{1 - xz} \, , \tag{30}$$

which can be solved for $D(x,z)$ to get

$$D(x,z) = \frac{f(x,z)}{g(x,z)}, \tag{31}$$

with

$$f(x,z) = \frac{z\left[B_{\text{green}}(xz) - A_{\text{green}}(xz)\right]}{1 - xz}; \tag{32}$$

$$g(x,z) = B_{\text{green}}(xz) - A_{\text{green}}(xz)R_{\text{red}}(z), \tag{33}$$

in view of the rational expression (4) for $R_{\text{green}}(z)$. The functions $f(x,z)$ and $g(x,z)$ are polynomial functions in x of degree $m_g - 1$ and m_g respectively. We can now do a partial fraction expansion of $D(x,z)$, based on its m_g poles in x, which we will denote x_p and we will assume to be distinct. Note that the x_p are functions of z but for notational simplicity the argument is omitted. We get

$$D(x,z) = \sum_{p=1}^{m_g} \frac{f(x_p,z)}{(x - x_p)g_x(x_p,z)}, \tag{34}$$

with

$$g_x(x,z) \triangleq \frac{\partial}{\partial x}g(x,z) = zB'_{\text{green}}(xz) - zA'_{\text{green}}(xz)R_{\text{red}}(z). \tag{35}$$

We can now obtain an expression for $D_k(z)$:

$$D_k(z) = \frac{1}{k!}\frac{\partial}{\partial x^k}D(x,z)\Big|_{(x=0)} = \sum_{p=1}^{m_g} \frac{-f(x_p,z)}{(x_p)^{k+1}g_x(x_p,z)}. \tag{36}$$

The delay of a vehicle arriving during a red slot with n more slots until the next green slot (excluding the current slot) and with k vehicles in the queue in front of the tagged vehicle can be described by the pgf $z^n D_k(z)$.

Now we consider the delay of an arbitrary vehicle V, arriving during the slot S. We denote with $\sigma_{\text{green}}\pi_{l|n}^{\text{green}}$ ($\sigma_{\text{red}}\pi_{l|n}^{\text{red}}$) the probability that S is the lth slot of a green (red) period of in total $(l+n)$ slots with

$$\pi_{l|n}^{\text{green}} \triangleq \frac{r_{\text{green}}(l+n)}{\bar{r}_{\text{green}}}; \qquad \pi_{l|n}^{\text{red}} \triangleq \frac{r_{\text{red}}(l+n)}{\bar{r}_{\text{red}}}. \tag{37}$$

The delay of a vehicle depends on k, the number of vehicles that are in the queue and that have priority over the tagged vehicle.

First we consider the case that S is a green slot. If the queue was empty at the beginning of S, the vehicle V crosses the intersection without delay and $k = 0$. Otherwise, k equals the number of vehicles present in the system at the beginning of S, minus 1 and plus those vehicles that arrive during S, but before V. Also now k can be equal to 0, but the delay of V will be equal to 1 slot or the length of a red period plus 1 slot. Even though the queue is empty, the vehicle V

cannot cross the intersection without slowing down, as at the moment of arrival of V the intersection is blocked by another vehicle which is still accelerating.

If S is a red slot, the number of vehicles that are in the queue and that have priority over V equals the number of vehicles that were present in the system at the beginning of S and those vehicles that arrive during S, but before V.

The pgf $F(z)$ of the number of vehicles that arrive during S but before V is known to be given by (see, e.g. [2])

$$F(z) = \frac{C(z) - 1}{\lambda(z - 1)}. \tag{38}$$

We will introduce t_l^{green} and t_l^{red} as the system content experienced by V given that S is the lth slot of a green period and a red period respectively. Their pgfs $T_l^{\text{green}}(z)$ and $T_l^{\text{red}}(z)$ can be obtained as

$$
\begin{aligned}
T_l^{\text{green}}(z) &= G_{l-1}^{\text{green}}(0) + \frac{F(z)}{z} \left[G_{l-1}^{\text{green}}(z) - G_{l-1}^{\text{green}}(0) \right] \\
&= \left[1 - \frac{F(z)}{z} \right] G_{l-1}^{\text{green}}(0) + \frac{F(z)}{z} \left(\frac{C(z)}{z} \right)^{l-1} G_0^{\text{green}}(z) \\
&\quad + \frac{F(z)}{z} \left[1 - \frac{C(z)}{z} \right] \sum_{j=1}^{l-1} \left(\frac{C(z)}{z} \right)^{j-1} G_{l-1-j}^{\text{green}}(0) .
\end{aligned}
\tag{39}
$$

$$T_l^{\text{red}}(z) = F(z) \left[C(z) \right]^{l-1} G_0^{\text{red}}(z) . \tag{40}$$

We will denote the inverse z-transform of the above pgfs as $t_l^{\text{green}}(k)$ and $t_l^{\text{red}}(k)$. Using the law of total expectation we can now develop the pgfs $W_{\text{green}}(z)$ and $W_{\text{red}}(z)$ of the delay of an arbitrary vehicle arriving during a green and red slot:

$$W_{\text{green}}(z) = \sum_{l=1}^{\infty} \sum_{n=0}^{\infty} \pi_{l|n}^{\text{green}} \left\{ G_{l-1}^{\text{green}}(0) \left[1 - D_0^n(z) \right] + \sum_{k=0}^{\infty} t_l^{\text{green}}(k) D_k^n(z) \right\} ; \tag{41}$$

$$W_{\text{red}}(z) = \sum_{l=1}^{\infty} \sum_{n=0}^{\infty} \pi_{l|n}^{\text{red}} \sum_{k=0}^{\infty} t_l^{\text{red}}(k) z^n D_k(z) . \tag{42}$$

The pgf $W(z)$ of the delay of an arbitrary customer is then given by

$$W(z) = \sigma_{\text{green}} W_{\text{green}}(z) + \sigma_{\text{red}} W_{\text{red}}(z) . \tag{43}$$

Using the expressions (36) for $D_k(z)$, (37) for $\pi_{l|n}^{\text{red}}$ and (40) for $T_l^{\text{red}}(z)$ we can work out $W_{\text{red}}(z)$ as

$$
\begin{aligned}
W_{\text{red}}(z) &= \sum_{l=1}^{\infty} \sum_{n=0}^{\infty} \frac{r_{\text{red}}(l+n)}{\bar{r}_{\text{red}}} \sum_{k=0}^{\infty} t_l^{\text{red}}(k) z^n \sum_{p=1}^{m_g} \frac{-f(x_p, z)}{(x_p)^{k+1} g_x(x_p, z)} \\
&= \sum_{p=1}^{m_g} \frac{-f(x_p, z) \left[R_{\text{red}} \left(C \left(\frac{1}{x_p} \right) \right) - R_{\text{red}}(z) \right]}{\bar{r}_{\text{red}} \left[C \left(\frac{1}{x_p} \right) - z \right] x_p g_x(x_p, z)} F \left(\frac{1}{x_p} \right) G_0^{\text{red}} \left(\frac{1}{x_p} \right) . \tag{44}
\end{aligned}
$$

In order to work out the expression for $W_{\text{green}}(z)$, we first look in more detail at the distribution of the green periods, which is characterized by the pgf $R_{\text{green}}(z)$. We restricted $R_{\text{green}}(z)$ to be a rational function of its argument and we will further assume that it has only poles of multiplicity 1. We can then rewrite it as

$$R_{\text{green}}(z) = \sum_{j=1}^{M_1} \gamma_j z^j + \sum_{j=1}^{M_2} \omega_j \frac{(1-\beta_j)z}{1-\beta_j z}, \tag{45}$$

where the summations do not necessarily appear both. In the remainder of this paper we will assume that they both appear, the results can be easily modified otherwise. We state that $\gamma_{M_1} \neq 0$. Note that some of the β_j can be complex-valued and the γ_j, ω_j can be negative or larger than 1. The corresponding probability mass function (pmf) $r_{\text{green}}(n)$ can then be written as

$$r_{\text{green}}(n) = \begin{cases} \gamma_n + \sum_{j=1}^{M_2} \omega_j (1-\beta_j)(\beta_j)^{n-1}, & \text{if } n \leq M_1; \\ \sum_{j=1}^{M_2} \omega_j (1-\beta_j)(\beta_j)^{n-1}, & \text{if } n > M_1. \end{cases} \tag{46}$$

Using the above form of the pmf $r_{\text{green}}(n)$ in the expression (37) for $\pi_{l|n}^{\text{green}}$ we can group the terms with γ_j and β_j in the expression for $W_{\text{green}}(z)$ and work it further out as $W_{\text{green}}(z) = W_1(z) + W_2(z)$ with

$$W_1(z) \triangleq \sum_{j=1}^{M_1} \frac{\gamma_j}{\overline{r}_{\text{green}}} \sum_{l=1}^{j} \left\{ G_{l-1}^{\text{green}}(0) \left[1 - D_0^{j-l}(z)\right] + \sum_{k=0}^{\infty} t_l^{\text{green}}(k) D_k^{j-l}(z) \right\}; \tag{47}$$

$$W_2(z) \triangleq \sum_{j=1}^{M_2} \frac{\omega_j}{\overline{r}_{\text{green}}} \sum_{n=0}^{\infty} \sum_{l=1}^{\infty} (1-\beta_j)\beta_j^{n+l-1}$$
$$\left\{ G_{l-1}^{\text{green}}(0) \left[1 - D_0^n(z)\right] + \sum_{k=0}^{\infty} t_l^{\text{green}}(k) D_k^n(z) \right\}. \tag{48}$$

We look first at $W_1(z)$ and introduce $D_k^n(z)$ according to (28). For the terms with $G_{l-1}^{\text{green}}(0)$ we distinguish between the situations where $l = j$ and where $l < j$ as in the former the delay includes a red period. Similarly we split the terms with $t_l^{\text{green}}(k)$ for $k < j - l$ and $k \geq j - l$. For the latter we can introduce $D_{k-(j-l)}(z)$ according to (36). This leads to

$$W_1(z) = \sum_{j=1}^{M_1} \frac{\gamma_j}{\overline{r}_{\text{green}}} \left\{ G_{j-1}^{\text{green}}(0) \left[1 - z R_{\text{red}}(z)\right] + \sum_{l=1}^{j-1} G_{l-1}^{\text{green}}(0) \left[1 - z\right] \right\}$$
$$+ \sum_{j=1}^{M_1} \frac{\gamma_j}{\overline{r}_{\text{green}}} \sum_{l=1}^{j} \sum_{k=0}^{j-l-1} t_l^{\text{green}}(k) \left[z^{k+1} + R_{\text{red}}(z) \sum_{p=1}^{m_g} \frac{z^{j-l}(x_p)^j f(x_p, z)}{(x_p)^{l+k+1} g_x(x_p, z)} \right]$$
$$- \sum_{j=1}^{M_1} \frac{\gamma_j}{\overline{r}_{\text{green}}} \sum_{l=1}^{j} R_{\text{red}}(z) \sum_{p=1}^{m_g} \frac{z^{j-l}(x_p)^j f(x_p, z)}{(x_p)^{l+1} g_x(x_p, z)} T_l^{\text{green}}\left(\frac{1}{x_p}\right). \tag{49}$$

In order to further work out the expression for $W_2(z)$, we first introduce the following auxiliary functions:

$$D_{k,j}(z) \triangleq \sum_{n=0}^{\infty} (1 - \beta_j)\beta_j^n D_k^n(z) , \; k \geq 0; \quad D_j(x,z) \triangleq \sum_{k=0}^{\infty} D_{k,j}(z)\, x^k . \quad (50)$$

We work out the definition of $D_j(x,z)$ making use of the expression (28) for $D_k^n(z)$ and obtain

$$D_j(x,z) = (1 - \beta_j)\sum_{k=0}^{\infty}\sum_{n=0}^{\infty} R_{\text{red}}(z)\, z^n \beta_j^n D_k(z)\, x^{n+k} + (1 - \beta_j)\sum_{k=0}^{\infty}\sum_{n=1}^{\infty} \beta_j^{n+k} z^k x^k$$

$$= (1 - \beta_j)\frac{R_{\text{red}}(z)\, D(x,z)}{1 - \beta_j zx} + \frac{\beta_j z}{1 - \beta_j zx}$$

$$= \frac{f^j(x,z)}{g(x,z)}, \quad (51)$$

with $g(x,z)$ as given in (33) and with

$$f^j(x,z) = \frac{(1 - \beta_j)R_{\text{red}}(z)\, f(x,z) + \beta_j zg(x,z)}{1 - \beta_j zx} . \quad (52)$$

It can be verified that $f^j(x,z)$ is a polynomial function in x of degree $m_g - 1$. We can then find an expression for $D_{k,j}(z)$ as

$$D_{k,j}(z) = \frac{1}{k!}\frac{\partial}{\partial x^k} D_j(x,z)\bigg|_{(x=0)} = \sum_{p=1}^{m_g} \frac{-f^j(x_p,z)}{(x_p)^{k+1} g_x(x_p,z)} . \quad (53)$$

Now we look at the function $W_2(z)$ as given in (48) and work it out further. We introduce the expression (28) for $D_k^n(z)$. For the terms with $G_{l-1}^{\text{green}}(0)$ we again make a distinction between the situations where the delay includes a red period (so when $n = 0$) or not (when $n > 0$). In the terms with $t_l^{\text{green}}(k)$ we can recognize the definition of the auxiliary function $D_{k,j}(z)$. We get

$$W_2(z) = \sum_{j=1}^{M_2} \frac{\omega_j}{\overline{r}_{\text{green}}}\left\{ \tilde{q}_j \left[1 - zR_{\text{red}}(z) + \frac{\beta_j(1 - z)}{1 - \beta_j}\right]\right.$$

$$\left. + \frac{1}{1 - \beta_j}\sum_{p=1}^{m_g} \frac{-f^j(x_p,z)}{x_p g_x(x_p,z)} T^{\text{green},j}\left(\frac{1}{x_p}\right)\right\}, \quad (54)$$

with

$$T^{\text{green},j}(z) \triangleq \sum_{l=1}^{\infty} (1 - \beta_j)\beta_j^{l-1} T_l^{\text{green}}(z)$$

$$= \frac{(1 - \beta_j)F(z)}{z - \beta_j C(z)} G_0^{\text{green}}(z) + \tilde{q}_j \left\{ 1 - \frac{F(z)}{z} + \frac{F(z)}{z}\frac{\beta_j\, [z - C(z)]}{z - \beta_j C(z)}\right\}, \quad (55)$$

and with

$$\widetilde{q}_j \triangleq \sum_{k=0}^{\infty} (1 - \beta_j)\beta_j^k G_k^{\text{green}}(0) \,. \tag{56}$$

We now have an expression for $W(z)$, but it is complex which makes obtaining the full pmf by means of inversion very difficult. It also still contains a (finite) number of unknowns. However we can obtain valuable information about the delay characteristics of a vehicle by means of a tail approximation.

We use the theory of the dominant singularity (see e.g. [3, 20]) in order to obtain the tail characteristics of the delay distribution. We need to find the dominant singularity z_0 of $W(z)$ and calculate the residue w_0 given by

$$w_0 = \lim_{z \to z_0} W(z)(z - z_0) \,. \tag{57}$$

For sufficiently large k the following approximation formula applies:

$$\text{Prob[delay} = k \text{ slots]} \approx -w_0 z_0^{-(k+1)} \,. \tag{58}$$

Note that z_0 is real-valued and larger than 1 and that w_0 is real-valued and negative. Looking at the expressions for $W_{\text{red}}(z)$, $W_1(z)$ and $W_2(z)$ in (44), (49) and (54) we can state that z_0 must be a pole of $G_0^{\text{green}}\left(\frac{1}{x_p}\right)$ and $G_0^{\text{red}}\left(\frac{1}{x_p}\right)$ (if z_0 exists). We have proven a similar conjecture in [17]. The pole z_0 is found for a specific value of p which we call ξ and we denote the value of $x_\xi(z_0)$ as x_0. We can obtain w_0 from (57) using L'Hôpital's rule as

$$
\begin{aligned}
w_0 = &- \frac{\left[R_{\text{red}}\left(C\left(\frac{1}{x_0}\right)\right) - R_{\text{red}}(z_0)\right] f(x_0, z_0) F\left(\frac{1}{x_0}\right) G_0^{\text{green},*}\left(\frac{1}{x_0}\right)}{(\overline{r}_{\text{green}} + \overline{r}_{\text{red}})\left[C\left(\frac{1}{x_0}\right) - z_0\right] x_0 g_x(x_0, z_0)} \\
&+ \sum_{j=1}^{M_1} \frac{\gamma_j}{\overline{r}_{\text{green}} + \overline{r}_{\text{red}}} \sum_{l=1}^{j} z_0^{j-l} R_{\text{red}}(z_0) \frac{-x_0^{j-l-1} f(x_0, z_0)}{g_x(x_0, z_0)} T_l^{\text{green},*}\left(\frac{1}{x_0}\right) \\
&+ \sum_{j=1}^{M_2} \frac{\omega_j}{(1 - \beta_j)(\overline{r}_{\text{green}} + \overline{r}_{\text{red}})} \frac{-f^j(x_0, z_0)}{x_0 g_x(x_0, z_0)} T^{\text{green},j,*}\left(\frac{1}{x_0}\right) \,, \tag{59}
\end{aligned}
$$

with

$$T_l^{\text{green},*}\left(\frac{1}{x_0}\right) \triangleq \left[C\left(\frac{1}{x_0}\right) x_0\right]^{l-1} F\left(\frac{1}{x_0}\right) x_0 G_0^{\text{green},*}\left(\frac{1}{x_0}\right) \,; \tag{60}$$

$$T^{\text{green},j,*}\left(\frac{1}{x_0}\right) \triangleq (1 - \beta_j) \frac{x_0 F\left(\frac{1}{x_0}\right)}{1 - \beta_j x_0 C\left(\frac{1}{x_0}\right)} G_0^{\text{green},*}\left(\frac{1}{x_0}\right) \,, \tag{61}$$

and with

$$G_0^{\text{green},*}\left(\frac{1}{x_0}\right) \triangleq \lim_{z \to z_0} G_0^{\text{green}}\left(\frac{1}{x_\xi(z)}\right)(z - z_0) \,. \tag{62}$$

Following an application of L'Hôpital's rule we can obtain $G_0^{\text{green},*}\left(\frac{1}{x_0}\right)$ from (13) by dividing the numerator of $G_0^{\text{green}}\left(\frac{1}{x_\xi(z)}\right)$ by the derivative with respect to z of its denominator and evaluating at $z = z_0$. This leads to

$$
G_0^{\text{green},*}\left(\frac{1}{x_0}\right) \triangleq \frac{R_{\text{red}}\left(C\left(\frac{1}{x_0}\right)\right)\left[\frac{x_0}{C\left(\frac{1}{x_0}\right)} - x_0^2\right]Q\left(x_0 C\left(\frac{1}{x_0}\right)\right)}{\left\{\begin{array}{l} R'_{\text{red}}\left(C\left(\frac{1}{x_0}\right)\right)C'\left(\frac{1}{x_0}\right)R_{\text{green}}\left(x_0 C\left(\frac{1}{x_0}\right)\right) \\ -R_{\text{red}}\left(C\left(\frac{1}{x_0}\right)\right)R'_{\text{green}}\left(x_0 C\left(\frac{1}{x_0}\right)\right)x_0^2 C\left(\frac{1}{x_0}\right) \\ +R_{\text{red}}\left(C\left(\frac{1}{x_0}\right)\right)R'_{\text{green}}\left(x_0 C\left(\frac{1}{x_0}\right)\right)x_0 C'\left(\frac{1}{x_0}\right) \end{array}\right\}\frac{dx_\xi}{dz}\bigg|_{z=z_0}}.
$$

$$(63)$$

In order to evaluate $\frac{dx_\xi}{dz}\big|_{z=z_0}$ we recall that x_ξ is a solution of $g(x,z) = 0$. We can then find

$$
\frac{dx_\xi}{dz}\bigg|_{z=z_0} \triangleq -\frac{R'_{\text{green}}(x_0 z_0)\,R_{\text{red}}(z_0)\,x_0 + R_{\text{green}}(x_0 z_0)\,R'_{\text{red}}(z_0)}{R'_{\text{green}}(x_0 z_0)\,R_{\text{red}}(z_0)\,z_0}. \tag{64}
$$

5   Numerical Example

The developed method can be used to quantify the effect of possible changes in the infrastructure or policy of a traffic light, as will be illustrated by a numerical example in this section.

A traffic light has two phases: when vehicles can and cannot cross the intersection. These phases are indicated with the effective green time and the effective red time. They do not fully overlap with the actual green and red phases of the traffic light. When the light turns green, there is a start-up loss time because drivers do not instantaneously react to the traffic light change [7,11]. A possible method to decrease the start-up loss time is the installation of countdown timers at a traffic light. The timers display the remaining time until the next green phase will start. This allows drivers to react quicker [8,19]. When the start-up loss can be reduced it results in an increased effective green time, without prolonging the cycle of the traffic light (and thus without decreasing the service capacity for other traffic directions).

We consider the following base situation (i.e. without countdown timers) of a variable-cycle traffic light:

$$
R_{\text{green}}(z) = \frac{z}{20 - 19z}; \quad R_{\text{red}}(z) = \frac{z}{40 - 39z}; \tag{65}
$$

$$
C(z) = e^{\lambda(z-1)}. \tag{66}
$$

We have a green phase with mean length 20 slots, a red phase with mean length 40 slots and a Poisson arrival process. Installing countdown timers increases the effective green times and decreases the effective red times. The following distributions lead to mean period lengths of 21 slots for the green phase and 39 slots for the red phase (resulting in an increase of the capacity with 5%):

$$R_{\text{green, cd}}(z) = \frac{z^2}{20 - 19z}; \quad R_{\text{red, cd}}(z) = \frac{z}{39 - 38z}. \quad (67)$$

In Fig. 1 we plot the mean system content $E[g]$ and the mean overflow queue content $E[g_0^{\text{red}}]$ in function of the arrival intensity λ. We can see that the overflow queue content is always smaller than the mean system content as can be expected. The system contents are lower when the service capacity is increased. The higher the load, the larger the impact of a small increase in capacity. In Fig. 2 we plot in blue the relative change of the mean system content in function of the load in the original traffic queue, i.e. for a given λ we calculate the mean system content for the traffic light with and without timers and plot the relative change in mean system content. At 50% load in the original traffic queue (i.e. with $\lambda = \frac{1}{6}$), the decrease in mean system content is 13% when the service capacity is increased with 5%. The larger the load, the bigger the effect.

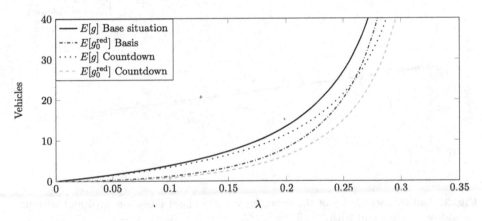

Fig. 1. Mean system content $E[g]$ and mean overflow queue content $E[g_0^{\text{red}}]$ in function of arrival intensity λ for a traffic light with and without countdown timers

It is often reported that an increase in service capacity attracts more traffic. With more service capacity available, the time cost for a certain trip reduces and thus more users are inclined to make this trip, up to the point that the time cost for that trip is equal to before the capacity increase [9]. Applying that mechanism to the current example, we can calculate how much the arrival intensity can increase to reach the point where the mean delay (obtained from the mean system content by applying Little's Law) is the same as before the introduction of countdown timers. The relative increase in traffic that the intersection can handle is plotted in red in Fig. 2. Here we can see that the effect is the strongest for low loads. When the original traffic queue is loaded at 50%, increasing the

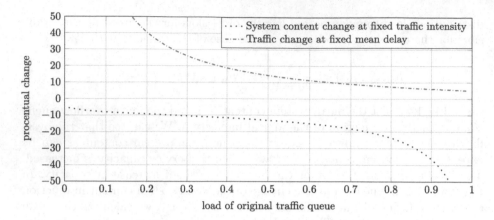

Fig. 2. Relative change in system content at fixed traffic intensity and relative change in capacity at fixed mean delay in function of traffic load of the original queue

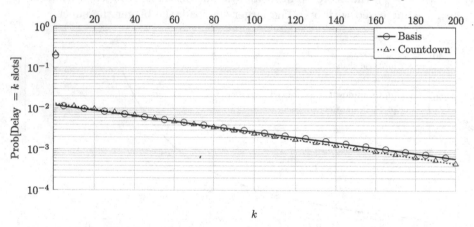

Fig. 3. Tail characteristics of the delay for a traffic light situation with and without countdown timers and with $\lambda = 0.1667$

capacity with 5% by installing countdown timers, can lead to an increase in 14% of traffic that can be handled for the same mean delay.

Let us now look at the delay distribution for the presented example and investigate the possible effect of countdown timers on the tail characteristics of the delay. In Fig. 3 we plot the delay distribution for the original situation and the situation with countdown timers for a traffic arrival intensity $\lambda = 0.1667$. We plot the results based on the tail approximation as discussed in this paper, as well as based on simulation (the markers). Using formula (27) we can obtain the probabilities that a vehicle can pass the intersection without slowing down, this is 20% for the original queue and 22% for the queue with countdown timers. These values are also confirmed by simulation. We can see that the delay characteristics for the original situation show a slightly heavier tail. This can also be concluded when looking at the delay percentiles for both situations. For the original traffic light, 90% of vehicles experience a delay smaller than 140 slots and 99% experience a delay smaller than 291 slots. With countdown timers, these percentiles are reduced to 123 and 259 slots.

References

1. Beckmann, M., McGuire, C., Winsten, C.: Studies in the Economics of Transportation. Yale University Press, New Haven (1956)
2. Bruneel, H., Kim, B.: Discrete-Time Models for Communication Systems Including ATM. Kluwer Academic Publishers Group, Dordrecht (1993)
3. Bruneel, H., Steyaert, B., Desmet, E., Petit, G.: Analytic derivation of tail probabilities for queue lengths and waiting times in ATM multiserver queues. Eur. J. Oper. Res. **76**(3), 563–572 (1994)
4. Bruneel, H., Wittevrongel, S.: Analysis of a discrete-time single-server queue with an occasional extra server. Perform. Eval. **116**, 119–142 (2017)
5. Darroch, J.: On the traffic-light queue. Ann. Math. Stat. **35**, 380–388 (1964)
6. van Leeuwaarden, J.: Delay analysis for the fixed-cycle traffic-light queue. Transp. Sci. **40**(2), 189–199 (2006)
7. Li, H., Prevedouros, P.: Detailed observations of saturation headways and start-up lost times. Transp. Res. Rec. **1802**, 44–53 (2002)
8. Limanond, T., Chookerd, S., Roubtonglang, N.: Effects of countdown timers on queue discharge characteristics of through movement at a signalized intersection. Transp. Res. Part C **17**(6), 662–671 (2009)
9. Litman, T.: Generated traffic and induced travel. Victoria Transport Policy Institute, Canada (2017)
10. Newell, G.: Queues for a fixed-cycle traffic light. Ann. Math. Stat. **31**, 589–597 (1960)
11. Oblakova, A.: Queueing models for urban traffic networks. Ph.D. thesis, University of Twente (2019)
12. Ohno, K.: Computational algorithm for a fixed cycle traffic signal and new approximate expressions for average delay. Transp. Sci. **12**, 29–47 (1978)
13. Tanner, J.: A theoretical analysis of delays at an uncontrolled intersection. Biometrika **49**, 163–170 (1962)
14. Tarko, A.: Random queues in signalized road networks. Transp. Sci. **34**, 415–425 (2000)
15. Verdonck, F., Bruneel, H., Wittevrongel, S.: Analysis of a 2-state discrete-time queue with stochastic state-period lengths and state-dependent server availability and arrivals. Perform. Eval. **135**, article no. 102026, 15p. (2019). https://doi.org/10.1016/j.peva.2019.102026
16. Verdonck, F., Bruneel, H., Wittevrongel, S.: Delay in a 2-state discrete-time queue with stochastic state-period lengths and state-dependent server availability and arrivals. Mathematics **9**(14), article no. 1709, 17p. (2021). https://doi.org/10.3390/math9141709
17. Verdonck, F., Bruneel, H., Wittevrongel, S.: Delay analysis of a discrete-time single-server queue with an occasional extra server. Ann. Oper. Res. **310**, 551–575 (2022)
18. Wardrop, J.: Some theoretical aspects of road traffic research. Proc. Inst. Civ. Eng. **1**, 325–362 (1952)

19. Webo, S., Zhaocheng, H., Xi, X., Feifei, X.: Exploring impacts of countdown timers on queue discharge characteristics of through movement at signalized intersections. Procedia Soc. Behav. Sci. **96**, 255–264 (2013)
20. Woodside, C., Ho, E.: Engineering calculation of overflow probabilities in buffers with Markov-interrupted service. IEEE Trans. Commun. **35**(12), 1272–1277 (1987)

$M/G/1$ Queue with State Dependent Service Times

Zsolt Saffer[1]([✉]) [ID], Karl Grill[1], and Miklós Telek[2,3]

[1] Institute of Statistics and Mathematical Methods in Economics, Vienna University
of Technology, Vienna, Austria
{zsolt.saffer,karl.grill}@tuwien.ac.at
[2] Department of Networked Systems and Services, Budapest University of
Technology and Economics, Budapest, Hungary
[3] ELKH-BME Informations Systems Research Group, Budapest, Hungary
telek@hit.bme.hu

Abstract. In this paper we study the state dependent $M/G/1$ queueing
system in which the service time can change at departure epochs. The
model is a special case of an already investigated model. As a result of
the narrowed scope we get numerically more effective and closed form
solutions.
We provide the steady-state distribution of the number of customers in
the system and the stability condition, both in terms of quantities computed by recursions.
We also study the model with finite number of state dependent service
time distributions. For this model variant, closed form expressions are
provided for the probability-generating function and the mean of the
steady-state number of customers, which are computed from a system of
linear equations.
Finally we also investigate the model with state dependent linear interpolation of two service times. For this model, we derive an explicit expression for the probability generating function of the steady-state number of
customers and establish a simple, explicit stability condition. This model
behaviour implements a control of number of customers in the system.

Keywords: queueing theory · state dependent service time
distribution · control of queues

1 Introduction

The requirement for controlling the behaviour of queueing systems is a natural
demand in the areas of their applications. One way of achieving this control is
to implement a state dependent behaviour of the server.

The authors thank the recommendations of the reviewers which helped to improve the
presentation of the paper. This work is partially supported by the Hungarian Scientific
Research Fund OTKA K-138208 project.

State dependent queueing systems appear in the literature since the 1960s. State dependency have been studied in the context of $M/M/$ systems as system with state dependent service rates. Such system has been analysed in the early work of [5], in which the service rate is specified to be proportional to the power of the number of customers. Harris [7] investigated a model, in which the dependency of service rate on the number of customers is linear. In the work [11] a two-state state dependent M/G/1 queue is investigated and the Laplace transform (LT) of the steady-state waiting time distribution is obtained. Gupta and Rao [6] studied a finite buffer queue with state dependent arrival rates and service times. They provided the distribution of the number of customers in the system. Kerner [8] considered an M/G/1 system with state dependent arrival rates and derived a closed form expression for the distribution of the customers in the system in terms of the idle probability.

M/G/1 queue with state dependent service times has been considered by many authors [3,10]. Abouee-Mehrizi and Baron [1] investigated an M/G/1 queue with state dependent arrival rates and service times. They provided expression for the steady-state distribution of the number of customers in the system in terms of quantities depending on the LTs of the conditional state dependent residual service times, given the state of the system. These LTs are computed recursively and the solution requires $\mathcal{O}(K^2)$ operational steps, where K is the highest state to be taken into account to get the solution in required accuracy. The follow-up work on [1] is [2], which extends the model to multiclass, multiserver queueing network, uses fluid approximation to provide stability condition and the equilibrium analysis of the system.

In this paper we study the state dependent M/G/1 queueing system in which the service time can change at departure epochs. This model is a special case of the model studied in [1], which is obtained by omitting the state dependency of arrival rates and at arrival epochs. As a result of the narrowed scope we get numerically more effective and for some cases simple closed form solutions.

We analyse the model at embedded departure epochs and use standard queueing arguments. We establish a forward recursion for computing the steady-state distribution of the number of customers in the system. This recursion requires also $\mathcal{O}(K^2)$ operational steps. We establish the stability condition of the model in terms of the quantities computed by the above mentioned forward recursion. We also study the special case of the model, in which only the first finite number of service times are state dependent. We provide closed form expressions for the probability-generating function (PGF) and the mean of the steady-state number of customers, which are computed from a system of linear equations. Additionally we also investigate the special case of the model with state dependent linear interpolation of two service times. For this model we derive an explicit expression for the PGF of the steady-state number of customers and establish a simple, closed form sufficient stability condition. The computation of the PGF and the moments of the steady-state number of customers requires $\mathcal{O}(K)$ operational steps, where K is a highest numerical index in infinite products and sums to be taken into account to get the PGF and the moments in required accuracy. This model is appropriate to implement a kind of control of number of customers in the system.

Compared to [1], there are significant differences in the applied analysis method in the current work. First of all, the state independent arrival process makes the Pasta property valid in our model, while it does not hold in [1]. The other significant simplification is that [1] is built on the computation of the steady-state residual service time distribution as a function of the customers in the system, while we compute the performance measures based on the number of Poisson arrivals during the state dependent service time.

The above $M/G/1$ queueing model with state dependent service times can be applied to queueing modelling in healthcare. An essential property of queues in healthcare is that their arrival processes, service times are state dependent [4]. This motivated us

1. to establish simple solvable state dependent queueing model, like the special case with finite number of different service times and
2. to provide simple control mechanism, like the special case with state dependent linear interpolation of two service times, in order to decrease the waiting time.

The rest of this paper is organized as follows. In Sect. 2 we describe the model and the notations. The steady-state analysis of the model is provided in Sect. 3. Section 4 is devoted to the model variant with finite number of state dependent service times. Finally the investigation of the model with state dependent linear interpolation of two service time is presented in Sect. 5.

2   Model Description

We consider an infinite buffer queue. The arrival process is Poisson with rate $0 < \lambda < \infty$. The customer service time depends on the number of customers in the system and it is set at the customer departure epoch (after the departure of the last served customer and before the start of the next customer service) in the server. B_n, $b_n(t)$, $\tilde{B}_n(s)$, b_n and $b_n^{(2)}$ denote the service time random variable, its probability density function (pdf), its LT, its mean and its second moment when the number of customers is $n \geq 0$ at the customer departure epoch, respectively. The customer service times are independent with finite means, i.e. $0 < b_n < \infty$ for $n \geq 0$. For notational convenience we introduce $B_0 = B_1$.

We impose the usual assumptions on the model. The arrival process and the customer service times are mutually independent. The customers are served in First-In-First-Out (FIFO) order as well as the service during the service period is work conserving and non-preemptive. We denote the above described $M/G/1$ queue as $M/G_n/1$ queue.

When $\hat{x}(z)$ is a PGF, $\hat{x}^{(k)}$ denotes its k-th derivative at $z = 1$ for $k \geq 1$, i.e., $\hat{x}^{(k)} = \frac{d^k}{dz^k}\hat{x}(z)|_{z=1}$. Similarly when $\tilde{y}(s)$ is a LT, then $\tilde{y}^{(k)}(s_0)$ denotes its k-th derivative at $s = s_0$ for $k \geq 1$, i.e., $\tilde{y}^{(k)}(s_0) = \frac{d^k}{ds^k}\tilde{y}(s)|_{s=s_0}$. Additionally $\tilde{y}^{(0)}(s_0)$ denotes $\tilde{y}(s_0)$.

3 General $M/G_n/1$ System

3.1 Relation for the PGF of the Steady-State Number of Customers

Let t_k^d stand for the epoch just after the departure of the k-th customer. Let $N(t)$ denote the number of customers in the system at time t for $t \geq 0$. Then the probabilities at arbitrary epoch, p_n, and at departure epochs, p_n^d, are defined as

$$p_n = \lim_{t \to \infty} P\{N(t) = n\}, n \geq 0,$$

$$p_n^d = \lim_{k \to \infty} P\{N(t_k^d) = n\}, n \geq 0.$$

The corresponding steady-state PGFs are defined as

$$\widehat{P}(z) = \lim_{t \to \infty} \sum_{n=0}^{\infty} P\{N(t) = n\} z^n, |z| \leq 1,$$

$$\widehat{P}^d(z) = \lim_{k \to \infty} \sum_{n=0}^{\infty} P\{N(t_k^d) = n\} z^n, |z| \leq 1.$$

We also define the joint transform $\bar{Q}(s, z)$ as

$$\bar{Q}(s, z) = \sum_{n=0}^{\infty} p_n \tilde{B}_n(s) z^n, Re(s) \geq 0 \text{ and } |z| \leq 1.$$

Theorem 1. *In the stable $M/G_n/1$ system*

$$\widehat{P}(z) = \frac{\bar{Q}(\lambda - \lambda z, z) - p_0 \tilde{B}_1(\lambda - \lambda z)}{z} + p_0 \tilde{B}_1(\lambda - \lambda z). \tag{1}$$

Proof. The PGF of the number of customers arriving during a service time with pdf $b_n(t)$ is $\tilde{B}_n(\lambda - \lambda z)$. If the number of customers present at the k-th departure epoch is $n \geq 1$ then it decreases by one at the next departure epoch due to the actual customer service and increases by the number of customers arriving during the service time B_n. This can be described on PGF level as multiplication by $\frac{\tilde{B}_n(\lambda - \lambda z)}{z}$. Assuming that the system is idle at the k-th departure epoch, the customers present at the next departure epoch are the ones arriving during the service time B_1. This means a multiplication by $\tilde{B}_1(\lambda - \lambda z)$ on PGF level. Putting it together gives

$$\sum_{n=0}^{\infty} P\{N(t_{k+1}^d) = n\} z^n = \sum_{n=1}^{\infty} P\{N(t_k^d) = n\} z^n \frac{\tilde{B}_n(\lambda - \lambda z)}{z}$$
$$+ P\{N(t_k^d) = 0\} \tilde{B}_1(\lambda - \lambda z).$$

Taking $\lim_{k\to\infty}$ we get

$$\lim_{k\to\infty} \sum_{n=0}^{\infty} P\left\{N(t_{k+1}^d) = n\right\} z^n = \lim_{k\to\infty} \sum_{n=1}^{\infty} P\left\{N(t_k^d) = n\right\} z^n \frac{\tilde{B}_n(\lambda - \lambda z)}{z}$$
$$+ \lim_{k\to\infty} P\left\{N(t_k^d) = 0\right\} \tilde{B}_1(\lambda - \lambda z).$$

Due to stability the limit and sum can be exchanged. Applying it on the first term on the r.h.s and using the notations for the steady-state quantities we get

$$\widehat{P}^d(z) = \sum_{n=1}^{\infty} p_n^d z^n \frac{\tilde{B}_n(\lambda - \lambda z)}{z} + p_0^d \tilde{B}_1(\lambda - \lambda z). \tag{2}$$

In this $M/G_n/1$ system the state of the system can be changed only in unit step and PASTA also holds due to Poisson arrivals. Thus the steady-state distribution of the number of customers at departure epochs, at arrival epochs as well as at arbitrary epochs are the same (see e.g. in [9]). Utilizing it we can rearrange (2) as

$$\widehat{P}(z) = \sum_{n=1}^{\infty} p_n z^n \frac{\tilde{B}_n(\lambda - \lambda z)}{z} + p_0 \tilde{B}_1(\lambda - \lambda z). \tag{3}$$

The statement comes by applying the definition of $\bar{Q}(s, z)$ in (3).   □

Corollary 1. *In the stable $M/G_n/1$ system, the steady-state probability that the system is in idle state, p_0 can be given as*

$$p_0 = 1 - \lambda \sum_{i=0}^{\infty} p_i b_i. \tag{4}$$

Proof. Expressing p_0 from (1) gives

$$p_0 = \frac{z\widehat{P}(z) - \bar{Q}(\lambda - \lambda z, z)}{(z - 1)\tilde{B}_1(\lambda - \lambda z)}. \tag{5}$$

Taking $\lim_{z\to 1}$ on (5) and using the L'Hospital rule we get

$$p_0 = \frac{1 + \widehat{P}^{(1)}(1) - \frac{\partial \bar{Q}(s,z)}{\partial s}\Big|_{s=\lambda-\lambda z, z=1}(-\lambda) - \frac{\partial \bar{Q}(s,z)}{\partial z}\Big|_{s=\lambda-\lambda z, z=1}}{1}$$

$$= 1 + \widehat{P}^{(1)}(1) - \lambda \sum_{i=0}^{\infty} p_i b_i - \widehat{P}^{(1)}(1) = 1 - \lambda \sum_{i=0}^{\infty} p_i b_i.$$

□

Remark 1. Probability that the queue is busy and the Little's law
Observe that the sum in (4) is exactly the mean service time, i.e. $E[B] = \sum_{i=0}^{\infty} p_i b_i$. Thus the steady-state probability of the system being busy is given as

$$P\{\text{busy}\} = 1 - p_0 = \lambda E[B],$$

which justifies the Little's law for this system due to $E[N_B] = P\{\text{busy}\}$, where N_B is the number of customers in the server.

Remark 2. Utilization and stability
The utilization, ρ of the system is defined as

$$\rho = \lambda E[B],$$

Thus the stability condition of the system can be given on two equivalent ways as

$$\rho < 1 \quad \Leftrightarrow \quad 0 < p_0.$$

3.2   The Steady-State Distribution of the Number of Customers

We define the following auxiliary quantities

$$c_{i,0} = \tilde{B}_i(\lambda), i \geq 1$$
$$c_{i,j} = \frac{(-\lambda)^j}{j!} \tilde{B}_i^{(j)}(\lambda), i, j \geq 1.$$

Remark 3. Interpretation of the quantities $c_{i,j}$
The quantity $c_{i,j}$, for $i \geq 1$ and $j \geq 0$ can be interpreted as the probability of arriving j customers during the service time B_i. This can be seen as

$$\frac{(-\lambda)^j}{j!} \tilde{B}_i^{(j)}(\lambda) = \frac{(-\lambda)^j}{j!} \int_{x=0}^{\infty} (-x)^j e^{-\lambda x} b_i(x) dx$$
$$= \int_{x=0}^{\infty} \frac{(\lambda x)^j}{j!} e^{-\lambda x} b_i(x) dx = P\{j \text{ arrivals during } B_i\}.$$

The above integral provides a way to compute the quantities $c_{i,j}$.

Theorem 2. *In the stable $M/G_n/1$ system the steady-state probabilities of the number of customers are given by*

$$p_0 = \frac{1}{\sum_{i=0}^{\infty} \alpha_i},$$
$$p_i = p_0 \alpha_i \quad i \geq 0, \tag{6}$$

where α_i-s can be determined recursively as

$$\alpha_0 = 1,$$

$$\alpha_1 = \frac{1 - c_{1,0}}{c_{1,0}},$$

$$\alpha_n = \alpha_{n-1}\frac{1 - c_{n-1,1}}{c_{n,0}} - \sum_{i=1}^{n-2}\alpha_i\frac{c_{i,n-i}}{c_{n,0}} - \frac{c_{1,n-1}}{c_{n,0}} \quad n \geq 2. \tag{7}$$

Proof. Let v_n be the steady-state probability that a departing customer leaves n customers in the system. The $c_{i,j}$ probabilities define the following relation of the v_n probabilities

$$v_n = \sum_{i=1}^{n+1} v_i c_{i,n-i+1} + v_0 c_{1,n}. \tag{8}$$

Utilizing that the number of customers in the queue can change by one at a time, a customer arriving in steady-state finds n customers in the queue with probability v_n. Additionally utilizing the PASTA property, we have $p_n = v_n$ for $n \geq 0$.

Using $p_n = v_n$ and expressing $p_{n+1}c_{n+1,0}$ from (8) gives

$$p_1 c_{1,0} = p_0(1 - c_{1,0})$$

$$p_{n+1}c_{n+1,0} = p_n(1 - c_{n,1}) - \sum_{i=1}^{n-1} p_i c_{i,n-i+1} - p_0 c_{1,n} \quad n \geq 1.$$

Changing the index $n + 1 \to n$ leads to the expression of p_n as

$$p_1 = p_0\frac{1 - c_{1,0}}{c_{1,0}}$$

$$p_n = p_{n-1}\frac{1 - c_{n-1,1}}{c_{n,0}} - \sum_{i=1}^{n-2} p_i\frac{c_{i,n-i}}{c_{n,0}} - p_0\frac{c_{1,n-1}}{c_{n,0}} \quad n \geq 2. \tag{9}$$

The recursive forms for determining α_i-s are coming from applying $p_i = p_0\alpha_i$ for $i \geq 0$, from (6), in (9). Finally p_0 can be determined from the normalization condition $1 = \sum_{i=0}^{\infty} p_i = p_0 \sum_{i=0}^{\infty} \alpha_i$. $\qquad\square$

Corollary 2. *The necessary and sufficient condition of the stability of the $M/G_n/1$ system is given by*

$$0 < \sum_{i=0}^{\infty} \alpha_i < \infty. \tag{10}$$

Proof. It follows directly from the expression of p_0 in (6). $\qquad\square$

4 $M/G_n/1$ System with a Finite Number of Different Service Time Distributions

Let $K \geq 0$, such that

$$B_i = B_\infty, i \geq K. \tag{11}$$

and $b_\infty(t)$, $\tilde{B}_\infty(s)$, b_∞ and $b_\infty^{(2)}$ stands for the related pdf, LT, first and second moment, respectively. That is, when the number of customers in the system is above K the service time pdf is always $b_\infty(t)$. This modelling restrictions allows to derive closed form expressions for the PGF of the steady-state number of customers and the mean steady-state number of customers.

4.1   The PGF of the Steady-State Number of Customers

Proposition 1. *In the stable $M/G_n/1$ system with finite number of state dependent service times, the steady-state probability that the system is idle is*

$$p_0 = 1 - \lambda \left(b_\infty + \sum_{i=0}^{K-1} p_i(b_i - b_\infty) \right). \tag{12}$$

Proof. For the system with finite number of state dependent service times, the sum $\sum_{i=0}^{\infty} p_i b_i$ can be rewritten as

$$\sum_{i=0}^{\infty} p_i b_i = \sum_{i=0}^{K-1} p_i b_i + \sum_{i=K}^{\infty} p_i b_\infty = \sum_{i=0}^{K-1} p_i b_i + \left(1 - \sum_{i=0}^{K-1} p_i\right) b_\infty$$

$$= b_\infty + \sum_{i=K}^{\infty} p_i(b_i - b_\infty). \tag{13}$$

Applying (13) in (4) gives the statement.   □

Let $d_0 = 1 + \lambda(b_1 - b_\infty)$ and for $i = 1, \ldots, K-1$ let $d_i = \lambda(b_i - b_\infty)$.

Theorem 3. *In the stable $M/G_n/1$ system with finite number of state dependent service times, the PGF of the steady-state number of customers in the system is*

$$\widehat{P}(z) = p_0 \frac{(1-z)\tilde{B}_1(\lambda - \lambda z)}{\tilde{B}_\infty(\lambda - \lambda z) - z} + \frac{\sum_{i=0}^{K-1} p_i(\tilde{B}_\infty(\lambda - \lambda z) - \tilde{B}_i(\lambda - \lambda z))z^i}{\tilde{B}_\infty(\lambda - \lambda z) - z}, \tag{14}$$

and the steady-state probabilities can be expressed as

$$(p_0, \ldots, p_{K-2}, p_{K-1}) = (0, \ldots, 0, 1 - \lambda b_\infty)\mathbf{M}^{-1}, \tag{15}$$

where the coefficient matrix $\mathbf{M}$ is given by

$$\mathbf{M} = \begin{pmatrix} c_{1,0} - 1 & c_{1,1} & c_{1,2} & \cdots & \cdots & c_{1,K-2} & d_0 \\ c_{1,0} & c_{1,1} - 1 & c_{1,2} & \cdots & \cdots & c_{1,K-2} & d_1 \\ & c_{2,0} & c_{2,1} - 1 & \cdots & \cdots & c_{2,K-3} & d_2 \\ & & c_{3,0} & \ddots & \cdots & c_{3,K-4} & d_3 \\ & & & \ddots & \ddots & \vdots & \vdots \\ & & & & c_{K-2,0} & c_{K-2,1} - 1 & d_{K-2} \\ & & & & & c_{K-1,0} & d_{K-1} \end{pmatrix}.$$

Proof. For the system with finite number of state dependent service times, the joint transform $\bar{Q}(s, z)$ can be rearranged as

$$\bar{Q}(s, z) = \sum_{i=0}^{\infty} p_i \tilde{B}_i(s) z^i = \sum_{i=0}^{K-1} p_i \tilde{B}_i(s) z^i + \sum_{i=K}^{\infty} p_i \tilde{B}_\infty(s) z^i$$

$$= \sum_{i=0}^{K-1} p_i \tilde{B}_i(s) z^i + \tilde{B}_\infty(s) \left(\widehat{P}(z) - \sum_{i=0}^{K-1} p_i z^i \right)$$

$$= \tilde{B}_\infty(s) \widehat{P}(z) + \sum_{i=0}^{K-1} p_i \left(\tilde{B}_i(s) - \tilde{B}_\infty(s) \right) z^i. \tag{16}$$

Applying (16) in (1) and rearranging it gives

$$\left(z - \tilde{B}_\infty(\lambda - \lambda z) \right) \widehat{P}(z) = p_0 (z - 1) \tilde{B}_1(\lambda - \lambda z)$$

$$+ \sum_{i=0}^{K-1} p_i \left(\tilde{B}_n(\lambda - \lambda z) - \tilde{B}_\infty(\lambda - \lambda z) \right) z^i. \tag{17}$$

Further rearranging of (17) results in the expression (14).
The relation (12) can be rearranged as

$$p_0 + \lambda \left(\sum_{i=0}^{K-1} p_i (b_i - b_\infty) \right) = 1 - \lambda b_\infty. \tag{18}$$

This provides a linear equation for $p_0, \ldots, p_{K-1}$, which is represented by the last column of matrix $\mathbf{M}$. The linear relations in Equation (8) for $n = 0, \ldots, K - 2$ are represented by the first $K - 1$ columns of matrix $\mathbf{M}$. $\qquad\square$

Equation (15) provides an explicit expression for $(p_0, \ldots, p_{K-2}, p_{K-1})$, which can be evaluated efficiently utilizing the quasi-triangular structure of matrix $\mathbf{M}$.

4.2 The Mean Steady-State Number of Customers

Corollary 3. *In the stable $M/G_n/1$ system with finite number of state dependent service times, the mean steady-state number of customers in the system,*

$p^{(1)}$, is given as

$$p^{(1)} = \frac{\lambda^2 b_\infty^{(2)}}{2(1 - \lambda b_\infty)} + \frac{(2\lambda b_1 + \lambda^2(b_1^{(2)} - b_\infty^{(2)}))p_0}{2(1 - \lambda b_\infty)}$$
$$+ \frac{\sum_{i=1}^{K-1} p_i(\lambda^2 b_i^{(2)} - \lambda^2 b_\infty^{(2)}) + 2\sum_{i=1}^{K-1} i \; p_i\lambda(b_i - b_\infty)}{2(1 - \lambda b_\infty)}. \tag{19}$$

Proof. The expression (14) can be rearranged as

$$\left(\tilde{B}_\infty(\lambda - \lambda z) - z\right)\hat{P}(z) = p_0(1 - z)\tilde{B}_1(\lambda - \lambda z)$$
$$+ \sum_{i=0}^{K-1} p_i(\tilde{B}_\infty(\lambda - \lambda z) - \tilde{B}_i(\lambda - \lambda z))z^i. \tag{20}$$

Taking the second derivative of (20) with respect to z and setting $z = 1$ gives

$$\lambda^2 b_\infty^{(2)} + 2(\lambda b_\infty - 1)p^{(1)} = -p_0 2\lambda b_1$$
$$+ \sum_{i=0}^{K-1} p_i \left((\lambda^2 b_\infty^{(2)} - \lambda^2 b_i^{(2)}) + 2\lambda(b_\infty - b_i)i\right).$$

This can be rearranged as

$$\lambda^2 b_\infty^{(2)} + 2(\lambda b_\infty - 1)p^{(1)} = -(2\lambda b_1 + \lambda^2(b_1^{(2)} - b_\infty^{(2)}))p_0$$
$$+ \sum_{i=1}^{K-1} p_i \left((\lambda^2 b_\infty^{(2)} - \lambda^2 b_i^{(2)}) + 2\lambda(b_\infty - b_i)i\right),$$

from which the statement comes by expressing $p^{(1)}$. $\qquad\square$

5 $M/G_n/1$ System with State Dependent Linear Interpolation of Two Service Times

In this section, we consider the special case when the state dependent service time is

$$B_n = (1 - Y_n)B_f + Y_n B_s, n \geq 1, \tag{21}$$

and Y_n is a Bernoulli distributed random variable with $P\{Y_n = 1\} = \eta(1 - \delta^{n-1})$, with $0 \leq \eta, \delta \leq 1$. That is, the state dependent service time is characterized by two service times B_f and B_s. Parameter η determines the maximum portion of the second service time, B_s, in B_n, while parameter δ controls the dependence on the number of customers in the system. Qualitatively, for small n values the service time is B_f with high probability and for large n values it is B_s with higher probability (i.e. closer to η).

In Laplace transform domain

$$\tilde{B}_n(s) = \left((1-\eta) + \eta\delta^{n-1}\right)\tilde{B}_f(s) + \eta(1-\delta^{n-1})\tilde{B}_s(s)$$
$$= (1-\eta)\tilde{B}_f(s) + \eta\left(\delta^{n-1}\tilde{B}_f(s) + (1-\delta^{n-1})\tilde{B}_s(s)\right), \quad (22)$$

where $\tilde{B}_f(s)$ and $\tilde{B}_s(s)$ are the LT of B_f and B_s, respectively, and $\tilde{B}_0(s) = \tilde{B}_1(s) = \tilde{B}_f(s)$ by definition.

This model enables also that the second service time is higher than the first one until keeping the queue stable. However in the usual application scenarios of this model, $B_s < B_f$ (meaning that $P\{B_s < t\} \geq P\{B_f < t\}$ for $\forall t > 0$) is implemented in order to realize a kind of control of number of customers in the system. This control mechanism makes the number of customers in the system and the waiting time lower.

5.1 The PGF of the Steady-State Number of Customers in the System

Theorem 4. *In the stable $M/G_n/1$ system with state dependent linear interpolation of two service times, $\widehat{P}(z)$ satisfies*

$$\widehat{P}(z) = \frac{\beta(\lambda - \lambda z)}{\alpha(\lambda - \lambda z) - z}\widehat{P}(\delta z) + \frac{\alpha(\lambda - \lambda z) - \beta(\lambda - \lambda z) - z\tilde{B}_f(\lambda - \lambda z)}{\alpha(\lambda - \lambda z) - z}p_0, \quad (23)$$

where $\alpha(s) = (1-\eta)\tilde{B}_f(s) + \eta\tilde{B}_s(s)$ and $\beta(s) = \frac{\eta}{\delta}\left(\tilde{B}_s(s) - \tilde{B}_f(s)\right)$.

Proof. For this system the joint transform $\bar{Q}(s,z)$ can be written as

$$\bar{Q}(s,z) = \sum_{n=0}^{\infty} p_n\tilde{B}_n(s)z^n = p_0\tilde{B}_0(s) + \sum_{n=1}^{\infty} p_n\tilde{B}_n(s)z^n$$

$$= p_0\tilde{B}_f(s) + \sum_{n=1}^{\infty} p_n\left((1-\eta)\tilde{B}_f(s) + \eta\left(\delta^{n-1}\tilde{B}_f(s) + (1-\delta^{n-1})\tilde{B}_s(s)\right)\right)z^n$$

$$= \tilde{B}_f(s)p_0 + \left((1-\eta)\tilde{B}_f(s) + \eta\tilde{B}_s(s)\right)\sum_{n=1}^{\infty} p_n z^n$$

$$+ \eta\left(\tilde{B}_f(s) - \tilde{B}_s(s)\right)\sum_{n=1}^{\infty} p_n\delta^{n-1}z^n$$

$$= \tilde{B}_f(s)p_0 + \left((1-\eta)\tilde{B}_f(s) + \eta\tilde{B}_s(s)\right)\left(\widehat{P}(z) - p_0\right)$$

$$+ \frac{\eta}{\delta}\left(\tilde{B}_f(s) - \tilde{B}_s(s)\right)\left(\widehat{P}(\delta z) - p_0\right).$$

Further rearrangement gives

$$\bar{Q}(s,z) = \Big((1-\eta)\tilde{B}_f(s) + \eta\tilde{B}_s(s) \Big) \widehat{P}(z) - \frac{\eta}{\delta} \Big(\tilde{B}_s(s) - \tilde{B}_f(s) \Big) \widehat{P}(\delta z)$$
$$- \bigg(\Big((1-\eta)\tilde{B}_f(s) + \eta\tilde{B}_s(s) \Big) - \frac{\eta}{\delta} \Big(\tilde{B}_s(s) - \tilde{B}_f(s) \Big) - \tilde{B}_f(s) \bigg) p_0.$$

This can be written by means of the functions $\alpha(s)$ and $\beta(s)$ as

$$\bar{Q}(s,z) = \alpha(s)\widehat{P}(z) - \beta(s)\widehat{P}(\delta z) - \Big(\alpha(s) - \beta(s) - \tilde{B}_f(s) \Big) p_0. \qquad (24)$$

Applying (24) in (1) and rearranging it gives

$$(z - \alpha(\lambda - \lambda z))\,\widehat{P}(z) = -\beta(\lambda - \lambda z)\widehat{P}(\delta z)$$
$$- \Big(\alpha(\lambda - \lambda z) - \beta(\lambda - \lambda z) - z\tilde{B}_f(\lambda - \lambda z) \Big) p_0. \quad (25)$$

The statement comes by further rearranging of (25). $\qquad\square$

Theorem 5. *In the stable $M/G_n/1$ system with state dependent linear interpolation of two service times, the PGF of the steady-state number of customers in the system, $\widehat{P}(z)$, is given as*

$$\widehat{P}(z) = \prod_{k=0}^{\infty} \frac{\beta(\lambda - \lambda\delta^k z)}{\alpha(\lambda - \lambda\delta^k z) - \delta^k z} p_0$$
$$+ \sum_{k=0}^{\infty} \frac{\alpha(\lambda - \lambda\delta^k z) - \beta(\lambda - \lambda\delta^k z) - \delta^k z\tilde{B}_f(\lambda - \lambda\delta^k z)}{\alpha(\lambda - \lambda\delta^k z) - \delta^k z}$$
$$\times \prod_{i=0}^{k-1} \frac{\beta(\lambda - \lambda\delta^i z)}{\alpha(\lambda - \lambda\delta^i z) - \delta^i z} p_0, \qquad (26)$$

where p_0 is given by

$$p_0 = \bigg(\prod_{k=0}^{\infty} \frac{\beta(\lambda - \lambda\delta^k)}{\alpha(\lambda - \lambda\delta^k) - \delta^k} \qquad\qquad\qquad (27)$$
$$+ \sum_{k=0}^{\infty} \frac{\alpha(\lambda - \lambda\delta^k) - \beta(\lambda - \lambda\delta^k) - \delta^k\tilde{B}_f(\lambda - \lambda\delta^k)}{\alpha(\lambda - \lambda\delta^k) - \delta^k} \prod_{i=0}^{k-1} \frac{\beta(\lambda - \lambda\delta^i)}{\alpha(\lambda - \lambda\delta^i) - \delta^i} \bigg)^{-1}.$$

Proof. Replacing z by $\delta^k z$ in (23) for $k \geq 0$ yields

$$\widehat{P}(\delta^k z) = \frac{\beta(\lambda - \lambda\delta^k z)}{\alpha(\lambda - \lambda\delta^k z) - \delta^k z} \widehat{P}(\delta^{k+1} z)$$
$$+ \frac{\alpha(\lambda - \lambda\delta^k z) - \beta(\lambda - \lambda\delta^k z) - \delta^k z\tilde{B}_f(\lambda - \lambda\delta^k z)}{\alpha(\lambda - \lambda\delta^k z) - \delta^k z} p_0.$$

Solving the above equation by recursive substitution for $k \geq 0$ leads to

$$
\widehat{P}(z) = \prod_{k=0}^{\infty} \frac{\beta(\lambda - \lambda \delta^k z)}{\alpha(\lambda - \lambda \delta^k z) - \delta^k z} \lim_{k \to \infty} \widehat{P}(\delta^k z)
$$
$$
+ \sum_{k=0}^{\infty} \frac{\alpha(\lambda - \lambda \delta^k z) - \beta(\lambda - \lambda \delta^k z) - \delta^k z \tilde{B}_f(\lambda - \lambda \delta^k z)}{\alpha(\lambda - \lambda \delta^k z) - \delta^k z}
$$
$$
\times \prod_{i=0}^{k-1} \frac{\beta(\lambda - \lambda \delta^i z)}{\alpha(\lambda - \lambda \delta^i z) - \delta^i z} p_0. \tag{28}
$$

Due to $\delta < 1$

$$
\lim_{k \to \infty} \widehat{P}(\delta^k z) = \widehat{P}(0) = p_0.
$$

Applying this limit in (28) gives the first part of the statement, the relation (26). The second relation, (27) comes by setting $z = 1$ in (26) and expressing p_0 from it. $\qquad\square$

Remark 4. Numerical complexity
The computation of $\widehat{P}(z)$ by means of (26) and the steady-state moments of the number of customers in the system requires $\mathcal{O}(K)$ operational steps, where K is the highest index in the infinite products and sums to be taken into account to get the PGF and the moments in required accuracy. This is because these computations require the computation of K points of the LTs $\tilde{B}_f(s)$, $\tilde{B}_s(s)$ and their derivatives.

5.2 Stability

Proposition 2. *The necessary and sufficient condition of the stability of $M/G_n/1$ system with state dependent linear interpolation of two service times is*

$$
\prod_{k=0}^{\infty} \frac{\beta(\lambda - \lambda \delta^k)}{\alpha(\lambda - \lambda \delta^k) - \delta^k} \tag{29}
$$
$$
+ \sum_{k=0}^{\infty} \frac{\alpha(\lambda - \lambda \delta^k) - \beta(\lambda - \lambda \delta^k) - \delta^k \tilde{B}_f(\lambda - \lambda \delta^k)}{\alpha(\lambda - \lambda \delta^k) - \delta^k} \prod_{i=0}^{k-1} \frac{\beta(\lambda - \lambda \delta^i)}{\alpha(\lambda - \lambda \delta^i) - \delta^i} < \infty.
$$

Proof. The necessary and sufficient condition of the stability is $p_0 > 0$, which is equivalent with the denominator of (27) being convergent. Thus this statement is a direct consequence of the expression (27). $\qquad\square$

Corollary 4. *A sufficient condition of the stability of $M/G_n/1$ system with state dependent linear interpolation of two service times is given as*

$$
\eta \left(\frac{1}{\delta} - 1 \right) \left(\tilde{B}_s(\lambda) - \tilde{B}_f(\lambda) \right) < \tilde{B}_f(\lambda), \quad if \;\; \tilde{B}_s(\lambda) \geq \tilde{B}_f(\lambda), \tag{30}
$$
$$
\eta \left(\frac{1}{\delta} + 1 \right) \left(\tilde{B}_f(\lambda) - \tilde{B}_s(\lambda) \right) < \tilde{B}_f(\lambda), \quad if \;\; \tilde{B}_s(\lambda) < \tilde{B}_f(\lambda).
$$

Proof. We evaluate the convergence of (29) under the condition

$$\left|\frac{\beta(\lambda)}{\alpha(\lambda)}\right| < 1. \tag{31}$$

Under this condition there exists an enough large K_1 for which $\left|\frac{\beta(\lambda-\lambda\delta^k)}{(\alpha(\lambda-\lambda\delta^k)-\delta^k)}\right| = r < 1$ for every $k \geq K_1$. Hence the first product term in (29) must vanish, in other words

$$\prod_{k=0}^{\infty} \frac{\beta(\lambda - \lambda\delta^k)}{\alpha(\lambda - \lambda\delta^k) - \delta^k} = 0. \tag{32}$$

The expression after the sum in the second term of (29) can be upper limited for enough large k as follows. According to the above argument the fraction $-\frac{\beta(\lambda-\lambda\delta^k)}{\alpha(\lambda-\lambda\delta^k)-\delta^k} < 1$ for any $k \geq K_1$. The functions $\tilde{B}_f(\lambda - \lambda z)$ and $\alpha(\lambda - \lambda z)$ are PGFs, since both $\tilde{B}_f(s)$ and $\alpha(s)$ are LTs of continuous random variables representing a durations and hence the above functions can be interpreted as the PGF of the number of arriving customers during these random durations. PGFs have positive values for $0 \leq z \leq 1$ and their value at z close to 0 are greater than z as far as the number of zero arrival in their above interpretations has positive probability. Thus there exists a K_2 for which for any $k \geq K_2$ $z = \delta^k$ is close enough to 0 to have $\frac{\tilde{B}_f(\lambda-\lambda\delta^k)}{\alpha(\lambda-\lambda\delta^k)-\delta^k} > 0$. The fraction $\frac{\alpha(\lambda-\lambda\delta^k)}{\alpha(\lambda-\lambda\delta^k)-\delta^k}$ is upper limited by $\frac{\alpha(\lambda-\lambda\delta^K)}{\alpha(\lambda-\lambda\delta^K)-\delta^K}$ for any $k \geq K$, where $K = max(K_1, K_2)$. Putting all these together

$$\frac{\alpha(\lambda - \lambda\delta^k) - \beta(\lambda - \lambda\delta^k) - \delta^k \tilde{B}_f(\lambda - \lambda\delta^k)}{\alpha(\lambda - \lambda\delta^k) - \delta^k} \leq \frac{\alpha(\lambda - \lambda\delta^K)}{\alpha(\lambda - \lambda\delta^K) - \delta^K} + 1 = U.$$

for any $k \geq K$.

This upper limit ensures that the infinite tail of the second term of (29) is convergent, which can be shown as

$$\sum_{k=K}^{\infty} \frac{\alpha(\lambda - \lambda\delta^k) - \beta(\lambda - \lambda\delta^k) - \delta^k \tilde{B}_f(\lambda - \lambda\delta^k)}{\alpha(\lambda - \lambda\delta^k) - \delta^k} \prod_{i=0}^{k-1} \frac{\beta(\lambda - \lambda\delta^i)}{\alpha(\lambda - \lambda\delta^i) - \delta^i}$$

$$\leq \sum_{k=K}^{\infty} U \prod_{i=0}^{k-1} \frac{\beta(\lambda - \lambda\delta^i)}{\alpha(\lambda - \lambda\delta^i) - \delta^i}$$

$$= U \prod_{j=0}^{K_1-1} \frac{\beta(\lambda - \lambda\delta^j)}{\alpha(\lambda - \lambda\delta^j) - \delta^j} \sum_{k=K}^{\infty} \prod_{i=K_1}^{k-1} \frac{\beta(\lambda - \lambda\delta^i)}{\alpha(\lambda - \lambda\delta^i) - \delta^i}$$

$$\leq U \prod_{j=0}^{K_1-1} \frac{\beta(\lambda - \lambda\delta^j)}{\alpha(\lambda - \lambda\delta^j) - \delta^j} \sum_{k-K_1=K-K_1}^{\infty} (r)^{k-K_1} < \infty. \tag{33}$$

It follows from (32) and (33) that the condition (31) ensures the convergence of (29), and hence it is sufficient for the stability.

Applying the expressions of the functions $\alpha(s)$ and $\beta(s)$ in (31) gives

$$\frac{\frac{\eta}{\delta}\left(\tilde{B}_s(\lambda) - \tilde{B}_f(\lambda)\right)}{(1-\eta)\tilde{B}_f(\lambda) + \eta\tilde{B}_s(\lambda)} < 1, \quad \text{if } \tilde{B}_s(\lambda) \geq \tilde{B}_f(\lambda),$$

$$\frac{\frac{\eta}{\delta}\left(\tilde{B}_f(\lambda) - \tilde{B}_s(\lambda)\right)}{(1-\eta)\tilde{B}_f(\lambda) + \eta\tilde{B}_s(\lambda)} < 1, \quad \text{if } \tilde{B}_s(\lambda) < \tilde{B}_f(\lambda). \tag{34}$$

The relations (34) can be rearranged as

$$\eta(\frac{1}{\delta} - 1)\tilde{B}_s(\lambda) < \tilde{B}_f(\lambda) + \eta(\frac{1}{\delta} - 1)\tilde{B}_f(\lambda), \quad \text{if } \tilde{B}_s(\lambda) \geq \tilde{B}_f(\lambda), \tag{35}$$

$$\eta(\frac{1}{\delta} + 1)\tilde{B}_f(\lambda) < \tilde{B}_f(\lambda) + \eta(\frac{1}{\delta} + 1)\tilde{B}_s(\lambda), \quad \text{if } \tilde{B}_s(\lambda) < \tilde{B}_f(\lambda).$$

The final form of the condition comes by rearranging (35). $\qquad\square$

References

1. Abouee-Mehrizi, H., Baron, O.: State-dependent M/G/1 queueing systems. Queueing Syst. **82**(1), 121–148 (2016)
2. Barjesteh, N., Abouee-Mehrizi, H.: Multiclass state-dependent service systems with returns. Naval Res. Logist., Special Issue: Serv. Oper. Manage. **68**(5), 631–662 (2021). https://doi.org/10.1002/nav.21908
3. Baron, O., Economou, A., Manou, A.: The state-dependent M / G / 1 queue with orbit. Queueing Syst. Theory Appl. **90**(1–2), 89–123 (2018). https://doi.org/10.1007/s11134-018-9582-1
4. Bekker, R.: Validating state-dependent queues in health care. Queueing Syst. **100**, 505–507 (2022). https://doi.org/10.1007/s11134-022-09827-x
5. Conway, R.W., Maxwell, W.L.: A queueing model with state dependent service rate. J. Ind. Eng. **12**, 132–136 (1961)
6. Gupta, U.C., Rao, T.S.: On the analysis of single server finite queue with state dependent arrival and service processes: $M_n/G_n/1/K$. OR Spektrum **20**, 83–89 (1998)
7. Harris, C.M.: Queues with state-dependent stochastic service rates. Oper. Res. **15**(1), 117–130 (1967)
8. Kerner, Y.: The conditional distribution of the residual service time in the $M_n/G/1$ queue. Stoch. Model. **24**, 364–375 (2008)
9. Kleinrock, L.: Queueing Systems. Volume 1: Theory. John Wiley, New York (1975)
10. Rodrigues, J., Prado, S.M., Balakrishnan, N., Louzada, F.: Flexible M/G/1 queueing system with state dependent service rate. Oper. Res. Lett. **44**(3), 383–389 (2016). https://doi.org/10.1016/j.orl.2016.03.011
11. Shanthikumar, J.G.: On a single-server queue with state-dependent service. Nav. Res. Logist. **26**(1), 305–309 (1979)

Sojourn Time in a Markov Driven Fluid Queue with Finite Buffer

Eleonora Deiana[1] , Guy Latouche[2], and Marie-Ange Remiche[1]([✉])

[1] Faculty of Computer Science, Namur University, Rue Grangagnage, 21,
5000 Namur, Belgium
{Eleonora.Deiana,Marie-Ange.Remiche}@unamur.be
[2] Faculty of Science, Computer Science Department, Université libre de Bruxelles,
Boulevard du Triomphe, 1050 Bruxelles, Belgium
Guy.Latouche@ulb.be

Abstract. We consider here a Markov-driven finite fluid queue: the buffer content is limited to the interval $[0, B]$, with $B < \infty$. This implies that at full capacity, entering fluid might be lost. We are interested in computing the sojourn time distribution. The lost of fluid at full capacity needs to be taken into account in order to extend the two step approach used in Deiana et al. in [3].

Keywords: Markov-driven fluid queue with finite capacity · sojourn time distribution · regenerative instants

1 Model and Problem Statement

In this work, we are interested in obtaining the sojourn time distribution at arrivals, when considering a fluid queue with finite capacity buffer. Such a model may be of crucial interest for particular systems where the capacity of the buffer matters. The results proposed here generalized results presented in Deiana et al. [3]. To study the finite buffer case will allow us to anticipate problems that may arise in a more general setting. Indeed, one possible goal is to determine sojourn time distribution in the case of a reactive fluid queue. Such models may serve to analyse energy-aware servers farms (see Deiana et al. [4]).

The fluid queue with finite buffer is not a new model, and it has been studied in the past with different techniques (see da Silva and Latouche [2], Bean *et al.* [1] or Van Foreest et al. in [6] and references therein). To allow the calculation of the sojourn time distribution, we proceed as in Deiana et al. [3], in two steps. The first step is to get the stationary distribution of the buffer level at arrival epochs (Sect. 2). However, due to the finite capacity of the buffer, some arriving fluid may now be lost. Therefore, to obtain the stationary distribution at arrivals, a different approach to that of Deiana et al. in [3] is provided here. The second step is to compute the distribution of the time needed by an arriving infinitesimal unit of fluid to leave the buffer (Sect. 3). This time, this distribution is not influenced by the finite buffer, since it only depends on the level of the buffer content when

M. Iacono et al. (Eds.): EPEW/ASMTA 2023, LNCS 14231, pp. 96–110, 2023.
https://doi.org/10.1007/978-3-031-43185-2_7

the unit of fluid arrives. These two steps are used in Sect. 4 to obtain the Laplace-Stieltjes transform of the sojourn time distribution. A numerical example is also provided. The paper ends up with a conclusion.

The finite buffer model we consider is now specified. Next we sketch the main lines to obtain the stationary distribution of the buffer content and in particular the regenerative instants we need to consider in this particular setting. Indeed, the fluid queue is studied as a semi-regenerative process in order to obtain the stationary distribution. This distribution is of crucial interest to get the stationary distribution at arrival times in Sect. 2.

1.1   The Finite Buffer Model

A buffer with finite capacity is of interest in this work. Accordingly the fluid queue $\{(X(t), \varphi(t)) : t \geq 0\}$ is such that $X(t) \in [0, B]$, with B finite. The phase process $\varphi(t)$ controls the evolution of the buffer content in the following way. When $\varphi(t) = i \in \mathcal{S}$, then

$$\frac{\mathrm{d}X(t)}{\mathrm{d}t} = \begin{cases} r_i & \text{if } 0 < X(t) < B, \\ \max\{0, r_i\} & \text{if } X(t) = 0, \\ \min\{0, r_i\} & \text{if } X(t) = B. \end{cases} \tag{1}$$

Matrix T is defined as the generator matrix of the phase process $\varphi(t)$, and R as the rate matrix. The state space $\mathcal{S}$ of the phase process is partitioned as $\mathcal{S} = \mathcal{S}_+ \cup \mathcal{S}_- \cup \mathcal{S}_0$ depending on the rate sign. Also matrices T and R follow this partition, it gives

$$T = \begin{bmatrix} T_{++} & T_{+-} & T_{+0} \\ T_{-+} & T_{--} & T_{-0} \\ T_{0+} & T_{0-} & T_{00} \end{bmatrix}, \quad R = \begin{bmatrix} R_+ & & \\ & R_- & \\ & & 0 \end{bmatrix}. \tag{2}$$

One also needs to define matrix $\widetilde{T}$

$$\widetilde{T} = \begin{bmatrix} T_{++} & T_{+-} \\ T_{-+} & T_{--} \end{bmatrix} + \begin{bmatrix} T_{+0} \\ T_{-0} \end{bmatrix} (-T_{00})^{-1} [T_{0+} \ T_{0-}]. \tag{3}$$

We assume in this context that the mean drift of the fluid, is strictly negative. In the finite buffer model, the stationary distribution exists for every value of the mean drift. However, when the mean drift is equal to 0, there are some technical difficulties (see Sect. 8 in [5]). This is not the type of mathematical questions we want to discuss in this work: so we do assume the mean drift to be strictly negative, to keep the buffer more often empty than full.

Without lack of generality, we assume in the following that $X(0) = 0$.

1.2   The Stationary Distribution of the Buffer Content at Any Time

The joint stationary distribution of the buffer content and the phase is obtained using a regenerative approach. We refer to Latouche and Nguyen [5] for a complete and step-by-step development of stationary distribution computation in the case of the infinite buffer model. Regenerative instants have to be defined in the context of this finite buffer model. Let $\{h_n\}_{n\geq 0}$ be the regenerative instants defined as

$$h_0 = 0, \tag{4}$$
$$h_{n+1} = \inf\{t > l_n : X(t) = 0 \vee X(t) = B\}. \tag{5}$$

with the help of $\{l_n\}_{n\geq 0}$

$$l_n = \inf\{t \geq h_n : (X(t) > 0 \wedge \varphi(t) \in \mathcal{S}_+) \vee (X(t) < B \wedge \varphi(t) \in \mathcal{S}_-)\}. \tag{6}$$

The joint stationary distribution of the buffer content and of the phase $\boldsymbol{\Pi}(x)$ is

$$\boldsymbol{\Pi}(x) = c\boldsymbol{\nu}M(x), \tag{7}$$

where primary, $\boldsymbol{\nu}$ is the stationary probability vector of the phases at regenerative epochs h_n. Secondary, $M(x)$ is the matrix of the expected times the buffer content $X(t)$ spends in $[0, x]$, during a *regenerative interval* $[h_n, h_{n+1})$ and tertiary, c is the normalizing constant.

Many quantities have to be specify to be able to compute $\boldsymbol{\Pi}(x)$. Let first Ψ (respectively $\widehat{\Psi}$) be the first return probability to the initial level from above (respectively from below) in the corresponding unrestrictive fluid queue. Both are minimal non-negative solution of quadratic Riccati equations, respectively

$$R_+^{-1}\widetilde{T}_{+-} + R_+^{-1}\widetilde{T}_{++}\Psi + \Psi|R_-|^{-1}\widetilde{T}_{--} + \Psi|R_-|^{-1}\widetilde{T}_{-+}\Psi = 0, \tag{8}$$
$$|R_-|^{-1}\widetilde{T}_{-+} + |R_-|^{-1}\widetilde{T}_{--}\widehat{\Psi} + \widehat{\Psi}R_+^{-1}\widetilde{T}_{++} + \widehat{\Psi}R_+^{-1}\widetilde{T}_{+-}\widehat{\Psi} = 0. \tag{9}$$

Let us define $\Gamma(x)$ ($\widehat{\Gamma}(x)$ respectively) as the matrix of the mean times the buffer content spends in $(0, x)$, between two successive instants when the buffer was empty (full respectively).

Theorem 1. *Let $x > 0$, then*

$$\Gamma_+(x) = \int_0^x e^{Ku}\,du\, R_+^{-1}, \tag{10}$$

$$\Gamma_-(x) = \int_0^x e^{Ku}\,du\, \Psi|R_-^{-1}|, \tag{11}$$

$$\Gamma_0(x) = \int_0^x e^{Ku}\,du\, (R_+^{-1}T_{+0} + \Psi|R_-^{-1}|T_{-0})(-T_{00})^{-1}. \tag{12}$$

and

$$\widehat{\Gamma}_+(x) = \int_{B-x}^B e^{\widehat{K}u}\,du\, \widehat{\Psi}R_+^{-1}, \tag{13}$$

$$\widehat{\Gamma}_-(x) = \int_{B-x}^B e^{\widehat{K}u}\,du\, |R_-^{-1}|, \tag{14}$$

$$\widehat{\Gamma}_0(x) = \int_{B-x}^B e^{\widehat{K}u}\,du\, \left(\widehat{\Psi}R_+^{-1}T_{+0} + |R_-|^{-1}T_{-0}\right)(-T_{00})^{-1}, \tag{15}$$

with K be defined as

$$K = R_+^{-1}\widetilde{T}_{++} + \Psi|R_-|^{-1}\widetilde{T}_{-+}. \tag{16}$$

and $\widehat{K}$ as

$$\widehat{K} = |R_-|^{-1}\widetilde{T}_{--} + \widehat{\Psi}R_+^{-1}\widetilde{T}_{+-}. \tag{17}$$

$\square$

Finally Φ_{+-} and Φ_{-+} are transition probability matrices from instants h_n to l_n. One may establish that matrices Φ_{-+} and Φ_{+-} are given by

$$\Phi_{-+} = \begin{bmatrix} I & 0 \end{bmatrix} \left(-\begin{bmatrix} T_{--} & T_{-0} \\ T_{0-} & T_{00} \end{bmatrix}\right)^{-1} \begin{bmatrix} T_{-+} \\ T_{0+} \end{bmatrix}, \tag{18}$$

$$\Phi_{+-} = \begin{bmatrix} I & 0 \end{bmatrix} \left(-\begin{bmatrix} T_{++} & T_{+0} \\ T_{0+} & T_{00} \end{bmatrix}\right)^{-1} \begin{bmatrix} T_{+-} \\ T_{0-} \end{bmatrix}. \tag{19}$$

Theorem 2. *In the case where the mean drift of the fluid is negative, we have for $0 \leq x \leq B$ that the stationary distribution $\Pi(x)$ has the following form:*

$$\Pi(x) = c \begin{bmatrix} 0 & p_-^0 & p_0^0 \end{bmatrix} + c \begin{bmatrix} \mu^0 & \mu^B \end{bmatrix} \begin{bmatrix} \Gamma(x) \\ \widehat{\Gamma}(x) \end{bmatrix} + c \begin{bmatrix} p_+^B & 0 & p_0^B \end{bmatrix} \mathbb{1}_{\{x=B\}}. \tag{20}$$

With ν partitioned as $\begin{bmatrix} \nu_+ & \nu_- \end{bmatrix}$, we have

$$\begin{bmatrix} p_-^0 & p_0^0 \end{bmatrix} = \begin{bmatrix} \nu_- & 0 \end{bmatrix} \left(-\begin{bmatrix} T_{--} & T_{-0} \\ T_{0-} & T_{00} \end{bmatrix}\right)^{-1}, \tag{21}$$

$$\begin{bmatrix} p_+^B & p_0^B \end{bmatrix} = \begin{bmatrix} \nu_+ & 0 \end{bmatrix} \left(-\begin{bmatrix} T_{++} & T_{+0} \\ T_{0+} & T_{00} \end{bmatrix}\right)^{-1}, \tag{22}$$

$$\begin{bmatrix} \mu^0 & \mu^B \end{bmatrix} = \begin{bmatrix} \nu_+ & \nu_- \end{bmatrix} \begin{bmatrix} 0 & \Phi_{+-} \\ \Phi_{-+} & 0 \end{bmatrix} \begin{bmatrix} I & e^{KB}\Psi \\ e^{\widehat{K}B}\widehat{\Psi} & I \end{bmatrix}^{-1} \tag{23}$$

$$\Theta = \begin{bmatrix} R_+^{-1} & \Psi|R_-|^{-1} & \left(R_+^{-1}T_{+0} + \Psi|R_-|^{-1}T_{-0}\right)(-T_{00})^{-1} \end{bmatrix} \tag{24}$$

$$\widehat{\Theta} = \begin{bmatrix} \widehat{\Psi}R_+^{-1} & |R_-|^{-1} & \left(\widehat{\Psi}R_+^{-1}T_{+0} + |R_-|^{-1}T_{-0}\right)(-T_{00})^{-1} \end{bmatrix}. \tag{25}$$

The normalizing constant is given by

$$c^{-1} = \left(p_-^0 \mathbf{1} + p_0^0 \mathbf{1}\right) + \left(\mu^0 \Gamma(B)\mathbf{1} + \mu^B \widehat{\Gamma}(B)\mathbf{1}\right) + \left(p_+^B \mathbf{1} + p_0^B \mathbf{1}\right). \tag{26}$$

$\square$

2 Stationary Distribution at Arrival Time

To obtain the sojourn time distribution requires two steps. The first step is the calculation of $\pi^*(x)$, that is the stationary density of the fluid at arrivals. Conditioning on the level x found by an arriving infinitesimal unit of fluid, the second step will be the calculation of $G(x, t)$, that is the distribution of the time needed by this infinitesimal unit of fluid to leave the buffer. This is the subject of Sect. 3. These two quantities, together with an adequate permutation matrix P^{out}, are used to obtain the sojourn time distribution, which in the finite buffer case is given by the following equation

$$V(t) = \boldsymbol{p}_0^* \boldsymbol{1} + \int_0^B \boldsymbol{\pi}^*(x) P^{out} G(x, t) \boldsymbol{1} \mathrm{d}x + \boldsymbol{p}_B^* G(B, t) \boldsymbol{1}. \tag{27}$$

To tackle this calculation, one first needs to specify a new clock design.

2.1 New Clock Principle

Let the *input process* be defined as the fluid queue $\{(Y(t), \varphi(t)) : t \geq 0\}$, where $Y(t) \in \mathbb{R}^+$ represents the total amount of fluid which entered the buffer up to time t.

As described in (1), as soon as $X(t) = B$, the rate of evolution of $X(t)$ changes to $\min\{0, r_i\}$. Accordingly, a part of the arriving fluid may be lost. In order to take this into account, we rewrite the input rate as

$$r_{in,i} = r_{join,i} + r_{lost,i}, \tag{28}$$

where $r_{join,i}$ is the arrival rate of the fluid that actually joins the buffer, while $r_{lost,i}$ is the rate of the lost fluid.

When $X(t) < B$, the arriving fluid may join the buffer. Accordingly, there is no lost of fluid, so $r_{lost,i} = 0$ and $r_{join,i} = r_{in,i}$. When the buffer is full, only the joining fluid contributes to the net rate, which is now equal to zero, and thus $r_{join,i} - r_{out,i} = 0$. As a consequence, the fluid actually joining the buffer does it at the same rate as the servers are working, that is $r_{join,i} = r_{out,i}$. Therefore, one may write that, when $\varphi(t) = i$, the input process $Y(t)$ evolves as

$$\frac{\mathrm{d}Y(t)}{\mathrm{d}t} = \begin{cases} r_{in,i} & \text{if } 0 \leq X(t) < B, \\ r_{out,i} & \text{if } X(t) = B, \end{cases} \tag{29}$$

and the rate matrices associated with it, are

$$\begin{cases} R_{in} & \text{when } 0 \leq X(t) < B, \\ R_{out} & \text{when } X(t) = B. \end{cases} \tag{30}$$

Accordingly, the evolution of $Y(t)$ does not only depend on the phase process $\varphi(t)$, but also on the actual level reached by the buffer content $X(t)$.

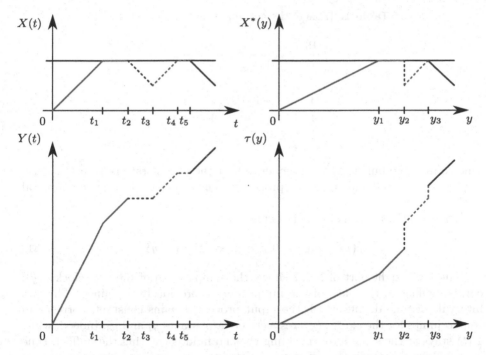

Fig. 1. An illustration of the different processes. On the upper left, a sample path of the original fluid $X(t)$, and below the corresponding input process $Y(t)$. On the right, the corresponding new clock $\tau(y)$ below, and the new process $X^*(y)$ above. Paths corresponding to the net rate 1, are in red: phase 1 with solid line, phase 2 with dotted line. Paths corresponding to the net rate -1 are in blue: solid line for phase 3 and dotted line for phase 4. (Color figure online)

The left side of Fig. 1 exhibits a sample path of the buffer content $X(t)$ above, and of its corresponding input process $Y(t)$ below. We consider a simple phase process $\varphi(t)$ with only four phases. The net rates only take values 1 and -1, but input and output rates take different values depending on the phase. These values are summarized in Table 1.

Consider the left side of Fig. 1. The buffer content $X(t)$ reaches bound B for the first time at instant t_1. Between 0 and t_1, the phase is 1, so that $X(t)$ is increasing at net rate $r_1 = 1$, while the input process $Y(t)$ is increasing at rate $r_{in,1} = 2$. At instant t_1 the buffer reaches its full capacity, and it stays so until instant t_2. During this interval of time the rate of the original process is thus 0. However, the input process continues increasing, but only at the rate of the fluid that may actually join the buffer. This rate is $r_{join,1} = r_{out,1} = 1$.

At time t_4, the buffer is full again and the phase is 2. At first, until t_4, both the buffer content $X(t)$ and the input process increase at rate 1, as $r_2 = 1$ and $r_{in,2} = 1$. However, at t_4, the buffer becomes full and the original fluid $X(t)$ remains equal to B until t_5. During this interval of time there is no more

Table 1. Table of the net, input and output rates

Phases	r_i	$r_{in,i}$	$r_{out,i}$
1	1	2	1
2		1	0
3	−1	1	2
4		0	1

fluid joining the buffer, as a consequence that the output rate is 0. In this case, $r_{join,2} = r_{out,2} = 0$ and the input process remains constant during this interval of time.

The *new clock* $\tau(y)$ is formally defined as

$$\tau(y) = \min\{u, 0 < u < \infty : Y(u) = y\}. \tag{31}$$

The lower-right part of Fig. 1 shows the sample path of the new clock $\tau(y)$, corresponding to the path of the input process previously explained. The two intervals of time during which the input process remains constant, correspond to two jumps for the clock $\tau(y)$, respectively at instants y_2 and y_3. Indeed, while $Y(t)$ stays at the same level (meaning that no fluid is entering the buffer), time is still flying. Accordingly, as soon as $Y(t)$ increases again, a given amount of time has been lost and this is recorded by a jump in $\tau(y)$.

2.2 Regenerative Approach for the Process with New Clock

Finally, we use the new clock $\tau(y)$ in order to define the following process

$$\{(X^*(y), \varphi^*(y)) : y \geq 0\} = \{(X(\tau(y)), \varphi(\tau(y))) : y \geq 0\}. \tag{32}$$

We observe the original fluid queue only at those times where the input process is increasing. A period of time where $Y(t)$ is constant is not taken into account in the new clock, and thus $X^*(y)$ jumps instantaneously downwards. This happens at time y_2 in the upper right plot of Fig. 1. Let us also note that the speed at which $X^*(y)$ varies is different from that of $X(t)$. The new clock is defined on basis of the input rates $r_{in,i}$, so the greater $r_{in,i}$ is, the slower the clock is advancing, whatever the speed at which servers are working. When level B is reached, new clock time is defined on $r_{out,i}$ only as long as r_i is positive.

The objective now is to obtain the stationary distribution $\Pi^*(x)$, and the corresponding density $\pi^*(x)$ of this new defined process. Thanks to the definition of the new clock, these are respectively the distribution and the density of the buffer content observed when an infinitesimal amount of fluid enters the buffer. In Deiana et al. [3], the stationary distribution $\Pi^*(x)$ was obtained from the stationary distribution of the original process with simple manipulations, based on a change of variables involving the new clock. The same approach is not possible here. The reason is that, due to the finite size of the buffer, the evolution of the new clock does not depend on the phase process only, but it also depends on the values of the original process $X(t)$. To overcome this, we use a regenerative approach. This allows us to study separately the different paths between two regenerative instants. The first step for this approach is the definition of the regenerative instants for the new process $X^*(y)$, these are the *regenerative points* $\{y_n\}_{n \geq 0}$. They are based on a transformation of the original regenerative instants $\{h_n\}_{n \geq 0}$ in the following manner.

Let us first consider a regenerative instant h_n for the original fluid, such that $\varphi(h_n) = i$ and $r_{in,i} > 0$. In this case one may define without any ambiguity the regenerative point y_n for $X^*(y)$, and the corresponding instant $\tau(y_n)$ as respectively

$$y_n = Y(h_n), \tag{33}$$

$$\tau(y_n) = h_n. \tag{34}$$

As a consequence, level and phase are the same for both the original process and the process with the new clock. One may then write

$$X^*(y_n) = X(\tau(y_n)) = X(h_n), \tag{35}$$

$$\varphi^*(y_n) = \varphi(\tau(y_n)) = \varphi(h_n). \tag{36}$$

Let us now consider a regenerative instant h_n where the phase $\varphi(h_n) = i$ is such that $r_{in,i} = 0$. This can only happen if $X(h_n) = 0$. Again, the point y_n is well defined as

$$y_n = Y(h_n), \tag{37}$$

and it represents the amount of fluid which entered the buffer up to time h_n. The inconvenience is that $\tau(y_n)$ is defined as the first instant when the total amount of fluid which entered the buffer is equal to y_n. However, this may happen for the first time before h_n, as there is no input rate. We then define the regenerative point y_n as the first instant where the three following conditions are fullfilled

$$\begin{cases} y_n = Y(h_n), \\ X^*(y_n) = X(h_n), \\ \varphi^*(y_n) = \varphi(h_n). \end{cases} \tag{38}$$

In this way the regenerative point y_n is unequivocally defined, and so is the level and phase for the process with the new clock.

The regenerative points $\{y_n\}_{n \geq 0}$ are then well defined for every phase. The stationary distribution $\boldsymbol{\Pi}^*(x)$ of the process with the new clock $\{(X^*(y), \varphi^*(y)) : y \geq 0\}$, i.e.

$$\boldsymbol{\Pi}^*(x) = c^* \boldsymbol{\nu}^* M^*(x), \tag{39}$$

which has the same form as the stationary distribution of the original process, as given in (4). The different elements of (39) may be calculated separately, through a transformation of the corresponding quantities for the original process, based on the new clock $\tau(y)$. We have the following result.

Theorem 3. *The stationary distribution $\mathit{\Pi}^*(x)$ of the process with new clock is given by:*

$$\mathit{\Pi}^*(x) = c^*\boldsymbol{\nu} M^*(x), \tag{40}$$

where

$$M^*(x) = M^0 R_{in} + M(0,x)R_{in} + M^B R_{out}\mathbb{1}_{\{x=B\}}, \tag{41}$$

$$(c^*)^{-1} = \boldsymbol{\nu}\left(M^0 R_{in} + M(0,B)R_{in} + M^B R_{out}\right)\mathbf{1}. \tag{42}$$

The matrix M^0 (respectively M^B) contains expected times the buffer content is 0 (B respectively) during $[h_n, l_n)$, while $M(0,x)$ contains the expected times the level $X(t)$ spends in $(0,x)$ during $[h_n, h_{n+1})$. One may established that

$$M(0,x) = \begin{bmatrix} 0 & \varPhi_{+-} \\ \varPhi_{-+} & 0 \end{bmatrix} \begin{bmatrix} H_+^B(x) \\ H_-^B(x) \end{bmatrix}, \tag{43}$$

with matrices matrices $H_+^B(x)$ and $H_-^B(x)$ solutions of the following system

$$\begin{bmatrix} I & e^{KB}\Psi \\ e^{\widehat{K}B}\widehat{\Psi} & I \end{bmatrix} \begin{bmatrix} H_+^B(x) \\ H_-^B(x) \end{bmatrix} = \begin{bmatrix} \Gamma(x) \\ \widehat{\Gamma}(x) \end{bmatrix}.$$

with Ψ

$$\varPhi = \begin{bmatrix} I & 0 \end{bmatrix} \left(- \begin{bmatrix} T_{--} & T_{-0} \\ T_{0-} & T_{00} \end{bmatrix}\right)^{-1} \begin{bmatrix} T_{-+} \\ T_{0+} \end{bmatrix}. \tag{44}$$

and $\Gamma(x)$ specified in Theorem 1. Matrix H is the transition probability matrix of the phases process restricted to the regenerative instants. We have

$$H = \begin{bmatrix} 0 & \varPhi_{+-} \\ \varPhi_{-+} & 0 \end{bmatrix} \begin{bmatrix} \Lambda^B & \Psi^B \\ \widehat{\Psi}^B & \widehat{\Lambda}^B \end{bmatrix}, \tag{45}$$

where matrices Ψ^B, Λ^B, $\widehat{\Psi}^B$ and $\widehat{\Lambda}^B$ are solutions of the following system

$$\begin{bmatrix} \Lambda^B & \Psi^B \\ \widehat{\Psi}^B & \widehat{\Lambda}^B \end{bmatrix} \begin{bmatrix} I & \Psi e^{UB} \\ \widehat{\Psi}e^{\widehat{U}B} & I \end{bmatrix} = \begin{bmatrix} e^{\widehat{U}B} & \Psi \\ \widehat{\Psi} & e^{UB} \end{bmatrix}. \tag{46}$$

Proof. There are three elements in Eq. (39) that we need to compute, in order to obtain $\mathit{\Pi}^*(x)$. Let us start with vector $\boldsymbol{\nu}^*$, the unique solution of the following linear system

$$\begin{cases} \boldsymbol{\nu}^* H^* = \boldsymbol{\nu}^*, \\ \boldsymbol{\nu}^*\mathbf{1} = 1, \end{cases} \tag{47}$$

with

$$H_{ij}^* = \Pr\left[\varphi^*(y_{n+1}) = j | \varphi^*(y_n) = i\right], \tag{48}$$

$$= \Pr\left[\varphi(h_{n+1}) = j | \varphi(h_n) = i\right], \tag{49}$$

$$= H_{ij}, \tag{50}$$

where the second line comes from the definition of the regenerative points y_n and from Eqs. (36) and (38). The equality between matrix H and H^* also implies that the vectors $\boldsymbol{\nu}$ and $\boldsymbol{\nu}^*$ are the same.

Let us now consider matrix $M^*(x)$ for $x < B$, this is defined as

$$(M^*(x))_{ij} = \mathrm{E}\left[\int_{y_n}^{y_{n+1}} \mathbb{1}\{X^*(u) \in [0,x], \varphi^*(u) = j\}\,du\,\middle|\,X^*(y_n), \varphi^*(y_n) = i\right],$$
(51)

where, again from the definition of y_n and from Eqs. (36) and (38), we may write $X^*(y_n) = X(h_n)$ and $\varphi^*(y_n) = \varphi(h_n)$, and replace it in the expected value, obtaining

$$(M^*(x))_{ij} = \mathrm{E}\left[\int_{y_n}^{y_{n+1}} \mathbb{1}\{X^*(u) \in [0,x], \varphi^*(u) = j\}\,du\,\middle|\,X(h_n), \varphi(h_n) = i\right].$$
(52)

We then only consider the integral inside the expected value, and make a change of variable by taking $t = \tau(y)$. Let us first consider $j \in \mathcal{S}_+^{in}$, then one has that $y = Y(t)$, and so $dy = r_{in,j}dt$. In fact, if $x < B$, $Y(t)$ evolves following the input rates $r_{in,j}$, as it is stated in the definition (29). Using the definition of $X^*(\cdot)$ and $\varphi^*(\cdot)$, we may write for the integral

$$\int_{y_n}^{y_{n+1}} \mathbb{1}\{X^*(u) \in [0,x], \varphi^*(u) = j\}\,du$$

$$= \int_{y_n}^{y_{n+1}} \mathbb{1}\{X(\tau(u)) \in [0,x], \varphi(\tau(u)) = j\}\,du \qquad (53)$$

$$= \int_{\tau(y_n)}^{\tau(y_{n+1})} \mathbb{1}\{X(t) \in [0,x], \varphi(t) = j\}\,r_{in,j}\,dt \qquad (54)$$

$$= \int_{h_n}^{h_{n+1}} \mathbb{1}\{X(t) \in [0,x], \varphi(t) = j\}\,r_{in,j}\,dt, \qquad (55)$$

where in the last line we used (34), since $r_{in,j} > 0$.

Replacing again this integral into the expected value, one obtains

$$(M^*(x))_{ij} = \mathrm{E}\left[\int_{h_n}^{h_{n+1}} \mathbb{1}\{X(t) \in [0,x], \varphi(t) = j\}\,r_{in,j}dt\,\middle|\,X(h_n), \varphi(h_n) = i\right]$$
(56)

for $j \in \mathcal{S}_+^{in}$ and $i \in \mathcal{S}$. Otherwise, if $j \in \mathcal{S}_0^{in}$, then the process $X^*(t)$ makes a jump so that

$$(M^*(x))_{ij} = 0. \qquad (57)$$

One can summarise it by writing that if $x < B$, then

$$M^*(x) = M(x)R^{in}. \qquad (58)$$

In that case, $M(x)$ may be written as $M(x) = M^0 + M(0, x)$ and that justifies the first two terms in (41).

Finally, if $x = B$, one only has to add the time spent at level B, which for the process with the new clock $X^*(t)$ is registered in matrix M^{B*}. This is properly defined as

$$\left(M^{B*}\right)_{ij} = \mathrm{E}\left[\int_{y_n}^{y_{n+1}} \mathbb{1}\{X^*(u) = B, \varphi^*(u) = j\}\, du \,\middle|\, X^*(y_n), \varphi^*(y_n) = i\right].$$
(59)

The conditions in the expected value can be modified as before, and we may again apply a change of variables in the integral. This time, one has $dy = r_{out,j}dt$, because when $x = B$ then the fluid that may still join the buffer, does it at rate $r_{out,j}$, as explained in (29). With these two modifications, we have

$$\left(M^{B*}\right)_{ij} = \mathrm{E}\left[\int_{h_n}^{h_{n+1}} \mathbb{1}\{X(t) = B, \varphi(t) = j\}\, r_{out,j}dt \,\middle|\, X(h_n), \varphi(h_n) = i\right],$$
(60)

$$= \left(M^B\right)_{ij} r_{out,j},$$
(61)

or in matrix form $M^{B*} = M^B R^{out}$.

Finally, one only has to compute the normalizing constant c^*, which may be computed from

$$(c^*)^{-1} = \nu M^*(B)\mathbf{1}.$$
(62)

□

The stationary density $\pi^*(x)$ is obtained by taking the derivative of $\Pi^*(x)$. With some simple manipulations one can obtain the following result for $\pi^*(x)$ and for the probability masses in bounds 0 and B. We give the following corollary without proof as it is straightforward combining Theorem 3 with elements defined in Theorem 2.

Corollary 1. *The stationary density $\pi^*(x)$ may be written as the following sum*

$$\pi^*(x) = c^* \boldsymbol{\mu}^0 e^{Kx} \Theta^{in} + c^* \boldsymbol{\mu}^B e^{\widehat{K}(B-x)} \widehat{\Theta}^{in},$$
(63)

where

$$\Theta^{in} = \Theta R^{in},$$
(64)

$$\widehat{\Theta}^{in} = \widehat{\Theta} R^{in}.$$
(65)

The probabilities for a unit of fluid to find an empty buffer, or a full buffer at its arrivals are respectively given by

$$\boldsymbol{p}_0^* = c^* \left[\mathbf{0}\ \boldsymbol{p}_-^0\ \boldsymbol{p}_0^0\right] R^{in},$$
(66)

$$\boldsymbol{p}_B^* = c^* \left[\boldsymbol{p}_+^B\ \mathbf{0}\ \boldsymbol{p}_0^B\right] R^{out}.$$
(67)

□

3   Time to Leave the Buffer

We compute here the quantity $G(x, t)$ of Eq. (27). This is the distribution of the time an arriving unit of fluid needs to leave the buffer, while the buffer content level at its arrival was $x \in [0, B]$. In order to compute this distribution, we consider the so-called *output process*, denoted by $\{(Z(t), \varphi(t)) : t \geq 0\}$, and obtained from the original process by only considering the output rates. The only influence the finite buffer has on this distribution is on the initial level.

The LS transform $\widehat{G}(x, s)$ of the distribution $G(x, t)$, has the following form

$$\widehat{G}(x, s) = \begin{bmatrix} I \\ -(T_{00}^{out} - sI)^{-1} T_{0-}^{out} \end{bmatrix} e^{Q_{out}(s)x}, \tag{68}$$

where

$$Q_{out}(s) = |R_{-}^{out}|^{-1} \left[(T_{--}^{out} - sI) - T_{-0}^{out} (T_{00}^{out} - sI)^{-1} T_{0-}^{out} \right], \tag{69}$$

and matrices T_{--}^{out}, T_{-0}^{out}, T_{00}^{out} and T_{0-}^{out} are blocks of the generator matrix T, when partitioned following the output rates $r_{out,j}$.

4   Numerical Illustration

We compute in this section the LS transform of the sojourn time distribution $V(t)$ as defined in Eq. (27). Its LS transform $\widehat{V}(s)$ is given by

$$\widehat{V}(s) = \int_0^B \pi^*(x) P^{out} \widehat{G}(x, s) 1 \mathrm{d}x. \tag{70}$$

The density $\pi^*(x)$, obtained in Sect. 2, is a vector partitioned following $\mathcal{S} = \mathcal{S}_+ \cup \mathcal{S}_- \cup \mathcal{S}_0$. The LS transform $\widehat{G}(x, s)$, is instead partitioned following $\mathcal{S} = \mathcal{S}_-^{out} \cup \mathcal{S}_0^{out}$. For this reason, in order to be able to multiply vector $\pi^*(x)$ by matrix $\widehat{G}(x, s)$ one needs the permutation matrix P^{out}, which switches from the partition $\mathcal{S} = \mathcal{S}_+ \cup \mathcal{S}_- \cup \mathcal{S}_0$ to the partition $\mathcal{S} = \mathcal{S}_-^{out} \cup \mathcal{S}_0^{out}$.

We now have all the elements to obtain the LS transform of the sojourn time distribution $\widehat{V}(s)$. We replace in Eq. (70) the components previously defined, and obtain the following equation

$$\widehat{V}(s) = c^* \mu^0 \int_0^B e^{Kx} \Theta^{in} B^{out}(s) e^{Q^{out}(s)x} 1 \mathrm{d}x$$
$$+ c^* \mu^B \int_0^B e^{\widehat{K}(B-x)} \widehat{\Theta}^{in} B^{out}(s) e^{Q^{out}(s)x} 1 \mathrm{d}x, \tag{71}$$

where Θ^{in} and $\widehat{\Theta}^{in}$ are defined in Corollary 1, and

$$B^{out}(s) = P^{out} \begin{bmatrix} I \\ -(T_{00}^{out} - sI)^{-1} T_{0-}^{out} \end{bmatrix}. \tag{72}$$

As it is not possible to obtain a closed form solution for the integrals in Eq. (71), we give without any proof, in the following theorem an alternative way to compute the LS transform of the sojourn time.

Theorem 4. *The LS transform of the sojourn time distribution $V(t)$ is given by*

$$\widehat{V}(s) = c^* \mu^0 A(s)\mathbf{1} + c^* \mu^B \widehat{A}(s)\mathbf{1}, \tag{73}$$

where the integrals $A(s)$ and $\widehat{A}(s)$, defined as

$$A(s) = \int_0^B e^{Kx} \Theta^{in} B^{out}(s) e^{Q^{out}(s)x} dx, \tag{74}$$

$$\widehat{A}(s) = \int_0^B e^{\widehat{K}(B-x)} \widehat{\Theta}^{in} B^{out}(s) e^{Q^{out}(s)x} dx, \tag{75}$$

are, respectively, the non-singular solutions of the following Sylvester equations

$$KA(s) + A(s)Q^{out}(s) + \Theta^{in} B^{out}(s) - e^{KB}\Theta^{in} B^{out}(s) e^{Q^{out}(s)B} = 0, \tag{76}$$

and

$$\widehat{K}\widehat{A}(s) - \widehat{A}(s)Q^{out}(s) + \widehat{\Theta}^{in} B^{out}(s) e^{Q^{out}(s)B} - e^{\widehat{K}B}\widehat{\Theta}^{in} B^{out}(s) = 0. \tag{77}$$

□

Let us consider the following example. The phase process cycles through 8 phases, spending on average one unit of time in every phase. The generator matrix of the process is given by

$$T = \begin{bmatrix} -1 & 1 & & \\ & -1 & 1 & \\ & & \ddots & \\ 1 & & & -1 \end{bmatrix}. \tag{78}$$

We fix the value of the mean input drift to be equal to $\lambda^{in} = 0.8$. For the input process we consider the following case: only one phase (phase 8) brings 3/4 of the total charge arriving into the buffer during a cycle. The vector of the input rates r_{in}, is given by

$$r_{in} = \begin{bmatrix} 2/7, 2/7, 2/7, 2/7, 2/7, 2/7, 2/7, 6 \end{bmatrix} \lambda^{in}. \tag{79}$$

For the output rates, we consider three different cases which are corresponding to the following vectors of output rates:

$$r^a_{out} = \begin{bmatrix} 1, 1, 1, 1, 1, 1, 1, 1 \end{bmatrix} \tag{80}$$

$$r^b_{out} = \begin{bmatrix} 2/3, 2/3, 2/3, 2/3, 2/3, 2/3, 2, 2 \end{bmatrix} \tag{81}$$

$$r^c_{out} = \begin{bmatrix} 4/7, 4/7, 4/7, 4/7, 4/7, 4/7, 4/7, 4 \end{bmatrix} \tag{82}$$

For the three cases, the mean output drift λ^{out} is equal to one, in order to have a negative total mean drift $\lambda^{in} - \lambda^{out} = -0.2$.

In Fig. 2 the sojourn time of the different cases is depicted. We consider the three cases for the output rates, and different types of buffer: infinite buffer, maximum capacity $B = 5$ and maximum capacity $B = 10$. The bounds for Case c do not make any impact on the sojourn time. Let us consider now Case a, where the output rates are constants and equal to 1. We first put our attention in the buffer limited to $B = 5$, but the same observation can be made for $B = 10$. Both the stationary distribution of the buffer content and the sojourn time of this case present a positive probability mass at $B = 5$. The reason is that the output rates are equal to one, so all the fluid that may enter at full capacity have the same sojourn time. Moreover, no fluid may spend more time into the buffer, so the value B in this case is also the maximum possible value for the sojourn time.

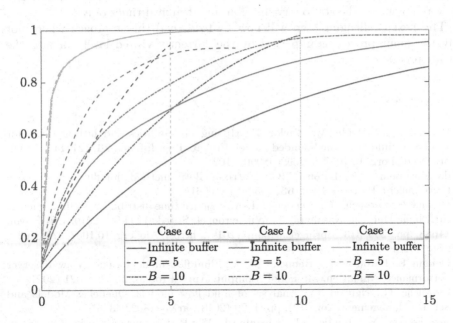

Fig. 2. Sojourn Time distribution with an infinite buffer, and with a finite buffer when $B = 5$ and $B = 10$ for the three different input environments, corresponding to Cases a, b and c.

In general, the maximum possible value for the sojourn time is reached by the units of fluid arriving when the buffer is completely full. These units of fluid have a sojourn time at most equal to

$$t_{MAX} = \frac{B}{\min_{i \in \mathcal{S}^{out}_-}\{r_{out,i}\}}. \qquad (83)$$

Of course, this value in Case a is equal to B. Let us now take Case b, with maximum capacity $B = 5$. In this case, the maximum possible value for the sojourn time is $t_{MAX} = 7.5$, and there is a positive mass corresponding to this value.

5   Conclusion

While the finite buffer case seems to be a natural extension of the infinite buffer model to study, challenging problems must be solved when dealing with the sojourn time distribution. In this latter case, the quantity of fluid that could not enter the finite buffer has indeed an incidence on the proposed technique in [3]. We overcome this problem extending the regenerative approach to the process where we only consider the original fluid queue at those times fluid enters the system. The new regenerative points we proposed, are univocally detailed and allow us to obtain the stationary distribution at arrival times only.

The next challenging step will now be to extend this approach where not only the input process is impacted by the currently visited level but also the output process.

References

1. Bean, N.G., O'Reilly, M., Taylor, P.: Hitting probabilities and hitting times for stochastic fluid flows the bounded model. Probab. Eng. Inf. Sci. **23**, 121–147 (2009). https://doi.org/10.1017/S0269964809000102
2. da Silva Soares, A., Latouche, G.: Matrix-analytic methods for fluid queues with finite buffers. Perform. Eval. **63**, 295–314 (2006)
3. Deiana, E., Latouche, G., Remiche, M.-A.: Sojourn time distribution in fluid queues. In: Phung-Duc, T., Kasahara, S., Wittevrongel, S. (eds.) QTNA 2019. LNCS, vol. 11688, pp. 295–313. Springer, Cham (2019). https://doi.org/10.1007/978-3-030-27181-7_18
4. Deiana, E., Latouche, G., Remiche, M.-A.: Fluid flow model for energy-aware server performance evaluation. Methodol. Comput. Appl. Probab. **23**, 801–821 (2021)
5. Latouche, G., Nguyen, G.: Analysis of fluid flow models. Queueing Models and Service Management, vol. 1(2), pp. 1–29 (2018). arXiv:1802.04355
6. Van Foreest, N., Mandjes, M., Scheinhardt, W.: Performance analysis of heterogeneous interacting TCP sources. Siam J. Control Optim. **01** (2001)

Performance Analysis of Multi-server Queueing Systems with Batch Services and Setup Times

Thu Le-Anh[1]([⊠])[iD], Kai Ishiguro[2], and Tuan Phung-Duc[3][iD]

[1] Graduate School of Science and Technology, University of Tsukuba,
Tsukuba, Japan
s2265261@u.tsukuba.ac.jp
[2] School of Science and Technology, University of Tsukuba, Tsukuba, Japan
s1711222@u.tsukuba.ac.jp
[3] Institute of Systems and Information Engineering, University of Tsukuba, 1-1-1,
Tennoudai, Tsukuba, Ibaraki 305-8573, Japan
tuan@sk.tsukuba.ac.jp

Abstract. Multi-server systems with setup time have been extensively investigated due to their applications in various contexts, especially in data center modeling. However, little attention has been paid to the effect of setup delays on the performance of server-sharing systems that allow multiple jobs to utilize a server simultaneously. This paper considers an $M/M/c/K$/Setup queueing model with a batching policy for resource-sharing systems. We present a simple analysis of the steady-state probabilities and performance measures under a Markovian setting. In addition, we show the effect of setup time distributions on the system performance by simulations where the setup time follows a deterministic, Erlang, exponential, and hyperexponential distribution. Our simulations reveal that the performance of the model is better when the setup time has a larger coefficient of variation.

Keywords: Multi-server queue · Setup time · Batching policy

1 Introduction

In queueing systems, operating servers consume a considerable amount of resources, whether they involve employees in a call center, medical professionals in a hospital, machines in a manufacturing system, or servers in a data center. The energy consumption of machines accounts for 67.5% of industrial electricity consumption, of which idle machines consume a significant amount [13]. In data centers, servers consume a lot of energy, and each idle server takes around 60% of its peak processing workload [2]. As a result, turning off idle servers becomes an effective solution to reduce resource consumption. Nevertheless, 'off' servers require a considerable setup time to back 'on' during that, they consume energy

M. Iacono et al. (Eds.): EPEW/ASMTA 2023, LNCS 14231, pp. 111–122, 2023.
https://doi.org/10.1007/978-3-031-43185-2_8

without processing jobs. Therefore, turning idle servers off makes a tradeoff between energy conservation and system performance.

Recently, many authors have studied similar or related setup systems driven by applications in various contexts (healthcare systems, manufacturing systems, call centers, data centers, etc.). Different conclusions might be derived due to the variety of model settings, including the cost function, setup times policy, and model details. Thus, multi-server systems with setup times have been extensively studied [4,7,8]; however, no theoretical study has analyzed the effect of setup time in queueing systems with group processing policies. The group processing or batching policy allows a server to process a batch (a set of jobs) instead of a single job at a time. It is practical in testing centers, freight companies, or manufacturing companies where items (samples, packages, etc.) are processed in batches.

Motivated by these applications, we propose to study an $M/M/c/K/$Setup queueing model with the batching policy. We assume that jobs arrive at the system according to the Poisson process, then they wait until a certain number of jobs in the queue are reached before requesting a server to set up. A server processes multiple jobs in a fixed-size batch at the same time. We assume that the service speed is independent of the batch size and the service time follows an exponential distribution, while the power consumption is directly proportional to the batch size. Compared to the $M/M/c/K/$Setup queueing system, our model may give better results with respect to the waiting time and blocking probability while it may consume more energy.

Recently, multi-server queues with setup time have received considerable attention. Chen et al. [3] were the first group to solve the problem of energy-aware resource provision using queueing theory. Since then, many variations of setup time models have been established. Gandhi et al. [5] analyze a multi-server system with a staggered setup policy. They obtain approximations for the setup systems where the number of setup servers can be equal to the number of jobs awaiting processing, and a server is turned off immediately when it is idle. Gandhi et al. [4] employ the recursive renewal reward approach to extend the setup model where servers are idle for a while before being shutdown. Phung-Duc [8] comprehensively compares the matrix analytic approach and the generating function method for a similar model and gives exact solutions for the queue-length distribution and other performance measures. Although the infinite capacity models with setup time have been widely studied, the finite buffer multi-server queues have received less attention. Phung-Duc et al. [9,10] analyze the finite buffer multi-server queue with the s-staggered setup policy and derive the stationary queue length distribution. They extend the standard policy in Artalejo et al. [1] from a single server to an arbitrary number of servers allowed in setup mode. Since a setup server requires a large amount of power, the staggered setup policy, which restricts the number of setup servers at once, saves energy. To reduce the number of setup servers for resource-sharing systems, we propose the $M/M/c/K/$Setup model with the batching policy. By comparing our model with the previous $M/M/c/K/$Setup model, we discover new insight into the effects of the setup conditions via batching policy that has not been observed in the literature.

In addition to the setup condition based on batch policies, it is important to consider the setup time distributions, especially in multi-server systems. According to Williams et al. [12], setup times in many systems are highly variable, while deterministic setup times are also typical in practice. For instance, the setup times of virtual machines (VM), as in [11], can be highly variable since their boot procedure includes steps that rely on the job's request. Meanwhile, the setup time for servers in the data center is constant in [6]. To understand the effect of setup time distribution on the system performance and power consumption, we compare our initial model with three other cases by simulation: hyperexponential distribution, Erlang distribution, and deterministic setup time.

The remainder of the paper is structured as follows. Section 2 describes the model in detail, while Sect. 3 presents the analysis of the joint stationary distribution and the performance of our system. In Sect. 4, we carry out numerical experiments and simulations to show insights into the system's performance. Finally, the conclusion and directions for future work are given in Sect. 5.

2 Model and Markov Chain

2.1 Model Description

In this paper, we analyze a c-server queueing system with a capacity of K, i.e., the system can simultaneously possess up to K jobs. We assume that jobs arrive in this system according to a Poisson process with rate λ; they are grouped and processed in fixed-size batches. The service time of batches follows an exponential distribution with mean $1/\mu$. Since a server requires some random setup time to go from "off" to "on," we assume the setup times are exponentially distributed with mean $1/\alpha$. Upon service completion, a server will shut down if no batches request the service. Otherwise, it instantly takes the waiting batch at the head of the queue to process. At that time, the server with the largest remaining setup time is turned off if the number of batches left in the queue is less than the number of setup servers, i.e., we are setting up more servers than needed.

Let i, j, and b denote the number of active servers, the number of jobs in the system, and the batch size, respectively. Then, the number of setup servers is $\min(\lfloor \frac{j-ib}{b} \rfloor, c - i)$. In our system, the number of active servers cannot exceed the number of batches in the system. Hence, a server has three states "setup," "busy," and "off." We assume that servers in this system operate under the First Come First Served (FCFS) discipline. We name our system the M/M/c/K/SET-BATCH queue, where SET and BATCH stand for setup and batching policy, respectively.

2.2 Markov Chain

Let $B(t)$ and $Q(t)$ be the number of busy servers and the number of jobs in the system at time t, respectively. In the current setting, the two-dimensional process $\{X(t) = (B(t), Q(t)); t \geq 0\}$ is a Markov chain in the state space

$$S = \{(i,j); i = 0, 1, \ldots, c, j = ib, ib + 1, \ldots, K\}.$$

Figure 1 shows the state transition diagram for the case $b = 2$, $c = 2$, and $K = 6$.

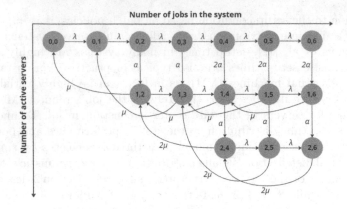

Fig. 1. State-transition diagram ($b = 2$, $c = 2$, $K = 6$).

3   Analysis

This section analyzes the joint stationary distribution to obtain the corresponding performance measures. The system can be formulated as a continuous time Markov chain with the infinitesimal generator Q whose entries are denoted by $(q_{(i,j),(i',j')}; (i,j), (i',j') \in S)$. Let $\mu_{i,j}$ and $\alpha_{i,j}$ denote the service rate and setup rate for state (i,j). We have

$$\mu_{i,j} = i\mu,$$
$$\alpha_{i,j} = \min(\lfloor \frac{j - ib}{b} \rfloor, c - i)\alpha.$$

The entries of the transition-rate matrix Q are given as follows

$$q_{(i,j),(i,j)} = -\Sigma_{(i',j') \neq (i,j)} q_{(i,j),(i',j')},$$
$$q_{(0,m),(0,m+1)} = \lambda, \quad m = 0, 1, \dots, K - 1,$$
$$q_{(0,m),(1,m)} = \alpha_{0,m}, \quad m = b, b+1, \dots, K.$$

The transition-rates for states with $0 < i < c$ and $j < bl$ are given by

$$q_{(l,m),(l,m+1)} = \lambda,$$
$$q_{(l,m),(l-1,m-b)} = \mu_{l,m}, \quad \text{for } l, m \in \mathbb{N},$$

where $\mathbb{N} = \{1, 2, 3, \dots\}$.
For states with $0 < i < c$ and $bl \leq j < K$, we have

$$q_{(l,m),(l,m+1)} = \lambda,$$
$$q_{(l,m),(l,m-b)} = \mu_{l,m},$$
$$q_{(l,m),(l+1,m)} = \alpha_{l,m}, \quad \text{for } l, m \in \mathbb{N}.$$

For $0 < i < c$ and $j = K$, we have

$$q_{(l,K),(l,K-b)} = \mu_{l,K},$$
$$q_{(l,K),(l+1,K)} = \alpha_{l,K}, \quad \text{for } l \in \mathbb{N}.$$

For $K = cb$, the transition-rates for states with $i = c$ are given as follows

$$q_{(c,K),(c,K)} = -\lambda - \mu_{c,K},$$
$$q_{(c,K),(c-1,K-b)} = \mu_{c,K}.$$

For $K > cb$ and $i = c$, we have

$$q_{(c,K),(c,K)} = -\mu_{c,K},$$
$$q_{(c,K),(c,K-b)} = \mu_{c,K}.$$

The transition-rates for states with $i = c$ and $j \neq K$ are given by

$$q_{(c,m),(c,m+1)} = \lambda,$$
$$q_{(c,m),(c,m-b)} = \mu_{c,m}, \quad \text{for } m \in \mathbb{N}.$$

For $i = c$ and $j < (c+1)b$, we have

$$q_{(c,m),(c,m+1)} = \lambda,$$
$$q_{(c,m),(c-1,m-b)} = \mu_{c,m}, \quad \text{for } m \in \mathbb{N}.$$

For $i = c$ and $(c+1)b \leq j < K$, we have

$$q_{(c,m),(c,m+1)} = \lambda,$$
$$q_{(c,m),(c,m-b)} = \mu_{c,m}, \quad \text{for } m \in \mathbb{N}.$$

The remaining entries of the transition-rate matrix Q are 0.

Let $\pi_{i,j} = P(B(t) = i, Q(t) = j)((i,j) \in S)$ denote the stationary distribution of $X(t)$. We define

$$\pi_i = (\pi_{i,ib}, \pi_{i,ib+1}, \ldots, \pi_{i,K}), \quad \text{for } i = 0, 1, 2, \ldots, c,$$

and

$$\pi = (\pi_1, \pi_2, \ldots, \pi_c).$$

Then, π is the solution of the following equations

$$\pi e = 1,$$
$$\pi Q = 0,$$

where $\mathbf{0}$ is a row vector of zeros and e is a column vector of ones with an appropriate dimension.

Let P_B be the blocking probability, i.e.,

$$P_B = \sum_{i=0}^{c} \pi_{i,K}.$$

Denote by $E[B]$ and $E[S]$ the average number of busy servers and setup servers, respectively. Then

$$E[B] = \sum_{(i,j) \in S} i\pi_{i,j},$$

$$E[S] = \sum_{i=0}^{c} \sum_{j=ib}^{K} \min(\lfloor \frac{j-ib}{b} \rfloor, c-i)\pi_{i,j}.$$

The power consumption cost per unit of time is given as follows

$$C = b(C_b E[B] + C_s E[S]),$$

where C_b and C_s represent the cost of processing a job for a busy server and a setup server per unit of time, respectively.

Denote by $E[L]$ the mean number of jobs in the system, then we have

$$E[L] = \sum_{i=0}^{c} \sum_{j=ib}^{K} j\pi_{i,j}.$$

Then from Litle's law, the mean response time can be obtained by

$$E[R] = \frac{E[L]}{\lambda(1 - P_B)}.$$

Furthermore, we define the integration cost by the product of the power consumption cost per unit of time and the mean response time. The integration cost is given by

$$P = E[R] \times C.$$

4   Numerical Examples and Simulation

This section presents several numerical experiments for the analysis in Sect. 3 to provide insights into the performance of our system. We also verify and compare our model with many cases with different setup time distributions using simulation.

4.1   Effect of the Setup Time Distribution

To analyze the effect of setup time distribution on the performance of our M/M/c/K/SET-BATCH system, we set $\mu = 1$, $\alpha = 0.5$, $b = 2$, $c = 10$, and the cost for a setup and an active server as $C_b = C_s = 1$. Then, we present some analytical results and simulations in Figs. 2, 3, 4, 5, 6 and 7 to show the effect of deterministic setup times, Erlang distribution for setup times with the shape parameter $k = 10$, exponentially distributed setup times, and hyperexponential distribution for setup times with the coefficient of variation CV $= 5$, which is a mixture of two exponential distributions. Parameters of all the setup time distributions are selected so that the mean remains the same as in the case of the exponential distribution.

Figures 2, 4, and 6 illustrate the blocking probability, the average response time, and the power usage against arrival rate λ for $K = cb = 20$, while that performance for $K = 35$ is presented in Figs. 3, 5, and 7. As observed, the blocking probability P_B and the average response time $E[R]$ decrease with the coefficient of variation CV and are bounded from below by that of the Setup-Hyperexponential curve. The shape of P_B curve does not change much, while that of $E[R]$ is very different when $K = 20$ and $K = 35$. When K is large, the blocking probability is almost unaffected by the setup time distribution. As expected, the blocking probability decreases with the capacity K and increases with the arrival rate λ.

We see in Figs. 4 and 5 that for all settings of the setup time distributions, the mean delay decreases with λ and converges to a fixed value for $K = 20$, and the changes corresponding to the case $K > cb$ are more complicated. In particular, Fig. 5 shows that when the arrival rate λ is small, the average response time decreases with λ, while it increases with λ as λ is large and then remains unchanged at high values of λ. This occurs because the setup time has a significant effect at low arrival rates. Specifically, increasing the arrival rate will increase the number of setup servers, thereby reducing the average response time. This is especially true for our model, where jobs wait in the queue and are processed in batches. On the other hand, when the arrival rate is large enough, all servers are always "on," and then the mean delay increases with λ, as in $M/M/c/K$ systems without setup time. Furthermore, when λ reaches a certain value, the mean response time remains unchanged due to finite buffer capacity.

The $M/M/c/K/$SET-BATCH system, where setup times follow Erlang distribution with a large shape parameter value, converges to the system with deterministic setup times. There is a significant difference in the blocking and delay in the system with deterministic and Erlang setup times and the systems with Setup-Exponential and Setup-Hyperexponential. Those values of the $M/M/c/K/$SET-BATCH system with a fixed setup time are larger than that of the corresponding system with a variable setup time. Thus, the highly variable setup times models are more effective than the deterministic setup model.

Figures 6 and 7 show that the Setup-deterministic and the Setup-Erlang systems are more power-consuming than the exponential and hyperexponential setup times systems. In other words, the larger the coefficient of variation CV is, the smaller the power consumption is. The power consumption in both cases of K converges at the value of cb when λ is large enough; it converges very quickly in the case $K > cb$. Although the difference in the power consumption is not as significant as the blocking probability and the mean delay when the setup time distribution changes, in general, variable setup times give better results for the $M/M/c/K/$SET-BATCH system.

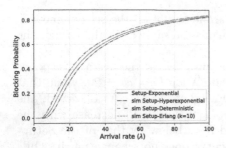

Fig. 2. Blocking probability against λ ($K = 20$).

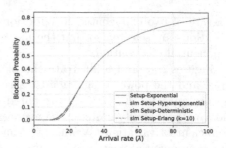

Fig. 3. Blocking probability against λ ($K = 35$).

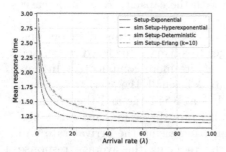

Fig. 4. Mean response time against λ ($K = 20$).

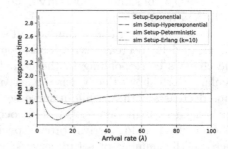

Fig. 5. Mean response time against λ ($K = 35$).

4.2   Effect of the Batching Policy

In this section, we consider the effect of the batching policy on the performance of our $M/M/c/K$/SET-BATCH system where $b = 2$, $K = 20, 30$, and 50. Under the same settings, for fixed $K = 50$ and $b = 2, 3$, we show how the performance measures change according to batch size b. We also derive these measures in the case without batching policy ($b = 1$) and use simulation to verify our analytical results.

Figures 8, 10, and 12 reflect the effect of the capacity K on the system performance. The capacity has no significant effect on the system performance when the arrival rate is low. The power consumption and the mean response time are relatively small, while the blocking probability is large in the case $K = cb$. Figures 9, 11, and 13 represent the effect of batch size b on the system performance. The blocking probability and the average response time decrease with the batch size while the power usage is directly proportional to the batch size. The average response time in the $M/M/c/K$/Setup without the batch policy ($b = 1$) is smaller than that of our model when λ is small while applying the batching policy reduces delay when λ is large enough.

Figures 8 and 9 show that the blocking probability increases with the arrival rate and decreases with the capacity and the batch size as expected. We observe in Fig. 11 that when the arrival rate λ is small, the average response time reduces

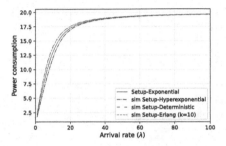

Fig. 6. Power consumption against λ ($K = 20$).

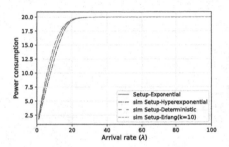

Fig. 7. Power consumption against λ ($K = 35$).

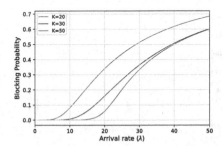

Fig. 8. Blocking probability against λ ($b = 2$).

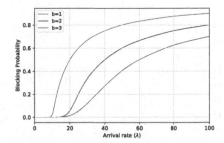

Fig. 9. Blocking probability against λ ($K = 50$).

with λ, while it increases with λ as λ is large and then remains unchanged at very high values of λ. This is because the setup time has a significant effect at low arrival rates; however, when the arrival rate is large enough, all servers are always "on," and then setup time has almost no effect. Moreover, when the arrival rate λ reaches a certain value, the mean system response time remains unchanged since the buffer capacity is finite.

Figures 14 and 15 reflect the changes in the integration cost against the arrival rate with different capacity and batch size values. Figures 14 shows that when the arrival rate is low, the integration cost increases with λ, and then remains unchanged at high values of λ. The integration cost is highly sensitive to the capacity K when λ is large. We observe in Fig. 15 that the integration cost increases with the batch size when λ is small while applying the batching policy reduces integration cost when λ is large enough. The integration cost remains unchanged when λ reaches a certain value. Thus, a batching policy with the appropriate batch size can balance the delay and energy trade-offs as the arrival rate λ is not too large.

4.3   Effect of the Capacity

In this section, we demonstrate how the system performance is affected by the capacity K. We consider the cases where $b = 2$ and $\alpha = 0.01, 0.5, 1$. Figures 16

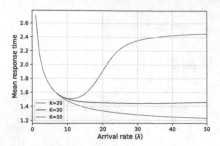

Fig. 10. Mean response time against λ ($b = 2$).

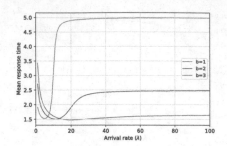

Fig. 11. Mean response time against λ ($K = 50$).

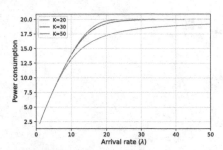

Fig. 12. Power consumption against λ ($b = 2$).

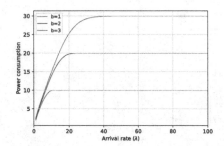

Fig. 13. Power consumption against λ ($K = 50$).

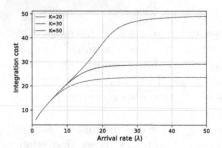

Fig. 14. Integration cost against λ ($b = 2$).

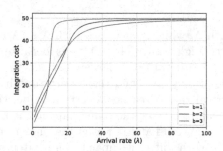

Fig. 15. Integration cost against λ ($K = 50$).

and 17 show that the blocking probability decreases with the capacity K while the average response time increases with K. When $\alpha = 1, 0.5$, the blocking probability and the average response time are unchanged for $K \geq 100$, i.e., our system converges to the corresponding M/M/c/∞/Setup system. Meanwhile, the curves for the case $\alpha = 0.01$ do not converge when $K < 200$. This shows that the mean delay is huge when $\alpha = 0.01$, and many jobs are lost due to the finite buffer restriction.

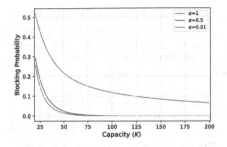

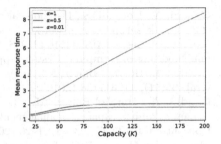

Fig. 16. Blocking probability against K $(b = 2)$.

Fig. 17. Mean response time against K $(b = 2)$.

5 Conclusions

This paper presents a simple analysis of the $M/M/c/K/$Setup model with batching policy for sharing server systems. We employ simulation to show that the adverse effect of setup times on delay in the deterministic setup time model is much more significant than that in exponential and hyperexponential setup time models. However, in many real-world systems, the setup time is fixed, so extending the current model to the deterministic setup time model may be worth investigating. The batch policy in this paper reduces the number of servers in the setup mode for resource-sharing systems, thereby reducing delays and blocking. However, this policy can significantly increase energy consumption when the batch size is large. Thus, a batch policy with the appropriate batch size can balance the trade-off between energy savings and performance. One direction of future work is analyzing the batching policy for natural systems such as testing centers, freight agencies, manufacturing firms, etc., where the energy cost function is much more complicated. In addition, it might be interesting to investigate how the performance of these systems is affected by batch policy and different setup time distributions.

Acknowledgments. The research of Tuan Phung-Duc was supported in part by JSPS KAKENHI, Grant Number 21K11765 and by F-MIRAI: R&D Center for Frontiers of MIRAI in Policy and Technology, the University of Tsukuba and Toyota Motor Corporation collaborative R&D center.

References

1. Artalejo, J.R., Economou, A., Lopez-Herrero, M.J.: Analysis of a multiserver queue with setup times. Queueing Syst. **51**, 53–76 (2005)
2. Barroso, L.A., Hölzle, U.: The case for energy-proportional computing. Computer **40**(12), 33–37 (2007)
3. Chen, Y., Das, A., Qin, W., Sivasubramaniam, A., Wang, Q., Gautam, N.: Managing server energy and operational costs in hosting centers. In: Proceedings of the 2005 ACM SIGMETRICS International Conference on Measurement and Modeling of Computer Systems, pp. 303–314 (2005)

4. Gandhi, A., Doroudi, S., Harchol-Balter, M., Scheller-Wolf, A.: Exact analysis of the M/M/k/setup class of Markov chains via recursive renewal reward. In: Proceedings of the ACM SIGMETRICS/International Conference on Measurement and Modeling of Computer Systems, pp. 153–166 (2013)
5. Gandhi, A., Harchol-Balter, M., Adan, I.: Server farms with setup costs. Perform. Eval. **67**(11), 1123–1138 (2010)
6. Gandhi, A., Harchol-Balter, M., Raghunathan, R., Kozuch, M.A.: Autoscale: dynamic, robust capacity management for multi-tier data centers. ACM Trans. Comput. Syst. (TOCS) **30**(4), 1–26 (2012)
7. Maccio, V.J., Down, D.G.: Exact analysis of energy-aware multiserver queueing systems with setup times. In: 2016 IEEE 24th International Symposium on Modeling, Analysis and Simulation of Computer and Telecommunication Systems (MASCOTS), pp. 11–20. IEEE (2016)
8. Phung-Duc, T.: Exact solutions for M/M/c/setup queues. Telecommun. Syst. **64**, 309–324 (2017)
9. Phung-Duc, T., Kawanishi, K.: Energy-aware data centers with s-staggered setup and abandonment. In: Wittevrongel, S., Phung-Duc, T. (eds.) ASMTA 2016. LNCS, vol. 9845, pp. 269–283. Springer, Cham (2016). https://doi.org/10.1007/978-3-319-43904-4_19
10. Phung-Duc, T., Kawanishi, K.: Delay performance of data-center queue with setup policy and abandonment. Ann. Oper. Res. **293**(1), 269–293 (2020)
11. Rzadca, K., et al.: Autopilot: workload autoscaling at google. In: Proceedings of the Fifteenth European Conference on Computer Systems, pp. 1–16 (2020)
12. Williams, J.K., Harchol-Balter, M., Wang, W.: The M/M/k with deterministic setup times. In: Proceedings of the ACM on Measurement and Analysis of Computing Systems, vol. 6, no. 3, pp. 1–45 (2022)
13. Zhou, L., Li, J., Li, F., Meng, Q., Li, J., Xu, X.: Energy consumption model and energy efficiency of machine tools: a comprehensive literature review. J. Clean. Prod. **112**, 3721–3734 (2016)

On a Stochastic Epidemic Model with Limited Number of Hospital Beds

A. Gómez-Corral[1]([⊠]) [ID], M. J. Lopez-Herrero[2] [ID], and D. Taipe[2] [ID]

[1] Department of Statistics and Operations Research, Faculty of Mathematical
Sciences, Complutense University of Madrid, 28040 Madrid, Spain
[2] Department of Statistics and Data Science, Faculty of Statistical Studies,
Complutense University of Madrid, 28040 Madrid, Spain
{agcorral,lherrero,dtaipe}@ucm.es

Abstract. The aim of this paper is to complement the work of Gómez-
Corral et al. [11] on a stochastic epidemic model where healthcare facilities, specifically the number of hospital beds for infectives, are limited.
We characterize the probability law of the total time that the hospital
ward remains functionally full during an outbreak of the disease. Our
methodology is mainly based on the use of sojourn and first-passage
times, and related hitting probabilities in a suitably defined absorbing
quasi-birth-death process. Numerical examples are presented to illustrate the effects of the person-to-person contact processes and exogenous
streams of infection on the dynamics of the epidemic when the occupancy
of the hospital ward is maximum.

Keywords: Epidemic model · Limited resources · Quasi-birth-death
process

1 Introduction

This paper deals with a stochastic epidemic model operating in a resource-limited
environment. The underlying resource-limited assumption yields a model for
which the dynamics of epidemics of SIS-type are combined with those of SI-like
models.

1.1 An Epidemic Model in a Resource-Limited Environment

Specifically, the focus is on the Markov chain model recently introduced in Ref.
[11], which is related to the spread of a pathogen within a finite population,

This work was supported by the Ministry of Science and Innovation (Government of
Spain), Project PID2021-125871NB-I00. D Taipe acknowledges the support of Banco
Santander and Complutense University of Madrid, Pre-doctoral Researcher Contract
CT63/19-CT64/19.

M. Iacono et al. (Eds.): EPEW/ASMTA 2023, LNCS 14231, pp. 123–138, 2023.
https://doi.org/10.1007/978-3-031-43185-2_9

with constant size N, under the assumption that infected individuals recover only after receiving medical treatment in a hospital ward where at most C individuals can be accommodated in beds, with $1 < C < N$. From now on, we use the term *inpatient* or simply *patient* to refer to hospital ward hosted individuals, and the term *individual* to those who are not. The epidemic evolving within the hospital ward is of the SIS-type with patient-to-patient contact rate $\beta > 0$, rate $\gamma > 0$ for a medical treatment to take effect on an infected patient, and rate $\nu > 0$ for a susceptible patient to be discharged after his healthy state is confirmed after recovering. The potential transmission of the pathogen outside the hospital ward is only possible when all the beds reserved to accommodate inpatients are simultaneously occupied. Therefore, an SI-like model with individual-to-individual contact rate $\beta_0 > 0$ describes how the pathogen spreads between susceptible and infected individuals. Two Poisson flows of rates $\lambda > 0$ and $\lambda_0 > 0$ are assumed in [11] to represent exogenous infections on the subpopulations of inpatients and individuals, respectively. In particular, the flow of rate λ_0 represents infections to individuals caused by persons who are not represented within the population, and the flow of rate λ corresponds to infections to inpatients caused by non-hospitalized persons, such as visitors and staff entering the ward. According to this idea, lowering the rate λ means that hospital authorities apply stricter isolation measures between the hospital ward and the outside, while stricter isolation measures between inpatients are related to smaller values of β.

The above description results in a compartmental epidemic model, formulated in [11, Section 2] as a finite quasi-birth-death (QBD) process. Concretely, the population is divided into four compartments or subpopulations of *susceptible* (S_0) and *infected* (I_0) individuals, and *susceptible* (S) and *infected* (I) inpatients, with respective sizes denoted by $S_0(t)$, $I_0(t)$, $S(t)$ and $I(t)$ at time t. First-passage times and hitting probabilities in finite QBD processes are used in [11], in a unified manner, to study the random length of an outbreak and related stochastic measures, such as the time to reach the maximum occupancy of the hospital ward, the number of critical events, and the number of admissions and therapeutic treatments occurring during an outbreak.

This paper aims to complement our work in [11] by characterizing the probability law of the total time that the hospital ward remains functionally full during an outbreak of the disease. This total sojourn time, which is inherently linked to the *resource-limited* condition $S(t) + I(t) = C$ shall be seen here to be a key index to prevent the spread of the pathogen among individuals through intra-hospital actions and measures to isolate inpatients from the outside.

1.2   Related Literature

There exists an abundant literature on epidemic models with limited resources, mainly from the deterministic perspective. Without intending to make an exhaustive enumeration, we mention here the paper by Zhou and Fan [24] on SIR-like models with varying amount of medical resources, the work of Jiang and Zhou [13–15] on two-layer epidemic networks where each subnetwork is of

SIS-type, and the nonlinear model with a limited number of hospital beds analyzed by Misra and Maurya [18], among others. In the setting of optimal control, some of the earlier work, mainly in the framework of SIR-like models, is done by Abakuks [1], Morton and Wickwire [19], Sethi [20], Sethi and Staats [21], and Wickwire [22]. For more recent work, see also Bolzoni et al. [4], Clancy [6], Lin et al. [17], and Zhang et al. [23], among others.

   From the stochastic perspective, the authors are only aware of the work of Amador and Lopez-Herrero [2] on the SEIR model with limited resources for treatment and nonlinear incidence function, and the SIQS model with limited carrying capacity for the quarantine compartment in Ref. [3], which are both analyzed by using finite QBD processes and matrix-analytic techniques, including phase-type random variables and censored structured Markov chains. For a summary of these techniques, we refer the reader to the books of He [12], and Latouche and Ramaswami [16], and the papers by Gaver et al. [7], and Gómez-Corral et al. [10].

1.3   Organization of the Paper

This paper proceeds as follows. To begin with, we present in Sect. 2 the finite QBD process used in Ref. [11] to study outbreaks of a disease when healthcare resources, more precisely the number of beds in the hospital ward to accommodate infectives, are assumed to be limited. In Sect. 3, we characterize the probability law of the total time that the hospital ward remains with its beds simultaneously occupied by inpatients during an outbreak of the disease by analyzing the dynamics of a suitably defined QBD process under a *taboo* subset of states. Our algorithmic solution is exemplified in Sect. 4 with some numerical experiments. A brief discussion is finally presented in Sect. 5.

2   The Epidemic Model as a Finite QBD Process

We consider the finite QBD process introduced in Ref. [11], which is defined as

$$\mathcal{X} = \{(I_0(t), J(t), I(t)) : t \geq 0\},$$

where $J(t) = S(t) + I(t)$, and takes values on the set $\mathcal{S} = \mathcal{S}_a \cup \mathcal{S}_l$, where the subset $\mathcal{S}_a$ consists of $s_a = 2^{-1}C(C+1)$ states of the form $(0, j, i)$, for $i \in \{0, ..., j\}$ and $j \in \{0, ..., C-1\}$, and the subset $\mathcal{S}_l$ contains $s_l = (C+1)(N-C+1)$ states of the form (n, C, i), where $i \in \{0, ..., C\}$ and $n \in \{0, ..., N-C\}$.

   The non-vanishing transition rates of $\mathcal{X}$ are specified (see [11, Table 1]) by

$$q_{(n,j,i),(n',j',i')} = \begin{cases} (N-j)\lambda_0, & \text{if } n'=n=0,\ j'=j+1 \text{ and } i'=i+1, \\ (N-C-n)\left(\frac{n\beta_0}{N-C}+\lambda_0\right), & \text{if } n'=n+1,\ j'=j=C \text{ and } i'=i, \\ (j-i)\left(\frac{i\beta}{j}+\lambda\right), & \text{if } n'=n,\ j'=j \text{ and } i'=i+1, \\ i\gamma, & \text{if } n'=n,\ j'=j \text{ and } i'=i-1, \\ (j-i)\nu, & \text{if } n'=n=0,\ j'=j-1 \text{ and } i'=i, \\ (j-i)\nu, & \text{if } n'=n-1,\ j'=j \text{ and } i'=i+1, \end{cases}$$

for states $(n,j,i),(n',j',i') \in \mathcal{S}$ with $(n',j',i') \neq (n,j,i)$, and $q_{(n,j,i),(n,j,i)} = -q_{(n,j,i)}$ with

$$q_{(n,j,i)} = (N - n - j)\left(\frac{n\beta_0}{N - j} + \lambda_0\right) + (j - i)\left(\frac{i\beta}{j} + \lambda + \nu\right) + i\gamma.$$

We adopt the notation of our work in [11], which will allow the interested reader to observe better the similarities and differences between the results in [11] and our approach in Sect. 3. For instance, we express subsets $\mathcal{S}_a$ and $\mathcal{S}_l$ in terms of *levels* as

$$\mathcal{S}_a = \bigcup_{j=0}^{C-1} \mathcal{L}_j^a \quad \text{and} \quad \mathcal{S}_l = \bigcup_{n=0}^{N-C} \mathcal{L}_n^l,$$

where $\mathcal{L}_j^a = \{(0,j,i) : i \in \{0,...,j\}\}$ and $\mathcal{L}_n^l = \{(n,C,i) : i \in \{0,...,C\}\}$, in such a way that Q, the q-matrix of process $\mathcal{X}$, can be written in terms of submatrices $A_{j,j'}$ and $B_{n,n'}$ (see [11, Section 2 and Appendix A]), where

(a) Submatrices $A_{j,j'}$ have dimension $(j + 1) \times (j' + 1)$ and consist of rates $q_{(0,j,i),(0,j',i')}$ associated with those jumps of $\mathcal{X}$ from states $(0,j,i) \in \mathcal{L}_j^a$ to states $(0,j',i') \in \mathcal{L}_{j'}^a$ with $j' \in \{\max\{0,j-1\}, j, \min\{C-1,j+1\}\}$ if $j \in \{0,...,C-1\}$, and to states $(0,j',i') \in \mathcal{L}_0^l$ with $j' = C$ if $j = C - 1$; and $A_{C,C-1}$ is a matrix, of dimension $(C+1) \times C$, recording rates $q_{(0,C,i),(0,C-1,i')}$, which are related to jumps from states $(0,C,i) \in \mathcal{L}_0^l$ to states $(0,C-1,i') \in \mathcal{L}_{C-1}^a$.

(b) Submatrices $B_{n,n'}$ are square matrices of order $C + 1$ that contain rates $q_{(n,C,i),(n',C,i')}$ associated with jumps from states $(n,C,i) \in \mathcal{L}_n^l$ to states $(n',C,i') \in \mathcal{L}_{n'}^l$, for integers $n \in \{0,...,N-C\}$ and $n' \in \{\max\{0,n-1\}, n, \min\{N-C,n+1\}\}$.

In a similar manner to [11, Section 3], we also let $\tau_{(n,j,i)}$ denote the first-passage time from (n,j,i) to state $(0,0,0)$, for states $(n,j,i) \in \mathcal{S} \setminus \{(0,0,0)\}$, so that $\tau_{(0,1,1)}$ is equivalent to the random length of an outbreak.

Throughout the article, $0_{b \times b'}$ denotes the null matrix of dimension $b \times b'$, I_b is the identity matrix of order b, 0_b is the column null vector of order b, 1_b is the column vector of order b of 1s, $e_b(b')$ is the row null vector of order b with a single 1 in its b'-th entry, superscript ^{T} denotes transposition, and $1_{\{E\}}$ is the indicator function of the set $\{E\}$.

3   The Time While the Hospital Ward Is Funtionally Full

We are concerned here with the total sojourn time $\tau_{(0,1,1)}(C)$ in states of the subset $\mathcal{S}_l$ before the first visit of process $\mathcal{X}$ to state $(0,0,0)$, starting from state $(0,1,1)$. Note that $\tau_{(0,1,1)}(C)$ equals zero when the hospital ward is always running with available beds during the outbreak. This is related to the sample paths of process $\mathcal{X}$ that satisfy $\{\tau_{(0,1,1)} < \tau'_{(0,1,1)}(C)\}$, where $\tau'_{(0,1,1)}(C)$ is the first-passage time from state $(0,1,1)$ to the subset $\mathcal{S}_l$. Thus, we have that

$$E\left[e^{-s\tau_{(0,1,1)}(C)}\right] = P\left(\tau_{(0,1,1)}(C) = 0\right) + \psi_{(0,1,1)}(s;C), \tag{1}$$

for $\Re(s) \geq 0$, with $P(\tau_{(0,1,1)}(C) = 0) = P(\tau_{(0,1,1)} < \tau'_{(0,1,1)}(C))$ and

$$\psi_{(0,1,1)}(s;C) = E\left[e^{-s\tau_{(0,1,1)}(C)} 1_{\{\tau'_{(0,1,1)}(C) < \tau_{(0,1,1)}\}}\right].$$

We shall derive briefly in Subsect. 3.1 the probability $P\left(\tau_{(0,1,1)}(C) = 0\right)$ by using a similar approach to that of [8] for extinction times in a two-species competition process. To derive the restricted Laplace-Stieltjes transform $\psi_{(0,1,1)}(s;C)$, we shall study in Subsect. 3.2 the dynamics of process $\mathcal{X}$ under a *taboo* subset of states in a suitable defined QBD process.

For later use, we define the more general random variables $\tau_{(n,j,i)}(C)$, for states $(n,j,i) \in \mathcal{S} \setminus \{(0,0,0)\}$, as the total sojourn times of process $\mathcal{X}$ in the subset $\mathcal{S}_l$ before the first visit to state $(0,0,0)$, provided that the *current* state of $\mathcal{X}$ is (n,j,i). For states $(0,j,i) \in \mathcal{S}_a \setminus \{(0,0,0)\}$, we also define $\tau'_{(0,j,i)}(C)$ as the first-passage time from state $(0,j,i)$ to $\mathcal{S}_l$, and $\psi_{(0,j,i)}(s;C)$ as the restricted Laplace-Stieltjes transform of $\tau_{(0,j,i)}(C)$ on the set $\{\tau'_{(0,j,i)}(C) < \tau_{(0,j,i)}\}$ of sample paths.

3.1   The Discrete Contribution of $\tau_{(0,1,1)}(C)$

For states $(0,j,i) \in \mathcal{S}_a \setminus \{(0,0,0)\}$, the probability $P(\tau_{(0,j,i)}(C) = 0)$ is related to the dynamics of process $\mathcal{X}$ under taboo of the subset $\mathcal{S}_l$ and, consequently, can be derived by using an auxiliary absorbing QBD process $\mathcal{X}'(C) = \{X'(t;C) : t \geq 0\}$ defined on the state space

$$\mathcal{S}'(C) = \{0'\} \cup \bigcup_{j=1}^{C-1} \mathcal{L}_j^a \cup \{C'\},$$

where states $0'$ and C' are assumed to be absorbing states, and states in the subset $\cup_{j=1}^{C-1} \mathcal{L}_j^a$ are transient. The q-matrix of $\mathcal{X}'(C)$ has the structured form

$$Q'(C) = \begin{pmatrix} 0 & 0_{s_a-1}^T & 0 \\ t_{0'}(C) & T(C) & t_{C'}(C) \\ 0 & 0_{s_a-1}^T & 0 \end{pmatrix},$$

where submatrix $T(C)$ is obtained from Q by removing rows and columns associated with state $(0,0,0)$ and states in $\mathcal{S}_l$, and column vectors $t_{0'}(C)$ and $t_{C'}(C)$ are given by $t_{0'}(C) = \nu e_{s_a-1}(1)$ and

$$t_{C'}(C) = (N - C + 1)\lambda_0 \begin{pmatrix} 0_{2^{-1}(C-2)(C+1)} \\ 1_C \end{pmatrix}.$$

In terms of process $\mathcal{X}'(C)$, the probability $P(\tau_{(0,j,i)} < \tau'_{(0,j,i)}(C))$ can be determined as the limit

$$P(\tau_{(0,j,i)} < \tau'_{(0,j,i)}(C)) = \lim_{t \to \infty} P\left(X'(t;C) = 0'|X'(0;C) = (0,j,i)\right),$$

for states $(0,j,i) \in \mathcal{S}_a \setminus \{(0,0,0)\}$, where

$$P\left(X'(t;C) = 0'|X'(0;C) = (0,j,i)\right)$$
$$= e_{s_a-1}(f_j(i))\left(I_{s_a-1} - e^{T(C)t}\right)(-T^{-1}(C))t_{0'}(C),$$

with $f_j(i) = 2^{-1}(j-1)(j+2)+i+1$, for $i \in \{0,...,j\}$ and $j \in \{1,...,C-1\}$. Since the subset $\bigcup_{j=1}^{C-1} \mathcal{L}_j^a$ is a finite class of transient states, the absorption of $\mathcal{X}(C)$ is certain and occurs in a finite expected time, regardless of the initial transient state $(0,j,i)$. This implies that submatrix $T(C)$ is a non-singular stable matrix with all its eigenvalues having strictly negative real part, from which it follows that

$$P(\tau_{(0,j,i)} < \tau'_{(0,j,i)}(C)) = \nu e_{s_a-1}(f_j(i))(-T^{-1}(C))e_{s_a-1}^T(1), \qquad (2)$$

for states $(0,j,i) \in \mathcal{S}_a \setminus \{(0,0,0)\}$.

Remark 1. Theorem 2.4.3 of [16] allows us to interpret the $(f_j(i), f_{j'}(i'))$-th entry of $-T^{-1}(C)$ as the expected total time that process $\mathcal{X}$ spends at state $(0,j',i')$ before its first visit to the subset $\{(0,0,0)\} \cup \mathcal{S}_l$, provided that $(I_0(0), J(0), I(0)) = (0,j,i)$, for states $(0,j,i), (0,j',i') \in \mathcal{S}_a \setminus \{(0,0,0)\}$.

The following theorem is an immediate consequence of Eq. (2).

Theorem 1. *The discrete contribution of $\tau_{(0,1,1)}(C)$ is given by*

$$P(\tau_{(0,1,1)}(C) = 0) = \nu e_{2^{-1}(C-2)(C+1)}(2)D_{1,1}(C)e_{2^{-1}(C-2)(C+1)}^T(1), \qquad (3)$$

where the square matrix $D_{1,1}(C)$ of order $2^{-1}(C-2)(C+1)$ is recursively determined, starting from $-T^{-1}(2) = -A_{1,1}^{-1}$, from the equations

$$D_{2,2}(k+1) = \left(-A_{k,k} - A_2^*(k+1)(-T^{-1}(k))A_1^*(k+1)\right)^{-1},$$
$$D_{2,1}(k+1) = D_{2,2}(k+1)A_2^*(k+1)(-T^{-1}(k)),$$
$$D_{1,2}(k+1) = -T^{-1}(k)A_1^*(k+1)D_{2,2}(k+1),$$
$$D_{1,1}(k+1) = -T^{-1}(k) - T^{-1}(k)A_1^*(k+1)D_{2,1}(k+1),$$

for integers $k \in \{2,...,C-1\}$. In these expressions, matrices $A_1^*(k+1)$, $A_2^*(k+1)$ and $-T^{-1}(k)$ are specified as

$$A_1^*(k+1) = \begin{pmatrix} 0_{(2^{-1}(k-2)(k+1))\times(k+1)} \\ A_{k-1,k} \end{pmatrix},$$
$$A_2^*(k+1) = \begin{pmatrix} 0_{(k+1)\times(2^{-1}(k-2)(k+1))}, & A_{k,k-1} \end{pmatrix},$$
$$-T^{-1}(k) = \begin{pmatrix} D_{1,1}(k) & D_{1,2}(k) \\ D_{2,1}(k) & D_{2,2}(k) \end{pmatrix}.$$

Proof. The proof is based on the fact that submatrix $-T^{-1}(C)$ can be determined from the inverse matrix $-T^{-1}(C-1)$. Indeed, we may write down

$$-T(k+1) = \begin{pmatrix} -T(k) & -A_1^*(k+1) \\ -A_2^*(k+1) & -A_{k,k} \end{pmatrix},$$

for integers $k \in \{2, ..., C-1\}$, with $-T(2) = -A_{1,1}$, so that straightforward algebra yields

$$-T^{-1}(k+1) = \begin{pmatrix} D_{1,1}(k+1) & D_{1,2}(k+1) \\ D_{2,1}(k+1) & D_{2,2}(k+1) \end{pmatrix}.$$

Then, Eq. (3) follows from (2) using the equality $P(\tau_{(0,1,1)}(C) = 0) = P(\tau_{(0,1,1)} < \tau'_{(0,1,1)}(C))$.

A first important observation is that, by defining the rate matrix $T_{\mathcal{L}_0}^l(C)$ as

$$T_{\mathcal{L}_0}^l(C) = \begin{pmatrix} 0_{(2^{-1}(C-2)(C+1)) \times (C+1)} \\ A_{C-1,C} \end{pmatrix},$$

the $(f_j(i), i'+1)$-th entry of $-T^{-1}(C)T_{\mathcal{L}_0}^l(C)$ records the hitting probability $P_{(0,j,i)}(0, C, i')$ that the first access of process $\mathcal{X}$ from state $(0, j, i)$ to the subset $\mathcal{L}_0^l$ (and, consequently, to $\mathcal{S}_l$) occurs via state $(0, C, i')$, for integers $j \in \{1, ..., C-1\}$, $i \in \{0, ..., j\}$ and $i' \in \{0, ..., C\}$. As a result, it is found that the probability $P_{(0,1,1)}(0, C, i)$ is given by

$$P_{(0,1,1)}(0, C, i) = e_{2^{-1}(C-2)(C+1)}(2) D_{1,2}(C) A_{C-1,C} e_{C+1}^T (i+1), \qquad (4)$$

for $i \in \{0, ..., C\}$, and the probability $P_{(0,C-1,i)}(0, C, i')$ has the form

$$P_{(0,C-1,i)}(0, C, i') = e_{\cup}(i+1) D_{2,2}(C) A_{C-1,C} e_{C+1}^T (i'+1), \qquad (5)$$

for $i \in \{0, ..., C-1\}$ and $i' \in \{0, ..., C\}$.

It must also be observed that, for states $(0, j, i) \in \mathcal{S}_a \setminus \{(0,0,0)\}$, the probability $P(\tau_{(0,j,i)}(C) = 0)$ can also be easily obtained from (2) and is given by the $f_j(i)$-entry of the column vector $\nu(-T^{-1}(C))e_{s_a-1}^T(1)$, with

$$\nu(-T^{-1}(C))e_{s_a-1}^T(1) = \nu \begin{pmatrix} D_{1,1}(C)e_{2^{-1}(C-2)(C+1)}^T(1) \\ D_{2,1}(C)e_{2^{-1}(C-2)(C+1)}^T(1) \end{pmatrix}.$$

In particular, it is seen that

$$P(\tau_{(0,C-1,i)}(C) = 0) = \nu e_C(i+1) D_{2,1}(C) e_{2^{-1}(C-2)(C+1)}^T(1), \qquad (6)$$

for $i \in \{0, ..., C-1\}$, which we shall use in Subsect. 3.2.

We conclude this subsection with a simple algorithm to compute the probabilities $P(\tau_{(0,j,i)}(C) = 0)$, for states $(0, j, i) \in \{(0, 1, 1)\} \cup \mathcal{L}_{C-1}^a$.

Algorithm 1: Iterative computation of $P(\tau_{(0,1,1)}(C) = 0)$ and $P(\tau_{(0,C-1,i)}(C) = 0)$, for $i \in \{0, ..., C-1\}$.

Step 1: $k := 2$;

$-T^{-1}(k) := -A_{k-1,k-1}^{-1}$;

while $k \leq C - 1$, do

$k := k + 1$;

$A_1^* := \begin{pmatrix} 0_{(2^{-1}(k-3)k) \times k} \\ A_{k-2,k-1} \end{pmatrix}$;

$A_2^* := \begin{pmatrix} 0_{k \times (2^{-1}(k-3)k)}, A_{k-1,k-2} \end{pmatrix}$;

$D_{2,2} := \left(-A_{k-1,k-1} - A_2^*(-T^{-1}(k-1))A_1^*\right)^{-1}$;

$D_{2,1} := D_{2,2}A_2^*(-T^{-1}(k-1))$;

$D_{1,2} := (-T^{-1}(k-1))A_1^* D_{2,2}$;

$D_{1,1} := -T^{-1}(k-1) - T^{-1}(k-1)A_1^* D_{2,1}$;

$-T^{-1}(k) := \begin{pmatrix} D_{1,1} & D_{1,2} \\ D_{2,1} & D_{2,2} \end{pmatrix}$.

Step 2: $P(\tau_{(0,1,1)}(C) = 0) := \nu e_{2^{-1}(C-2)(C+1)}(2) D_{1,1} e_{2^{-1}(C-2)(C+1)}^T(1)$.

Step 3: $i := 0$;

while $i \leq C - 1$, do

$P(\tau_{(0,C-1,i)}(C) = 0) := \nu e_C(i+1) D_{2,1} e_{2^{-1}(C-2)(C+1)}^T(1)$;

$i := i + 1$.

3.2 The Continuous Contribution of $\tau_{(0,1,1)}(C)$

In order to determine the restricted Laplace-Stieltjes transform $\psi_{(0,1,1)}(s; C)$ in (1), we observe that the dynamics of process $\mathcal{X}$ on the set of sample paths $\{\tau'_{(0,1,1)}(C) < \tau_{(0,1,1)}\}$ are related to the first access of $\mathcal{X}$ from state $(0, 1, 1)$ to the subset $\mathcal{L}_0^l$ via state $(0, C, i)$, for $i \in \{0, ..., C\}$, which occurs with probability $P_{(0,1,1)}(0, C, i)$; the first-passage time $U_{(0,C,i)}(0, C-1, i')$ from state $(0, C, i)$ to the subset $\mathcal{L}_{C-1}^a$ via state $(0, C-1, i')$, for $i' \in \{0, ..., C-1\}$; and the total sojourn time $\tau_{(0,C-1,i')}(C)$ in the subset $\mathcal{S}_l$ before the first visit to state $(0, 0, 0)$. This results in

$$\psi_{(0,1,1)}(s; C) = \sum_{i=0}^{C} P_{(0,1,1)}(0, C, i) \sum_{i'=0}^{C-1} \phi_{(0,C,i)}(s; (0, C-1, i'))$$

$$\times \left(P(\tau_{(0,C-1,i')}(C) = 0) + \psi_{(0,C-1,i')}(s; C)\right), \quad (7)$$

for $\Re(s) \geq 0$, where $\phi_{(0,C,i)}(s; (0, C-1, i')) = E[e^{-sU_{(0,C,i)}(0,C-1,i')}]$, and hitting probabilities $P_{(0,1,1)}(0, C, i)$ and discrete contributions $P(\tau_{(0,C-1,i')}(C) = 0)$ are given by Eqs. (4) and (6), respectively.

In deriving the Laplace-Stieltjes transforms $\phi_{(0,C,i)}(s; (0, C-1, i'))$ in (7), we first define the Laplace-Stieltjes transform $\phi_{(n,C,i)}(s; (0, C-1, i'))$ of the first-passage time $U_{(n,C,i)}(0, C-1, i')$ from state (n, C, i) to the subset $\mathcal{L}_{C-1}^a$ via state $(0, C-1, i')$, for $n \in \{1, ..., N-C\}$, $i \in \{0, ..., C\}$ and $i' \in \{0, ..., C-1\}$, and then introduce matrices $\Phi_n(s; C-1)$, for $n \in \{0, ..., N-C\}$, of dimension $(C+1) \times C$ with $(i+1, i'+1)$-th entry $\phi_{(n,C,i)}(s; (0, C-1, i'))$, for $i \in \{0, ..., C\}$ and $i' \in \{0, ..., C-1\}$.

A simple and often efficient approach is based on first-step analysis, which is here applied to process $\mathcal{X}$ and leads to the matrix equations

$$(sI_{C+1} - B_{0,0})\,\Phi_0(s; C-1) = A_{C,C-1} + B_{0,1}\Phi_1(s; C-1), \tag{8}$$

$$(sI_{C+1} - B_{n,n})\,\Phi_n(s; C-1) = B_{n,n-1}\Phi_{n-1}(s; C-1)$$
$$+(1 - \delta_{n,N-C})B_{n,n+1}\Phi_{n+1}(s; C), \tag{9}$$

for $n \in \{1, ..., N-C\}$.

Then, for a fixed integer $k \in \mathbb{N}$, the following equations for matrices $\Phi_n^{(k)}(C-1)$, for $n \in \{0, ..., N-C\}$, of dimension $(C+1) \times C$ with $(i+1, i'+1)$-th entry $E[U_{(n,C,i)}^k(0, C-1, i')]$, for $i \in \{0, ..., C\}$ and $i' \in \{0, ..., C-1\}$, are derived from (8)–(9) by taking derivatives k times with respect to s, evaluating at point $s = 0$ and multiplying by $(-1)^k$:

$$- B_{n,n}\Phi_n^{(k)}(C-1) - k\Phi_n^{(k-1)}(C-1) = (1 - \delta_{n,0})B_{n,n-1}\Phi_{n-1}^{(k)}(C-1)$$
$$+(1 - \delta_{n,N-C})B_{n,n+1}\Phi_{n+1}^{(k)}(C-1), \tag{10}$$

for $n \in \{0, ..., N-C\}$. We may then apply block-Gaussian elimination to Eqs. (8)–(9) (respectively, (10)). This results in an iterative computation of matrices $\Phi_n(s; C-1)$ (respectively, $\Phi_n^{(k-1)}(C-1)$), for $n \in \{0, ..., N-C\}$, as we show in the next theorem.

Theorem 2. *Starting from $\Phi_{N-C}(s; C-1) = J_{N-C}(s)$, matrices $\Phi_n(s; C-1)$, for $n \in \{0, ..., N-C-1\}$, can be determined, in reverse order, from the equations*

$$\Phi_n(s; C-1) = J_n(s) + F_n^{-1}(s)B_{n,n+1}\Phi_{n+1}(s; C-1),$$

where $F_n(s) = sI_{C+1} - B_{n,n} - (1-\delta_{n,0})B_{n,n-1}F_{n-1}^{-1}(s)B_{n-1,n}$, for $n \in \{0, ..., N-C\}$, and $J_n(s) = F_0^{-1}(s)A_{C,C-1}$ if $n = 0$, and $F_n^{-1}(s)B_{n,n-1}J_{n-1}(s)$ if $n \in \{1, ..., N-C\}$. Moreover, starting from $\Phi_{N-C}^{(k)}(C-1) = F_{N-C}^{-1}f_{N-C}^{(k)}$, matrices $\Phi_n^{(k)}(C-1)$, for $n \in \{0, ..., N-C-1\}$, can be evaluated, in reverse order, from the equations

$$\Phi_n^{(k)}(C-1) = F_n^{-1}\left(f_n^{(k)} + B_{n,n+1}\Phi_{n+1}^{(k)}(C-1)\right),$$

where $F_n = F_n(0)$ and $f_n^{(k)} = k\Phi_n^{(k-1)}(C-1) + (1-\delta_{n,0})B_{n,n-1}F_{n-1}^{-1}f_{n-1}^{(k)}$, with $\Phi_n^{(0)}(C-1) = \Phi_n(0; C-1)$.

Algorithm 2 uses iterative equations in Theorem 2 and computes, for a fixed integer $k \in \mathbb{N}$, the kth moments of the first-passage times $U_{(n,C,i)}(0, C-1, i')$, for states $(n, C, i) \in S_l$ and $i' \in \{0, ..., C-1\}$, in terms of previously computed $(k-1)$-th moments.

Algorithm 2: Iterative computation of $\Phi_n^{(k)}(C-1)$, for $n \in \{0, ..., N-C\}$ and a fixed integer $k \in \mathbb{N}_0$.

Step 1: $l := 0$;
$\quad n := 0$;
$\quad F_n := -B_{n,n}$;
$\quad J_n := F_n^{-1}A_{C,C-1}$;
$\quad$ while $n < N - C$, do
$\quad\quad n := n + 1$;
$\quad\quad F_n := -B_{n,n} - B_{n,n-1}F_{n-1}^{-1}B_{n-1,n}$;
$\quad\quad J_n := F_n^{-1}B_{n,n-1}J_{n-1}$.
Step 2: $\Phi_n^{(l)}(C-1) := J_{N-C}$;
$\quad$ while $n > 0$, do
$\quad\quad n := n - 1$;
$\quad\quad \Phi_n^{(l)}(C-1) := J_n + F_n^{-1}B_{n,n+1}\Phi_{n+1}^{(l)}(C-1)$.
Step 3: While $l < k$, do
$\quad l := l + 1$;
$\quad n := 0$;
$\quad f_n^{(l)} := l\Phi_n^{(l-1)}(C-1)$;
$\quad$ while $n < N - C$, do
$\quad\quad n := n + 1$;
$\quad\quad f_n^{(l)} := l\Phi_n^{(l-1)}(C-1) + B_{n,n-1}F_{n-1}^{-1}f_{n-1}^{(l)}$;
$\quad \Phi_n^{(l)}(C-1) := F_{N-C}^{-1}f_{N-C}^{(l)}$;
$\quad$ while $n > 0$, do
$\quad\quad n := n - 1$;
$\quad\quad \Phi_n^{(l)}(C-1) := F_n^{-1}(f_n^{(l)} + B_{n,n+1}\Phi_{n+1}^{(l)}(C-1))$.

By repeating the argument which led to Eq. (7), it is found that

$$\psi_{(0,C-1,i)}(s;C) = \sum_{i'=0}^{C} P_{(0,C-1,i)}(0,C,i') \sum_{i''=0}^{C-1} \phi_{(0,C,i')}(s;(0,C-1,i''))$$
$$\times \left(P(\tau_{(0,C-1,i'')}(C) = 0) + \psi_{(0,C-1,i'')}(s;C)\right),$$

for $i \in \{0,...,C-1\}$, where $P_{(0,C-1,i)}(0,C,i')$ and $P(\tau_{(0,C-1,i'')}(C) = 0)$ are specified by (5) and (6), respectively. Indeed, this equation can be expressed in matrix form as

$$\Psi_{C-1}(s;C) = D_{2,2}(C)A_{C-1,C}\Phi_0(s;C-1)$$
$$\times \left(\nu D_{2,1}(C)e_{2^{-1}(C-2)(C+1)}^T(1) + \Psi_{C-1}(s;C)\right), \quad (11)$$

for $\Re(s) \geq 0$, where $\Psi_{C-1}(s;C)$ denotes a column vector of order C with $(i+1)$-th entry $\psi_{(0,C-1,i)}(s;C)$, for $i \in \{0,...,C-1\}$. Similarly, it is also noted that, by (4), Eq. (7) becomes

$$\Psi_{(0,1,1)}(s;C) = e_{2^{-1}(C-2)(C+1)}(2)D_{1,2}(C)A_{C-1,C}\Phi_0(s;C-1)$$
$$\times \left(\nu D_{2,1}(C)e_{2^{-1}(C-2)(C+1)}^T(1) + \Psi_{C-1}(s;C)\right), \quad (12)$$

for $\Re(s) \geq 0$.

The next theorem shows the recursive determination of the kth moment of $\tau_{(0,1,1)}(C)$ from (11)–(12).

Theorem 3. *For a fixed integer $k \in \mathbb{N}$, the kth moment of the total sojourn time $\tau_{(0,1,1)}(C)$ is given by*

$$
E\left[\tau_{(0,1,1)}^k(C)\right] = e_{2^{-1}(C-2)(C+1)}(2)D_{1,2}(C)A_{C-1,C}
$$

$$
\times \left(\Phi_0^{(k)}(C-1)1_C + \sum_{k'=0}^{k-1} \binom{k}{k'} \Phi_0^{(k')}(C-1)\Psi_{C-1}^{(k-k')}(C) \right), \quad (13)
$$

where

$$
\Psi_{C-1}^{(k)}(C) = \left(I_C - D_{2,2}(C)A_{C-1,C}\Phi_0^{(0)}(C-1) \right)^{-1}
$$

$$
\times D_{2,2}(C)A_{C-1,C}\left(\nu\Phi_0^{(k)}(C-1)D_{2,1}(C)e_{2^{-1}(C-2)(C+1)}^T(1) \right.
$$

$$
\left. + \sum_{k'=1}^{k} \binom{k}{k'} \Phi_0^{(k')}(C-1)\Psi_{C-1}^{(k-k')}(C) \right), \quad (14)
$$

with $\Psi_{C-1}^{(0)}(C) = 1_C - \nu D_{2,1}(C)e_{2^{-1}(C-2)(C+1)}^T(1)$.

Proof. To prove (13), it is first observed that, by (1), the kth moment of $\tau_{(0,1,1)}(C)$ is equivalent to its kth moment on the set $\{\tau'_{(0,1,1)}(C) < \tau_{(0,1,1)}\}$ of sample paths of process $\mathcal{X}$, for $k \in \mathbb{N}$. Therefore, the expected value $E[\tau_{(0,1,1)}^k(C)]$ can be obtained from (12) by differentiating k times with respect to s, setting $s = 0$ and multiplying by $(-1)^k$, and using the equality $\Psi_{C-1}^{(0)}(C) = \Psi_{C-1}(0; C)$. In a similar manner, Eq. (11) yields

$$
\Psi_{C-1}^{(k)}(C) = D_{2,2}(C)A_{C-1,C}\left(\nu\Phi_0^{(k)}(C-1)D_{2,1}(C)e_{2^{-1}(C-2)(C+1)}^T(1) \right.
$$

$$
\left. + \sum_{k'=0}^{k} \binom{k}{k'} \Phi_0^{(k')}(C-1)\Psi_{C-1}^{(k-k')}(C) \right),
$$

which is equivalent to (14). The expression for column vector $\Psi_{C-1}^{(0)}(C)$ is readily derived from (6) by noting that $\psi_{(0,C-1,i)}(0; C) = 1 - P(\tau_{(0,C-1,i)}(C) = 0)$.

4 Numerical Work

In this section, we present some selected numerical experiments to illustrate the influence of the contact rates β_0 and β, the rates λ_0 and λ of exogenous infection, and the number C of beds for inpatients on the probability law of the total sojourn time $\tau_{(0,1,1)}(C)$. We consider our scenarios in [11, Section 4], which are linked to a lower transmission of the pathogen via inpatient-to-inpatient contact than through individual-to-individual contact (out of hospital ward); i.e., $\beta < \beta_0$. The spread of the pathogen by individual-to-individual contact is more likely to occur than getting the infection by exogenous causes, in such a way that $\lambda_0 = 0.05\beta_0$. Various isolation levels for inpatients from the outside are analyzed with the selection $\lambda = \alpha\beta_0$, for values of $\alpha \in \{0.01, 0.5, 1.0, 1.5\}$. For the sake of brevity, in Figs. 1 and 2, we focus on the selection $\beta_0 = 1.3$, and the assumption that any infected patient recovers, on average, in $\gamma^{-1} = 1.0$ units of time, and is discharged after, on average, $\nu^{-1} = 0.1$ units of time after recovering.

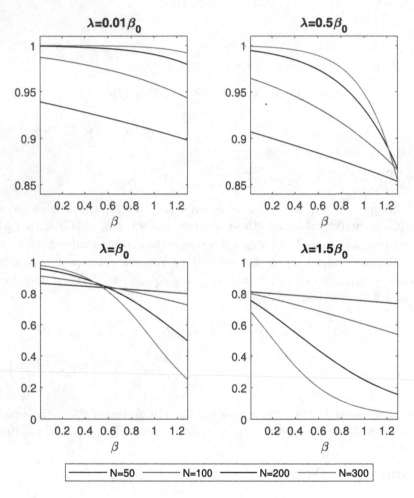

Fig. 1. Probability $P(\tau_{(0,1,1)}(C) = 0)$, with $C = 0.2N$, as a function of β, for values of $\lambda \in \{0.01\beta_0, 0.5\beta_0, \beta_0, 1.5\beta_0\}$ with $\beta_0 = 1.3$, and population size $N \in \{50, 100, 200, 300\}$.

Larger values of the probability $P(\tau_{(0,1,1)}(C) = 0)$ can be interpreted as indicators that the hospital ward will be working functionally full for a brief period of time during an outbreak. Figure 1 therefore shows that, as intuition tells us, the more relevant the interaction between hospital ward patients is, the more difficult it is to find available beds for newly infected individuals; i.e., the probability $P(\tau_{(0,1,1)}(C) = 0)$ decreases as a function of β, irrespectively of the fixed values of N and λ. It is also observed that, for fixed values of N and β, finding available beds for newly infected individuals becomes less likely to occur when isolation measures for inpatients are relaxed, which is related to increasing values of λ in Fig. 1.

From Fig. 1, we cannot make a general statement about the influence of the population size N and the associated bed limit $(C = 0.2N)$ on predicting the probability of finding available beds in the hospital ward. To be concrete, the scenarios with the most severe isolation measures for inpatients (i.e., scenario with $\lambda = 0.01\beta_0$) and the least severe ones ($\lambda = 1.5\beta_0$) exhibit two opposite behaviors for $P(\tau_{(0,1,1)}(C) = 0)$ as a function of N. The first behavior is related to the scenario with $\lambda = 0.01\beta_0$. Under the most severe isolation measures for inpatients, the probability of bed availability increases with the population size, indirectly implying hospital wards with a larger number of beds in absolute terms. The second behavior is linked to the most relaxed isolation measures ($\lambda = 1.5\beta_0$). In this case, finding available space to accommodate newly infected individuals will be less likely guaranteed in larger populations than in smaller ones, especially when interactions between inpatients increase. Moderate isolation levels for inpatients in Fig. 1 (scenarios with $\lambda = \alpha\beta_0$ and $\alpha \in \{0.5, 1.0\}$) exhibit the first behavior when the interaction between inpatients is *relatively* small (equivalently, β is smaller than a certain threshold), while a higher interaction leads to the second behavior. A suitable choice on the number of beds C, as a function of N, is therefore an important issue for practical use.

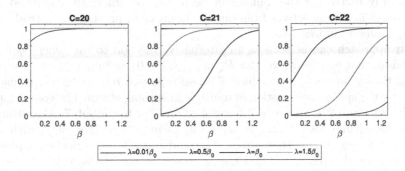

Fig. 2. Ratio $(E[\tau_{(0,1,1)}])^{-1}E[\tau_{(0,1,1)}(C)]$ as a function of β, for values of $\lambda \in \{0.01\beta_0, 0.5\beta_0, \beta_0, 1.5\beta_0\}$ with $\beta_0 = 1.3$, $N = 100$ and $C \in \{0.2N, 0.21N, 0.22N\}$.

The ratio of expectations $(E[\tau_{(0,1,1)}])^{-1}E[\tau_{(0,1,1)}(C)]$ represents the proportion of time that the hospital ward is working functionally full during an outbreak; specifically, larger values of this ratio would imply a more significant potential transmission of the pathogen outside the hospital ward, while smaller values would guarantee a greater availability of beds for inpatients and therefore less frequent transmission of the pathogen by individual-to-individual contact. For the above choices on λ, $N = 100$ and values of $C \in \{0.2N, 0.21N, 0.22N\}$, Fig. 2 shows how a minor variation in the number of beds can result in a hospital ward operating under its maximum capacity for longer periods of time when both measures of isolation (between inpatients, and between the hospital ward and the outside) are more severe. The choice of the number of beds as the twenty percent of N in Fig. 2 is related to scenarios with a high risk of potential transmission of the pathogen (specifically, by individual-to-individual contact) outside the hospital ward, even in the case of severe isolation measures and limited interaction between inpatients. As the number of beds is increased by one or two more units, the time that the hospital ward is expected to be working full will decrease compared to the length of the outbreak, therefore decreasing the risk of potential transmission of the pathogen between individuals.

5   Discussion and Conclusions

For a Markov chain model with limited resources in [11], we have characterized the probability law of the total time $\tau_{(0,1,1)}(C)$ that the hospital ward remains with its beds simultaneously occupied by inpatients during an outbreak of the disease. To be concrete, we have used first-step analysis and block-Gaussian elimination to derive an algorithmic solution for computing the probability $P(\tau_{(0,1,1)}(C) = 0)$, the restricted Laplace-Stieltjes transform $\psi_{(0,1,1)}(s; C)$ and related moments. Our numerical experiments in Figs. 1 and 2 are self-explanatory of how intra-hospital actions, based on limiting interaction between inpatients and slightly increasing the number of beds, can be efficiently combined with measures to isolate infectives from outside in order to prevent the spread of the pathogen among individuals.

Our approach can be seen as an alternative method to the general-purpose expressions that can be derived for $P(\tau_{(0,1,1)}(C) = 0)$ and $\psi_{(0,1,1)}(s; C)$ from the general theory on random variables of phase-type and related hitting probabilities under the special assumption of multiple absorbing states. The complexity of Algorithms 1 and 2 in Subsects. 3.1 and 3.2 is similar to that of [7, Algorithm A], and [9, Algorithms 1A-3A]. In the same way as in [11], we have dealt with state spaces consisting of more than fifteen thousand states in numerical examples not reported here. Therefore, we find the approach to be satisfactory for practical use in the setting of small communities (e.g., intensive care units, nursing homes, and prisons) where stochastic effects could generate important deviations from the deterministic solution. Large values of N will require to approximate the Markov chain model by analyzing, similarly to the paper [5] in the framework of SIR models, whether the large population limit of process $\mathcal{X}$ is a suitably defined first-order partial differential equation. An aspect that deserves further exploration is how the ideas from [5] could be adapted to derive the first-order partial differential equation arising in the dynamics of the deterministic version of the random time $\tau_{(0,1,1)}(C)$.

As an example of future work on the model, we mention the study of the waiting time of an infected individual before being admitted to the hospital ward, and the length of stay in the hospital ward, including the probability of reinfection. Another interesting problem is related to the multiple visits of process $\mathcal{X}$ to the subset $\{(0, j, 0) : j \in \{0, ..., C\}\}$ of disease-free states before the end of an outbreak, and the comparison of the random length of the outbreak with the first extinction of the pathogen.

References

1. Abakuks, A.: Optimal immunization policies for epidemics. Adv. Appl. Probab. **6**, 494–511 (1974)
2. Amador, J., Lopez-Herrero, M.J.: Cumulative and maximum epidemic sizes for a nonlinear SEIR stochastic model with limited resources. Discrete Cont. Dyn. B **23**, 3137 (2018)
3. Amador, J., Gómez-Corral, A.: A stochastic model with two quarantine states and a limited carrying capacity for quarantine. Phys. A **544**, 121899 (2020)
4. Bolzoni, L., Bonacini, E., Marca, R.D., Groppi, M.: Optimal control of epidemic size and duration with limited resources. Math. Biosci. **315**, 108232 (2019)
5. Chalub, F.A.C.C., Souza, M.O.: The SIR epidemic model from a PDE point of view. Math. Comput. Model. **53**, 1568–1574 (2011)
6. Clancy, D.: Optimal intervention for epidemic models with general infection and removal rate functions. J. Math. Biol. **39**, 309–331 (1999)
7. Gaver, D.P., Jacobs, P.A., Latouche, G.: Finite birth-and-death models in randomly changing environments. Adv. Appl. Probab. **16**, 715–731 (1984)
8. Gómez-Corral, A., López García, M.: Extinction times and size of the surviving species in a two-species competition process. J. Math. Biol. **64**, 255–289 (2012)
9. Gómez-Corral, A., López-García, M.: Perturbation analysis in finite LD-QBD processes and applications to epidemic models. Numer. Linear Algebr. **25**, 2160 (2018)
10. Gómez-Corral, A., López-García, M., Lopez-Herrero, M.J., Taipe, D.: On first-passage times and sojourn times in finite QBD processes and their applications in epidemics. Mathematics **8**, 1718 (2020)
11. Gómez-Corral, A., Lopez-Herrero, M.J., Taipe, D.: A Markovian epidemic model in a resource-limited environment. Appl. Math. Comput. **458**, 128252 (2023)
12. He, Q.M.: Fundamentals of Matrix-Analytic Methods. Springer, New York (2014)
13. Jiang, J., Zhou, T.: The influence of time delay on epidemic spreading under limited resources. Phys. A **508**, 414–423 (2018)
14. Jiang, J., Zhou, T.: Resource control of epidemic spreading through a multilayer network. Sci. Rep. **8**, 1629 (2018)
15. Jiang, J., Zhou, T.: Nontrivial effects of uncertainty on epidemic spreading under limited resources. Phys. A **532**, 121453 (2019)
16. Latouche, G., Ramaswami, V.: Introduction to Matrix Analytic Methods in Stochastic Modeling. ASA-SIAM Series on Statistics and Applied Probability, Philadelphia (1999)
17. Lin, F., Muthuraman, K., Lawley, M.: An optimal control theory approach to non-pharmaceutical interventions. BMC Infect. Dis. **10**, 1–13 (2010)
18. Misra, A.K., Maurya, J.: Bifurcation analysis and optimal control of an epidemic model with limited number of hospital beds. Int. J. Biomath. **16**, 2250101 (2023)

19. Morton, R., Wickwire, K.H.: On the optimal control of a deterministic epidemic. Adv. Appl. Probab. **6**, 622–635 (1974)
20. Sethi, S.P.: Optimal quarantine programmes for controlling an epidemic spread. J. Oper. Res. Soc. **29**, 265–268 (1978)
21. Sethi, S.P., Staats, P.W.: Optimal control of some simple deterministic epidemic models. J. Oper. Res. Soc. **29**, 129–136 (1978)
22. Wickwire, K.H.: Optimal isolation policies for deterministic and stochastic epidemics. Math. Biosci. **26**, 325–346 (1975)
23. Zhang, H., Yang, Z.W., Pawelek, K., Liu, S.: Optimal control strategies for a two-group epidemic model with vaccination-resource constraints. Appl. Math. Comput. **371**, 124956 (2020)
24. Zhou, L., Fan, M.: Dynamics of an SIR epidemic model with limited medical resources revisited. Nonlinear Anal. Real **13**, 312–324 (2012)

Pricing Models for Digital Renting Platforms

K. S. Ashok Krishnan[1]([✉])[ID], Samir M. Perlaza[1,2,3][ID], and Eitan Altman[1,4,5][ID]

[1] INRIA, Centre Inria d'Université Côte d'Azur, Sophia Antipolis, France
{ashok-krishnan.komalan-sindhu,samir.perlaza,eitan.altman}@inria.fr
[2] ECE Department, Princeton University, Princeton, NJ 08544, USA
[3] GAATI, Université de la Polynésie Française, Faaa, French Polynesia
[4] Laboratoire d'Informatique d'Avignon (LIA), Université d'Avignon,
Avignon, France
[5] Laboratory of Information, Network and Communication Sciences, Paris, France

Abstract. This paper considers different pricing models for a platform based rental system, such as Airbnb. A linear model is assumed for the demand response to price, and existence and uniqueness conditions for Nash equilibria are obtained. The Stackelberg equilibrium prices for the game are also obtained, and an iterative scheme is provided, which converges to the Nash equilibrium. Different cooperative pricing schemes are studied, and splitting of revenues based on the Shapley value is discussed. It is shown that a division of revenue based on the Shapley value gives a revenue to the platform proportional to its control of the market. The demand response function is modified to include user response to quality of service. It is shown that when the cost to provide quality of service is low, both renter and the platform will agree to maximize the quality of service. However, if this cost is high, they may not always be able to agree on what quality of service to provide.

Keywords: pricing · platforms · Airbnb · Nash equilibrium · Shapley value

1 Introduction

A platform is defined as "an entity that brings together economic agents and actively manages network effects between them" [7]. A platform enables economic interactions between users, while also profiting from them, by charging an appropriate fee from the users. A number of digital platforms have arisen in recent years, which cater to diverse requirements such as renting (Airbnb, Booking.com), transportation (Uber) and dating (Tinder). Depending on the nature of the interactions, the pricing model and the revenue generation mechanism of the platform can vary. In this article, we study the pricing and revenue

This research was partially supported in part by in part by the Agence Nationale de la Recherche (ANR) through the project PARFAIT; E. Altman was supported in part by the Indo-French project LION (Learning in Operations and Networks).

M. Iacono et al. (Eds.): EPEW/ASMTA 2023, LNCS 14231, pp. 139–153, 2023.
https://doi.org/10.1007/978-3-031-43185-2_10

of digital renting platforms such as Airbnb. A renting platform can be considered an example of a two sided market [4], with the two sides referring to the apartment owners (renters) and the users (consumers), who interact through the platform. A number of works study pricing of platforms in two sided markets. In [4], the authors consider two groups interacting through a platform, with utilities for one group being a linear function of the number of users of the other group. They find equilibrium prices for joining the platforms, for monopoly and multiple competing platforms. Another work [20], studies competition between platforms. In [6], the authors study how a gatekeeper, which controls the availability of price information of commodities, should price itself. They characterize the equilibrium in which gatekeeper profits are maximized.

There have been a large number of works looking at different aspects of Airbnb, which is a major player in short term renting [3]. Airbnb is often used by home owners to supplement their income by renting out of a part of their apartment full time or part time. From the perspective of customers, it represents the possibility of monetary savings as well as non monetary utility, that may come from sharing a living space with a renter [11]. In [22], the authors consider an online platform model for systems such as Airbnb. There are a finite number of sellers (of apartments), and a single representative user, interacting through an online platform. The platform can discriminate regarding how many sellers are to be displayed on its web interface. Under a *multinomial logit* model of demand, it is shown in [22] that social welfare maximization corresponds to displaying all sellers on the platform, while for maximizing the total revenue of the sellers, only the best sellers need to be displayed. However, they do not model the revenue generation mechanism for the platform.

Empirical studies have also been done to study different aspects of the Airbnb economic system. In [14], the authors study dynamic pricing in Airbnb, and conclude that hosts with more resources and experience are more likely to vary prices. In [5], the authors observe that ratings of quality, based on feedback from previous customers, lets hosts set higher prices for their apartments. There have also been studies on the impact of Airbnb on the hotel ecosystem [19]. A work in this direction has noticed an improved revenue for mainstream hotels owing to the presence of Airbnb [10], because Airbnb absorbs budget tourists, who are more price elastic, leaving higher paying customers to hotels. These remaining users are less price elastic. The hotels can now set higher prices for these remaining users. This increases the revenue of the hotels.

1.1   Contributions

The contributions of this work are as follows.

1. We propose a linear response model to capture the demand for apartments (or rooms). We first propose a competitive game formulation, where the platform and multiple renter classes find the Nash equilibrium as a pricing solution. By means of the diagonal strict concavity condition, we establish the uniqueness of the Nash equilibrium. We also study the Stackelberg or leader-follower

formulation, and demonstrate that being able to pick the price first gives a competitive advantage in terms of revenue.

2. We consider different cooperative scenarios, where the platform and the renters take a joint pricing decision. We obtain the Shapley value for the problem as a means to split the revenue in the cooperative case. With two platforms and multiple renters, we show that the revenue division using Shapley value results in the revenue of the platform being proportional to its market share.

3. We modify the model to introduce the notion of quality of service, and show that, depending on the cost of providing quality of service, the platform and the renter may have similar or differing perspectives on how much quality of service to provide to the customer. At low values of the cost, both the platform and the renter will agree to provide high quality of service. However, if this cost exceeds a threshold, they may not agree on the quality of service to be provided.

1.2 System Model

The renting game is played between three sets of users - the platform (for example, Airbnb), the renters (people who rent out their apartments through the platform) and end users or guests (people who are looking to rent out apartments). There are multiple variables of interest which come into play in this interaction. We model the end users by means of a *demand response* function, and we model the platform and renters as playing a game in response to the properties of this function. We will assume there are N classes of renters, each representing a particular category of apartments. Each class may represent, for example, a group of apartments that are preferred by a section of end users. High paying users prefer different apartments from low paying users. Since the purchasing power of these two groups are different, it will be reflected in the demand response of these two classes of apartments to the price charged as rent. The price charged to the guest has two components: the price paid to the platform, and the price paid to the renter. For class i, the price paid to the platform will be denoted by p_i^0, and the price paid to the renter will be denoted by p_i^1. Define the price vector

$$\mathbf{p} = (\mathbf{p}^0, \mathbf{p}^1) = (p_1^0, p_2^0, ..., p_N^0, p_1^1, p_2^1, ..., p_N^1). \qquad (1)$$

For renter class i, the demand response of the users, as a function of the price vector $\mathbf{p}$ charged, is given by

$$R_i(\mathbf{p}) = \alpha_i - \beta_i(p_i^0 + p_i^1), \qquad (2)$$

where $\alpha_i > 0$ and $\beta_i > 0$. This is a linear demand model, which is commonly used in the literature for modeling demand response [12,13]. The linear model can be interpreted as the mean behaviour of a market with stochastic demand. This models the condition that as price increases, demand for the apartments decreases. This also leads to a concave utility function, which is the commonly

142   K. S. Ashok Krishnan et al.

used form of utility for tractable optimization, in the literature [8]. Note that $R_i(\mathbf{p})$ represents the number of apartments that are rented out from all apartments of class i, when the price vector is $\mathbf{p}$. The parameter α_i represents the maximum demand for an apartment of class i. The parameter β_i models the rate of decrease of demand of apartments of class i, as price increases. Note that we do not assume any dependency between the prices of different classes of apartments. This is because we assume that demand for the different classes of apartments arise from disjoint sections of the population of customers.

For a price vector $\mathbf{p}$, the revenue (or utility) for the renter i is given by

$$U_i(\mathbf{p}) = R_i(\mathbf{p})p_i^1 \text{ for } i = 1, ..., N. \tag{3}$$

The revenue (or utility) for the platform is given by

$$U_0(\mathbf{p}) = \sum_{i=1}^{N} R_i(\mathbf{p})p_i^0. \tag{4}$$

Define the set

$$\mathcal{P} = \{\mathbf{p} \succeq \mathbf{0} : p_i^j \leq \frac{\alpha_i}{\beta_i}, \text{ for each } i = 1, ..., N \text{ and } j = 0, 1.\} \tag{5}$$

where $\mathbf{0}$ represents the all zero vector and $\succeq$ represents the element-wise "greater than or equal to" inequality. This represents a maximum constraint on the prices being set by the platform and the renter, separately. We will restrict our prices to take values from $\mathcal{P}$. This is a larger set than the set commonly used in the literature, which corresponds to the set of non negative prices with non negative demand. However, we use this more relaxed definition because the existence and uniqueness results in Sect. 2 become simpler to demonstrate. Nevertheless, with either set of constraints, we will obtain the same equilibrium points.

Remark 1. *The demand response, and consequently the utility, can be negative at some points in the constraint set $\mathcal{P}$. In economic scenarios, one expects utility to be a non negative quantity. In order to be consistent with this notion of non negative utility, one can assume that the actual utility received by the players is offset by a large enough positive value, i.e., of the form $C + U_j(\mathbf{p})$, so that it is always non negative.*

Remark 2. *Another approach to define the demand response would be to use a response function of the form $R_i(\mathbf{p}) = \max(\alpha_i - \beta_i(p_i^0 + p_i^1), 0)$. This ensures non negative demand over any set of prices. However, this can lead to multiple Nash equilibria, arising at price vectors that have zero demand.*

1.3   Global Optimum

The global optimal pricing for this model will be that which maximizes the sum of the revenue of the renters and the platform. The total revenue is

$$U(\mathbf{p}) = U_0(\mathbf{p}) + \sum_{i=1}^{N} U_i(\mathbf{p}).$$

Define the global optimal revenue,

$$U^* = \max_{\mathbf{p}\in\mathcal{P}} U(\mathbf{p}). \tag{6}$$

It is easy to see that the maximum total revenue is $U^* = \sum_{i=1}^{N} \frac{\alpha_i^2}{4\beta_i}$, achieved at any $\mathbf{p}$ for which $p_i^0 + p_i^1 = \frac{\alpha_i}{2\beta_i}$ for all renters i. If the revenue is split equally between the platform and the renter, renter i will receive a revenue of $\frac{\alpha_i^2}{8\beta_i}$, and the platform gets $\sum_i \frac{\alpha_i^2}{8\beta_i}$.

2 Non Cooperative Game

From Sect. 1.2, we can see that the pricing problem can be considered as a game, between the platform and the renters. Let us consider the case where there is no cooperation between the renters and the platform. Clearly the renters (indexed by i going from 1 through N) have strategies p_i^1, while the platform has a strategy vector $\mathbf{p}^0$. The corresponding utilities are $U_1, ..., U_N$ for the renters and U_0 for the platform. The price vector $\mathbf{p}$ takes values in the set $\mathcal{P}$. It is easy to see that $\mathcal{P}$ is closed, bounded and convex. We define a Nash equilibrium point for this game as follows.

Definition 1 (Nash equilibrium). *A Nash equilibrium point for the above pricing game is given by a point $\mathbf{p}^* = (\mathbf{p}^{0*}, \mathbf{p}^{1*}) \in \mathcal{P}$ where $\mathbf{p}^{0*} = (p_1^{0*}, p_2^{0*}, ..., p_N^{0*})$ and $\mathbf{p}^{1*} = (p_1^{1*}, p_2^{1*}, ..., p_N^{1*})$ such that*

$$U_0(\mathbf{p}^*) = \max_q \{U_0(\mathbf{q}, \mathbf{p}^{1*})\}, \tag{7}$$

$$U_i(\mathbf{p}^*) = \max_{q_i} \{U_i(\mathbf{p}^{0*}, p_1^{1*}, ..., q_i, ..., p_N^{1*})\} \text{ for } i = 1, ..., N. \tag{8}$$

We prove below that a Nash equilibrium exists for this game.

Lemma 1. *A Nash equilibrium exists for the above game.*

Proof. From the form of the utility function we can see that

1. $U_i(\mathbf{p})$ is continuous in $\mathbf{p}$ for all $\mathbf{p} \in \mathcal{P}$, for $i = 0, 1, 2, ..., N$
2. $U_i(\mathbf{p}^0, \mathbf{p}^1)$ is concave in p_i^1 for each fixed value of $(\mathbf{p}^0, p_1^1, ..., p_{i-1}^1, p_{i+1}^1, ..., p_N^1)$, for $i = 1, ..., N$,
3. $U_0(\mathbf{p}^0, \mathbf{p}^1)$ is concave in $\mathbf{p}^0$ for each fixed value of $\mathbf{p}^1$.

Since it is a concave game with finite players, a Nash equilibrium $\mathbf{p}^*$ exists [21].

We will now show that this Nash equilibrium is unique. For any non negative vector $\mathbf{s} = (s_0, s_1, ..., s_N)$, let us define the weighted sum of utilities,

$$V_\mathbf{s}(\mathbf{p}) = \sum_{i=0}^{N} s_i U_i(\mathbf{p}). \tag{9}$$

Define the *pseudogradient* $v_s(\mathbf{p})$ of $V_s(\mathbf{p})$ as the matrix

$$v_s(\mathbf{p}) = \begin{bmatrix} s_0 \nabla_0 U_0(\mathbf{p}) \\ s_1 \nabla_1 U_1(\mathbf{p}) \\ \vdots \\ s_N \nabla_N U_N(\mathbf{p}) \end{bmatrix}, \tag{10}$$

where $\nabla_1, ... \nabla_N$ denote gradient with respect to $p_1^1, ..., p_N^1$, and ∇_0 denotes gradient with respect to the vector $(p_1^0, ..., p_N^0)$. For two vectors x and y of the same dimension, let $\langle x, y \rangle$ denote their inner product. The notion of *diagonal strict concavity* is defined as follows.

Definition 2. *The function $V_s(\mathbf{p})$ is diagonally strict concave for fixed non negative $\mathbf{s}$ if for every $\mathbf{p} \neq \mathbf{q}$,*

$$\langle \mathbf{q} - \mathbf{p}, v_s(\mathbf{p}) - v_s(\mathbf{q}) \rangle > 0. \tag{11}$$

We are now ready to prove the uniqueness of the Nash equilibrium price $\mathbf{p}^*$ in Lemma 1. We will use the following result from [21].

Proposition 2 (Theorem 2 of [21]). *If $V_s(\mathbf{p})$ is diagonally strict concave for some non zero vector $\mathbf{s}$, the Nash equilibrium point $\mathbf{p}^*$ is unique.*

Note that the above result holds when the constraint set $\mathcal{P}$ is uncoupled, i.e., the player constraints do not interact with each other, as in our case.

Theorem 3. *The Nash equilibrium point $\mathbf{p}^*$ in Lemma 1 is unique.*

The proof follows by verifying that $V_s(\mathbf{p})$ is diagonally strict concave, and then using Proposition 3. The full proof is provided in [16].

2.1 Characterizing the Nash Equilibrium

We calculate the Nash equilibrium point $\mathbf{p}^*$ by solving the utility maximization conditions in Definition 1. From first order conditions for renter class i,

$$\frac{\partial U_i(\mathbf{p})}{\partial p_i^1} = \alpha_i - \beta_i p_i^{0*} - 2\beta_i p_i^1. \tag{12}$$

Equating this to zero, we obtain $p_i^{1*} = \frac{\alpha_i}{2\beta_i} - \frac{p_i^{0*}}{2}$. Similarly, we have

$$\frac{\partial U_0(\mathbf{p})}{\partial p_i^0} = \alpha_i - \beta_i p_i^{1*} - 2\beta_i p_i^0, \tag{13}$$

which is equal to zero at $p_i^{0*} = \frac{\alpha_i}{2\beta_i} - \frac{p_i^{1*}}{2}$. These are simultaneously true at $p_i^{0*} = p_i^{1*} = \frac{\alpha_i}{3\beta_i}$, which indicates a symmetric pricing between the renter and the platform. This yields the utilities at the Nash equilibrium point as

$$U_i(\mathbf{p}^*) = \frac{\alpha_i^2}{9\beta_i}, \quad U_0(\mathbf{p}^*) = \sum_{i=1}^{N} \frac{\alpha_i^2}{9\beta_i}. \tag{14}$$

At the Nash equilibrium, the price charged to a customer of class i is $\frac{2\alpha_i}{3\beta_i}$, which is higher than the price charged at global optimum, $\frac{\alpha_i}{2\beta_i}$. However, the total revenue obtained is $\sum_i \frac{2\alpha_i^2}{9\beta_i}$, which is smaller than the global optimal revenue U^*. The *price of anarchy* (*PoA*) [15], defined as the ratio of the total utility at the global optimum , to the total utility at the equilibrium, is given by

$$PoA = \frac{\sum_i \frac{\alpha_i^2}{4\beta_i}}{\sum_i \frac{2\alpha_i^2}{9\beta_i}}. \tag{15}$$

Note that $PoA \geq 1$. The PoA indicates how close the equilibrium is to the global optimum, in terms of total revenue. With $N = 1$, $PoA = \frac{9}{8}$.

2.2   An Iterative Pricing Scheme

We present a sequence of prices that converges to the Nash equilibrium price. Observe that the constraint region $\mathcal{P}$ can be written as $\mathcal{P} = \{\mathbf{p} : h(\mathbf{p}) \succeq \mathbf{0}\}$, where h is a vector of $4N$ linear constraints $h_k(\mathbf{p})$,

$$h_k(\mathbf{p}) = \begin{cases} p_k^0, & k = 1, ..., N, \\ p_{k-N}^1, & k = N+1, ..., 2N, \\ \frac{\alpha_{k-2N}}{\beta_{k-2N}} - p_{k-2N}^0, & k = 2N+1, ..., 3N, \\ \frac{\alpha_{k-3N}}{\beta_{k-3N}} - p_{k-3N}^1, & k = 3N+1, ..., 4N. \end{cases} \tag{16}$$

Define the index sets

$$\mathcal{I}_-(\mathbf{p}) = \{i : h_i(\mathbf{p}) \leq 0\}, \quad \mathcal{I}_+(\mathbf{p}) = \{i : h_i(\mathbf{p}) > 0\}. \tag{17}$$

Let $\mathbf{s} = (s, s..., s)$, $s > 0$. Define, for a $4N$ length vector $\mathbf{w} = (w_1, ..., w_{4N})$,

$$F_\mathbf{s}(\mathbf{p}, \mathbf{w}) = v_\mathbf{s}(\mathbf{p}) + \langle \nabla h(\mathbf{p}), \mathbf{w} \rangle, \tag{18}$$
$$\mathcal{W}(\mathbf{p}) = \{\mathbf{w} : w_i \geq 0 \text{ for } i \in \mathcal{I}_-(\mathbf{p}), w_i = 0 \text{ for } i \in \mathcal{I}_+(\mathbf{p})\} \tag{19}$$
$$W(\mathbf{p}) = \arg_{\mathbf{w} \in \mathcal{W}(\mathbf{p})} \min \|F_\mathbf{s}(\mathbf{p}, \mathbf{w})\|. \tag{20}$$

Consider the sequence of prices $\mathbf{p}(0), \mathbf{p}(1), ...$, where the components evolve as,

$$\mathbf{p}(t+1) = \mathbf{p}(t) + \Delta(t)[v_\mathbf{s}(\mathbf{p}(t)) + \langle \mathbf{w}(t), \nabla h(\mathbf{p}(t)) \rangle], \tag{21}$$
$$\mathbf{w}(t) \in W(\mathbf{p}(t)), \tag{22}$$
$$\Delta(t) = -\frac{F_\mathbf{s}(\mathbf{p}(t), \mathbf{w}(t))^T J_F F_\mathbf{s}(\mathbf{p}(t), \mathbf{w}(t))}{\|J_F F_\mathbf{s}(\mathbf{p}(t), \mathbf{w}(t))\|^2}, \tag{23}$$

where J_F is a mean value of the Jacobian of F. This sequence satisfies the following.

Theorem 4. *Let $p(t)$ satisfy the iterative scheme in (21)-(23). Then, for any starting point in $\mathcal{P}$, $p(t)$ converges to the Nash equilibrium.*

The proof is provided in [16].

2.3   Stackelberg Pricing

Now we consider the leader follower pricing, where one (set of) players declares their price first, and the other set follows. Let the renters be leaders (all choosing their prices simultaneously), followed by the platform. The price for the platform will be such that

$$\frac{\partial U_0(\mathbf{p})}{\partial p_i^0} = 0 \text{ for all } i. \tag{24}$$

This is equivalent to $p_i^0 = \frac{\alpha_i}{2\beta_i} - \frac{p_i^1}{2}$. The renter pricing will take this into account. Substituting this value for the renters, we get

$$U_i(\mathbf{p}) = R_i(\mathbf{p})p_i^1 = p_i^1(\frac{\alpha_i}{2} - \frac{\beta_i p_i^1}{2}).$$

For the renters, taking $\frac{\partial U_i}{\partial p_i^1} = 0$ gives

$$p_i^0 = \frac{\alpha_i}{4\beta_i}, \quad p_i^1 = \frac{\alpha_i}{2\beta_i}.$$

The renters obtain a total revenue of $\sum_i \frac{\alpha_i^2}{8\beta_i}$, and the platform gets $\sum_i \frac{\alpha_i^2}{16\beta_i}$, and the total revenue is $\sum_i \frac{3\alpha_i^2}{16\beta_i}$. Note that the leader, the renter, obtains the same revenue as in the globally optimal pricing (assuming equal division of revenues between renters and the platform). We see that the total price charged to a class i customer at the Stackelbeg equilibrium is higher than the price at Nash equilibrium as well as the global optimal price. The price of anarchy at the Stackelberg equilibrium is

$$PoA = \frac{\sum_i \frac{\alpha_i^2}{4\beta_i}}{\sum_i \frac{3\alpha_i^2}{16\beta_i}}.$$

For $N = 1$, $PoA = \frac{4}{3}$, which is greater than PoA at the Nash equilibrium. The Stackelberg equilibrium prices are hence costlier for the customer, while yielding lower revenues globally.

If we consider the inverse case, with the platform being leader and renters being followers, the platform now gets $\sum_i \frac{\alpha_i^2}{8\beta_i}$, and the renters obtain $\sum_i \frac{\alpha_i^2}{16\beta_i}$. Thus, getting to choose the price first lets you have a higher fraction of the revenue. In practice, who gets to choose first depends on the market. If the platform is a monopoly, and the renters have no option but to use the platform for their operation, then the platform may be able to take the lead in pricing, and getting an advantage. On the other hand, if there are multiple platforms that compete with in the market, or there is an open market where renters can easily get customers, renters may be in a position to choose to price first, and thus make a gain. Thus, renters who exclusively use one platform may stand to lose in revenue, whereas renters who use multiple platforms or have access to the real market, can make a gain.

3   Cooperative Pricing

In this section we consider cooperative pricing scenarios. The change from the preceding section is that the division of revenues between the platform and renter is fixed by arrangement. The renter pays the the platform from his revenue, at some previously agreed fixed rate f. We can think of two ways in which revenue can be shared between renter and the platform. In both these cases we will see that the total revenue extracted is the maximum possible (i.e., equal to the globally optimal revenue). However, the sharing of the revenues between renter and the platform is slightly different. The difference between these two ways is that in the first, the payment to the platform is viewed as a kind of 'tax' on the renter, and the end user is not aware of it, while in the second case the payment to the platform is presented as a 'tax' to the end user, directly on the price.

3.1   Case a

The renter i announces a price p_i, and then pays fp_i to the platform, where $f \in (0,1)$. The utilities of platform and renters i are

$$U_0(\mathbf{p}) = \sum_i (\alpha_i - \beta_i p_i) f p_i, \quad U_i(\mathbf{p}) = (\alpha_i - \beta_i p_i)(p_i - fp_i), \quad i = 1, ..., N.$$

At the optimal point given by equating the partial derivatives of U_i with respect to p_i to zero, we get $p_i = \dfrac{\alpha_i}{2\beta_i}$, $U_i(\mathbf{p}) = (1 - f)\dfrac{\alpha_i^2}{4\beta_i}$ and $U_0 = f \sum_i \dfrac{\alpha_i^2}{4\beta_i}$. Total revenue extracted is $\sum_i \dfrac{\alpha_i^2}{4\beta_i}$. The rental platform Booking.com uses this model, and charges an f between 10% and 20%, depending on the location of the property [2].

3.2   Case B

The renter i announces a price $p_i(1 + f)$, and then pays fp_i to the platform, for some $f > 0$. The utilities of platform and renters i are

$$U_0(\mathbf{p}) = \sum_i (\alpha_i - \beta_i(1 + f)p_i) f p_i, \quad U_i(\mathbf{p}) = (\alpha_i - \beta_i(1 + f)p_i)p_i, \quad i = 1, ..., N.$$

In this case the equilibrium prices and revenues are $p_i = \dfrac{\alpha_i}{2(1 + f)\beta_i}, U_i = \dfrac{\alpha_i^2}{4\beta_i(1 + f)}, U_0 = \dfrac{f}{1 + f}(\sum_i \dfrac{\alpha_i^2}{4\beta_i})$. Total revenue extracted is $\sum_i \dfrac{\alpha_i^2}{4\beta_i}$. This is close to the pricing model of Airbnb, which charges around 14.2% as f. However, they also charge 3% on the guest separately [1].

Compared to method A, under method B, for the same f, the revenue of the platform decreases, and that of the renters increases (because for $f > 0$,

$1 - f < \dfrac{1}{1+f}$). Under both methods A and B, the customer sees a price $\dfrac{\alpha_i}{2\beta_i}$. However, the splitting of revenues is different. Since $\lim_{f \to \infty} \dfrac{f}{1+f} = 1$, we see that by increasing f in method B, the platform can achieve a revenue as close to the total revenue as it wants. The customer sees the same price irrespective of the choice of f.

3.3 Revenue Division and Shapley Value

In the preceding discussion, we had assumed that there was a previously agreed, fixed number f, which decided the division of revenue between the platform and the renter. How can one obtain a "fair" value of f? One may also have a different value of f_i for each renter. How do we decide this division f_i? One solution is to consider the Shapley value [18] associated with the game. In a related context, Shapley value was used in the study of pricing of internet service prividers in [17]. The Shapley value is an indicator of the relative power that different players have in the coalition, when they cooperate. To obtain the Shapley value, we assign values to different coalitions between the players and the platform. We will also need to take into account the presence of other competitors to the platform in the market.

Let $P_l = \{A_1, A_2, R_1, ..., R_N\}$ denote the set of players, with A_1 and A_2 representing two platforms, and R_i representing renter class i. Let $\mu(C)$ represent the *worth* obtained by a coalition $C \subseteq P_l$. This denotes the total utility when all users in C work together as a coalition. We will assume that platform A_i controls a fraction ρ_i of the customer population, with $\rho_1 + \rho_2 = 1$. This can be considered as a generalization of our model with renter class j of the market share controlled by platform i, having a demand response of the form

$$R_{i,j}(\mathbf{p}) = \sqrt{\rho_i}\alpha_j - \beta_j(p_{i,j}^0 + p_{i,j}^1), \ i = 1, 2, \tag{25}$$

with the price vector $\mathbf{p}$ of appropriate dimensions. Using the results of the preceding discussions in Sect. 3, it is clear that the worth obtained by a coalition between both platforms and a subset of the renters $i_1, i_2, .., i_n \in \{1, ..., N\}$ is

$$\mu(\{A_1, A_2, R_{i_1}, R_{i_2}, ..., R_{i_n}\}) = \sum_{j=1}^{n} \frac{\alpha_{i_j}^2}{4\beta_{i_j}}. \tag{26}$$

If a single platform and a subset of renters collaborate, the revenue will be proportional to the market share of that platform. Thus, for any $i_1, i_2, .., i_n \in \{1, ..., N\}$ we have

$$\mu(\{A_n, R_{i_1}, R_{i_2}, ..., R_{i_n}\}) = \rho_n \sum_{j=1}^{n} \frac{\alpha_{i_j}^2}{4\beta_{i_j}}, \ n = 1, 2. \tag{27}$$

The apartments or the platforms have no worth in the absence of the other, i.e.,

$$\mu(\{A_1\}) = \mu(\{A_2\}) = \mu(C) = 0, \ \forall C \subseteq P_l \backslash \{A_1, A_2\}, \tag{28}$$

where for two sets A, B, $A \backslash B = A \cap B^c$ denotes the difference of the sets. We define the Shapley value for the game.

Definition 3. *The Shapley value η is the vector given by*

$$\eta(i) = \sum_{C \subseteq P_l \backslash \{i\}} \frac{|C|!(|P_l| - |C| - 1)!}{|P_l|!} \left(\mu(C \cup \{i\}) - \mu(C) \right), \ \forall i \in P_l. \qquad (29)$$

Here $\mu(C \cup \{i\}) - \mu(C)$ represents the increment in worth in the coalition C by the addition of some player $i \in P_l$. The Shapley value is a unique function, that provides a measure of each player's contribution to the coalition and satisfies certain desirable properties [18, Theorem 9.3]. With N renter classes and two platforms, and worth function μ defined as discussed, we obtain the following.

Lemma 5. *For the worth μ defined in (26)-(28), the Shapley value is given by*

$$\eta(A_i) = \frac{\rho_i}{2} \sum_{j=1}^{N} \frac{\alpha_j^2}{4\beta_i}, i = 1, 2, \quad \eta(R_i) = \frac{1}{2} \frac{\alpha_i^2}{4\beta_i}, \quad i = 1, ..., N. \qquad (30)$$

The proof of this lemma is provided in [16].

Thus, if $\rho_1 = 1$, there is one platform that has monopoly, and it will take away half of the renters' revenues. Since the total pooled revenue in a cooperative game between one platform and N renters yields a total utility of $\sum_i \frac{\alpha_i^2}{4\beta_i}$, we can see that the Shapley values of the players tells us how to divide this total utility. From the perspective of renter class i, the Shapley division represents a more than twofold improvement over what it gets in the Nash price. The platform also makes a similar gain.

In the case of two platforms, they will divide half of the total revenue between themselves. The fraction of total price taken as fee by the platform, is an indicator of how much of the market it controls, or how close it is to being a monopoly. Both Airbnb and Booking.com charge well below 50% of the total price as a service charge. As discussed before, the former charges around 17% and the latter, 15%, on average.

4   Quality of Service/Experience

In this section, we modify the demand response to incorporate the quality of service offered by the renter (alternatively, the quality of experience of the guest). In the sequel we will assume that there is just one renter class, i.e., $N = 1$. The demand response function is given by

$$R(p) - \alpha - \beta p + \gamma q. \qquad (31)$$

Here p represents the price. In this section we will remove the upper bound on price, and allow it to take values in the set $[0, \infty)$. The variable q represents a quality of service parameter, taking values in the set $[q_{min}, q_{max}]$, with $0 \leq$

$q_{min} \leq q_{max}$. Higher values of q correspond to the renters providing better services for the client (in terms of customer satisfaction). This can correspond to better behaviour, better services and creating a better ambience around the property. However, per apartment, maintaining this value of q comes at a price of cq for the renter. Let us also assume that of the price charged, the renter gets to keep a fraction $f \in (0,1]$, and the platform keeps $1 - f$. The utility of the renter and platform at a price p are,

$$U_r(p) = (fp - cq)R(p), \quad U_a(p) = (1 - f)pR(p). \qquad (32)$$

We are interested in whether the renter and the platform can agree on a value of q as the quality of service to provide to the customer. We first find the optimal price from the renter's perspective, and calculate the utilities for the renter and the platform at this optimal price. Then we examine how these utilities vary, as we vary the parameter q.

4.1   Maximizing the Renter's Utility

The partial derivatives of $U_r(p)$ with respect to p are,

$$\frac{\partial U_r(p)}{\partial p} = \alpha f + cq\beta + fq\gamma - 2pf\beta, \quad \frac{\partial^2 U_r(p)}{\partial p^2} = -2f\beta < 0. \qquad (33)$$

Since the function is strictly concave, it follows [9, Sec. 4.2.3] that the renter's utility is maximized when $\frac{\partial U_r(p)}{\partial p} = 0$, at $p = p_r^*$, where

$$p_r^* = \frac{\alpha + q\gamma}{2\beta} + \frac{cq}{2f}. \qquad (34)$$

The utility for the renter with price p_r^* is

$$U_r(p_r^*) = \frac{f}{4\beta}(\alpha + q\gamma - \frac{cq\beta}{f})^2. \qquad (35)$$

Since this is a function of q, we look at how it varies with q. We have

$$\frac{\partial U_r(p_r^*)}{\partial q} = \frac{f}{2\beta}(\alpha + q\gamma - \frac{cq\beta}{f})(\gamma - \frac{c\beta}{f}) = \frac{\alpha}{\beta}\frac{(f\gamma - c\beta)}{2} + q\frac{(f\gamma - c\beta)^2}{2f\beta}.$$

The utility for the platform at the price p_r^* is

$$U_a(p_r^*) = \frac{1-f}{4\beta}\left[(\alpha + q\gamma)^2 - \left(\frac{cq\beta}{f}\right)^2\right]. \qquad (36)$$

The rate of change of the utility of the platform at price p_r^* with respect to q is

$$\frac{\partial U_a(p_r^*)}{\partial q} = \frac{1-f}{2\beta}\left[\alpha\gamma + q\left(\gamma^2 - \frac{c^2\beta^2}{f^2}\right)\right]. \qquad (37)$$

The variation of the function $U_r(p_r^*)$ with q depends on the value of c, which is the cost per unit quality of service provided to the client by the renter. We consider the following cases, for different values of the cost c.

4.2   Case 1: $c = \frac{f\gamma}{\beta}$

In this case, we see that

$$U_r(p_r^*) = \frac{f\alpha^2}{4\beta}, \quad U_a(p_r^*) = \frac{(1-f)(\alpha^2 + 2q\alpha\gamma)}{4\beta}. \tag{38}$$

The renter is insensitive to q, while the platform can gain from a higher q.

4.3   Case 2: $c < \frac{f\gamma}{\beta}$

In this case, we have $f\gamma - c\beta > 0$. From (36),

$$\frac{\partial U_r(p_r^*)}{\partial q} > 0 \text{ for } q > -\frac{\alpha f}{(f\gamma - c\beta)}. \tag{39}$$

Since $q_{min} \geq 0$, and $-\frac{\alpha f}{(f\gamma - c\beta)} < 0$, it follows that $\frac{\partial U_r(p_r^*)}{\partial q}$ is positive for all $q \in [q_{min}, q_{max}]$. Hence $U_r(p_r^*)$ is always increasing, as we increase q from q_{min} to q_{max}. Hence, from the renter's perspective, it makes sense to choose the highest value $q = q_{max}$, to maximize revenue.

Similarly, from (37), observe that whenever $q > -\frac{f^2\alpha\gamma}{f^2\gamma^2 - c^2\beta^2}$, we see that $\frac{\partial U_a(p_r^*)}{\partial q} > 0$. Since $c < \frac{f\gamma}{\beta}$, it follows that

$$-\frac{f^2\alpha\gamma}{f^2\gamma^2 - c^2\beta^2} = -\frac{f^2\alpha\gamma}{(f\gamma - c\beta)(f\gamma + c\beta)} < 0,$$

and hence $U_a(p_r^*)$ is increasing, as we increase q from q_{min} to q_{max}. Hence, it is optimal for the platform to have $q = q_{max}$ as well. Both renter and the platform are in agreement about the quality of service to be provided. Thus, if the cost of providing quality of service is below a threshold, both parties can agree on providing the best quality of service possible.

4.4   Case 3: $c > \frac{f\gamma}{\beta}$

This is the case where the cost of providing quality of service is high. We have

$$\frac{\partial U_r(p_r^*)}{\partial q} \begin{cases} < 0, & \text{for } q < -\frac{f\alpha}{f\gamma - c\beta}, \\ > 0, & \text{for } q > -\frac{f\alpha}{f\gamma - c\beta}. \end{cases} \tag{40}$$

Since $\frac{\partial^2 U_r(p_r^*)}{\partial q^2} > 0$, $q = -\frac{f\alpha}{f\gamma - c\beta}$ is a minimizer of $U_r(p_r^*)$ in the variable q.

Also, since $f^2\gamma^2 - c^2\beta^2 = (f\gamma - c\beta)(f\gamma + c\beta) < 0$, we have that $\frac{\partial^2 U_a(p_r^*)}{\partial q^2} < 0$. Hence, for the platform, the value of q that maximizes its revenue is such that $\frac{\partial U_a(p_r^*)}{\partial q} = 0$, which is given by

$$q_a^* = -\frac{f^2\alpha\gamma}{f^2\gamma^2 - c^2\beta^2}. \tag{41}$$

Note that since $f\gamma < c\beta$, it follows that $\frac{-f\alpha}{f\gamma - c\beta} > 0$ and hence, $q_a^* = \frac{f\gamma}{f\gamma + c\beta}\frac{-f\alpha}{f\gamma - c\beta} < \frac{-f\alpha}{f\gamma - c\beta}$. For $q > q_a^*$, we see that $\frac{\partial U_a(p_r^*)}{\partial q}$ is negative, and when $q < q_a^*$, we have $\frac{\partial U_a(p_r^*)}{\partial q}$ positive. We have the following possibilities.

1. If $q_{min} > -\frac{f\alpha}{f\gamma - c\beta}$, then $\frac{\partial U_r(p_r^*)}{\partial q} > 0$ for $q \in [q_{min}, q_{max}]$. Therefore, the renter must choose $q = q_{max}$. However, since $q_{min} > q_a^*$, the utility of the platform will decrease as we increase q from q_{min} to q_{max}. Hence, the optimal q for the platform is q_{min}.

2. If $q_{max} < -\frac{f\alpha}{f\gamma - c\beta}$, then $\frac{\partial U_r(p_r^*)}{\partial q} < 0$ for $q \in [q_{min}, q_{max}]$. The renter must choose $q = q_{min}$. In this case, the platform has two possibilities. If $q_{max} \leq q_a^*$, it will prefer $q = q_{max}$. If $q_{max} > q_a^* \geq q_{min}$, the platform will prefer q_a^*. If $q_{min} > q_a^*$, the platform will prefer $q = q_{min}$.

3. If $-\frac{f\alpha}{f\gamma - c\beta} \in [q_{min}, q_{max}]$, the renter will prefer q_{min} or q_{max} depending on which one maximizes the value of $U_r(p_r^*)$. The preference of the platform will be dependent on the position of p_a^* relative to the interval $[q_{min}, q_{max}]$ as before.

The sharing of revenues between renter and the platform is through the fraction f. In the quantity $f\gamma - c\beta$, neither renter nor the platform can influence β, γ (determined by the population behaviour) and c (determined by service cost in the economy). By choosing a particular value of f, the renter and the platform enter into one of the cases listed above.

5   Conclusion and Future Directions

In this work we have presented different models to understand optimal pricing and revenue in a digital platform pricing problem and obtain different properties of non cooperative and cooperative solutions. We also study how providing quality of service impacts price. While these models help quantify questions of optimality and equilibrium behaviour in the prices, they do not fully capture all the dynamics of the rental market. Even though platforms such as Airbnb and Booking.com have a cooperative pricing mechanism, it is not clear if the Shapley value by itself fully explains their revenue structure. The next step is to model different externalities and how they impact prices, revenue and optimality.

References

1. Airbnb service fees - Airbnb help center. https://www.airbnb.com/help/article/1857
2. Booking.com service fees. https://your.rentals/blog/booking-com-fees-how-are-they-calculated/
3. Adamiak, C.: Current state and development of Airbnb accommodation offer in 167 countries. Curr. Issue Tour. **25**(19), 3131–3149 (2022)
4. Armstrong, M.: Competition in two-sided markets. Rand J. Econ. **37**(3), 668–691 (2006)

5. Aznar, P., Maspera, J.M.S., Segarra, G., Claveria, J.: Airbnb competition and hotels' response: the importance of online reputation (2018)
6. Baye, M.R., Morgan, J.: Information gatekeepers on the internet and the competitiveness of homogeneous product markets. Am. Econ. Rev. **91**(3), 454–474 (2001)
7. Belleflamme, P., Peitz, M.: The Economics of Platforms. Cambridge University Press, Cambridge (2021)
8. Bernstein, F., Federgruen, A.: Decentralized supply chains with competing retailers under demand uncertainty. Manage. Sci. **51**(1), 18–29 (2005)
9. Boyd, S.P., Vandenberghe, L.: Convex Optimization. Cambridge University Press, Cambridge (2004)
10. Coyle, D., Yeung, T.: Understanding Airbnb in fourteen European cities. Jean-Jacques Laffont Digit. Chair Work. Pap. **7088**, 1–33 (2016)
11. Cui, Y., Orhun, A.Y., Hu, M.: Shared lodging and customer preference: theory and empirical evidence from Airbnb. SSRN 3136138 (2020)
12. El Azouzi, R., Altman, E., Wynter, L.: Telecommunications network equilibrium with price and quality-of-service characteristics. In: Teletraffic science and engineering, vol. 5, pp. 369–378. Elsevier (2003)
13. Gallego, G., Huh, W.T., Kang, W., Phillips, R.: Price competition with the attraction demand model: existence of unique equilibrium and its stability. Manuf. Serv. Oper. Manage. **8**(4), 359–375 (2006)
14. Gibbs, C., Guttentag, D., Gretzel, U., Yao, L., Morton, J.: Use of dynamic pricing strategies by Airbnb hosts. Int. J. Contemp. Hospitality Manag. **30**, 2–20 (2018)
15. Guo, X., Yang, H.: The price of anarchy of cournot oligopoly. In: Deng, X., Ye, Y. (eds.) WINE 2005. LNCS, vol. 3828, pp. 246–257. Springer, Heidelberg (2005). https://doi.org/10.1007/11600930_24
16. Krishnan.K.S., A., Perlaza, S., Altman, E.: Pricing for platforms: games, equilibria and cooperation. INRIA, Centre Inria d'Université Côte d'Azur, Sophia Antipolis, France. Technical report RR. 9510 (2023)
17. Ma, R.T., Chiu, D.M., Lui, J.C., Misra, V., Rubenstein, D.: Interconnecting eyeballs to content: a Shapley value perspective on ISP peering and settlement. In: Proceedings of the 3rd International Workshop on Economics of Networked Systems, pp. 61–66 (2008)
18. Myerson, R.B.: Game Theory: Analysis of Conflict. Harvard University Press, Cambridge (1997)
19. Nguyen, Q.: A study of Airbnb as a potential competitor of the hotel industry. Masters thesis at the University of Nevada, Las Vegas (2014)
20. Rochet, J.C., Tirole, J.: Platform competition in two-sided markets. J. Eur. Econ. Assoc. **1**(4), 990–1029 (2003)
21. Rosen, J.B.: Existence and uniqueness of equilibrium points for concave n-person games. Econometrica J. Econometric Soc. **33**, 520–534 (1965)
22. Zheng, Z., Srikant, R.: Optimal search segmentation mechanisms for online platform markets. In: Caragiannis, I., Mirrokni, V., Nikolova, E. (eds.) WINE 2019. LNCS, vol. 11920, pp. 301–315. Springer, Cham (2019). https://doi.org/10.1007/978-3-030-35389-6_22

Decision Analysis in Stochastic Sociocultural Systems

Amy Sliva[1]([✉]), Emanuele Borgonovo[2], Alexander Levis[3],
Christopher Pawlenok[1], and Nathaniel Plaspohl[1]

[1] King's College, Wilkes-Barre, PA, USA
amysliva@kings.edu, christopherpawlenok@kings.edu,
nathanielplaspohl@kings.edu
[2] Bocconi University, Milan, Italy
emanuele.borgonovo@unibocconi.it
[3] George Mason University, Fairfax, VA, USA
alevis@gmu.edu
http://www.kings.edu/ , http://www.unibocconi.eu , https://www.gmu.edu

Abstract. Decision makers often leverage causal and predictive scientific models, enabling them to better estimate the behavior of a physical or social system and explore several possible outcomes of a policy before actually enacting it. However, these models are only useful insofar as decision makers can effectively interpret the model behaviors, their outputs, and their inherent limitations. In this paper, we describe a methodological approach for uncertainty quantification in stochastic models of sociocultural systems via application of sensitivity analysis techniques, enabling decision makers to explicitly account for uncertainties in model validity or stability when using them as decision support tools. To demonstrate the efficacy of this approach, we present a case study analysis of food security challenges in Gambella, Ethiopia. Using a Bayesian modeling approach, we represent the stochastic interaction of meteorological, agroeconomic, political, and social factors influencing the likelihood of famine in the region and examine the impact of possible mitigating courses of action. Leveraging several sensitivity analysis approaches, we provide evidence of improvements to the decision-making process by making our models more transparent, highlighting the impact that model uncertainty may have when determining the optimal course of action and indicating to the decision maker what questions they need to ask, what data they need to gather, or where to focus their attention to identify and implement the best policy to prevent famine conditions.

Keywords: Sensitivity Analysis · Decision Analysis · Complex Systems

1 Introduction

Decision makers in a wide variety of complex, uncertain domains, ranging from business administration and finance to engineering and socioeconomic policies—

often leverage causal and predictive scientific models, enabling them to better estimate the behavior of a physical or social system and explore several possible outcomes of a policy before actually enacting it. Ideally, these scientific models are derived from reliable data or an overarching theoretical framework, which determines the conditions and hypotheses under which the model is valid or "true", mathematically characterizing the behavior or system of interest. However, these models are only useful insofar as decision makers can effectively interpret the model behaviors, their outputs, and their inherent limitations.

A major source of such limits on a model's efficacy for decision making is uncertainty; unknown or uncertain inputs yield challenges in producing an output mapping that is accurate and reliable, a problem that is further exacerbated by a stochastic response observed in the system under analysis. This is especially true for models of sociocultural domains and phenomena, which attempt to capture the behavioral dynamics caused by a confluence of interacting and interdependent social, cultural, economic, and political systems. A wide variety of modeling frameworks exist to address the issues of stochasticrelationships and influences—probabilistic graphical models (e.g., Bayesian networks [11], influence diagrams [8], probabilistic relational models [5]), probabilistic soft logic [9], Markov logic networks [12], and many others—but these cannot necessarily account for all of the uncertainty in a model's behavior and results. A model may have incorrect parameters, the chosen values may not be good approximations of reality, or the models may be learned or derived from noisy or incomplete data.

Researchers often refer to these challenges as "The Black Box Menace", where the behavior and impact of model inputs or parameters are unknown and can propagate uncertainty during analysis. Begoli, Bhattacharya, and Kusnezov [1] highlight the issue that the "absence of theory" in purely data-driven models often makes a proper uncertainty quantification an essential ingredient in the use of black-box machine learning algorithms—even if the model behavior cannot be explicitly understood, the propagation of uncertainty and sources of uncertainty in the model inputs and outputs can be. In [6] Guidotti et al. suggest that interpretability and explainability issues impact the use of artificial intelligence methods, and without proper uncertainty quantification, modeling quality and validity is at stake, especially as model complexity increases [13].

In this paper, we describe a methodological approach for uncertainty quantification in stochastic models of sociocultural systems via application of sensitivity analysis techniques, enabling decision makers to explicitly account for uncertainties in model validity or stability when using them as decision support tools. Sensitivity analysis provides a mathematically rigorous approach to model introspection and explanation, increasing transparency, enhancing interpretability, and increasing awareness of the model behavior [6,13]. Applying sensitivity analysis techniques, especially to a complex probabilistic model, enables (1) increased awareness of model behavior; (2) understanding of model limitations; (3) understanding of areas requiring additional modeling efforts; (4) redirection of data collection to more important or impactful features; (5) proper communication of results; and (6) proper assessment of outcomes and effects.

To demonstrate the efficacy of this approach, we present a case study of food security challenges in Gambella, Ethiopia in 2019 (before the COVID-19 pandemic, which had substantial impact on global social, economic, and political dynamics). Using a Bayesian modeling approach, we represent the stochastic interaction of meteorological, agroeconomic, political, and social factors influencing the likelihood of famine and examine the impact of possible mitigating courses of action. This case study model is an abstract approximation of a very complex sociocultural process, but still successfully illustrates the challenges posed by uncertainty in the model parameters, values, and interactions. Leveraging several sensitivity analysis approaches, we provide evidence of improvements to the decision-making process by making our models more transparent, highlighting the impact that model uncertainty may have when determining the optimal course of action and indicating to the decision maker what questions they need to ask, what data they need to gather, or where to focus their attention to identify and implement the best policy to prevent famine conditions.

The remainder of this paper is organized as follows. In Sect. 2, we describe the Bayesian modeling approach used to capture the food security system in Gambella, Ethiopia. Section 3 describes a baseline approach to using this type of model for decision support, which considers the stochastic system dynamics and some minimal parameter variation, but does not attempt to quantify model uncertainty or its propagation effect. Then in Sect. 4 we introduce our sensitivity analysis for uncertainty quantification and illustrate how this leads to more robust decision making. Finally, in Sect. 5 we discuss the results of our proposed methodology and our plans for further research.

2   Modeling Food Security in Gambella, Ethiopia

To demonstrate the technical approach, we present a use case based on the food security and migration situation in the Gambella region of Ethiopia in 2019, where displaced persons from various civil conflicts in adjacent regions have caused a massive increase in population, and hence a possible food security crisis as the number of mouths to feed has also increased. This use case has many features that illustrate the complexity and stochastic nature of sociocultural systems modeling, including interdependent subsystems, temporal dynamics with seasonal variations based on the agricultural cycle, and several possible decision points for interventions or policy analysis. The model was developed from data describing actual events [3], but it is a simplified representation of the actual conditions in Gambella; while it is realistic and plausible, the goal of the model is not to provide accurate predictions for the purposes of making real-world decisions for the region, but to illustrate the value of uncertainty quantification in improving upon baseline decision analysis.

This model focuses on agroeconomic processes (e.g., agricultural inputs, crop yields, food transport), the temporal effects of precipitation, (e.g., drought and flood conditions during different parts of the agricultural cycle), sociopolitical events (e.g., civil conflict and human migration), and the impact of various *interventions*, including:

1. Increase in food imports (current purchases)
2. Provision of Food Aid
3. Provision of Direct Aid
4. Conflict Resolution

We define a *course of action (COA)* to be a combination of these policy interventions. For n possible interventions, we can construct 2^n COAs. In order to examine the impact of these COAs under a variety of conditions, we also designate several aspects of the model to be *scenario inputs*; these factors will be fixed for any particular analysis:

1. Precipitation
2. Flood conditions
3. Drought conditions
4. Civil conflict
5. Human migration

2.1 Model Implementation

Our approach was to capture all scenarios with a single causal model that has dependencies on temporal events (weather, labor, displaced persons) and on external decisions (investments, aid, conflict resolution). The model captures external events in a timeline (monthly) modeled on historical and scientific observations, and models external decisions in a timeline based on reasonable expectations, given a humanitarian goal.

The model was implemented as a timed influence net (TIN) [7], a probabilistic graphical model that is a variation on a standard influence diagram [8] that captures probabilistic and uncertain influences in dynamic environments. Like classic influence diagrams, a TIN contains different types of nodes—those that represent random variables and those that represent decisions—and can be transformed to a Bayesian network [11] for analysis and inference. However, in TINs, the model parameters (i.e., the influence parameters) are described in the *Causal Strength (CAST)* [7] format indicating the conditional influence of the parent nodes on their children. The CAST representation was originally developed to provide domain experts with a more intuitive method for specifying probabilistic relationships, requiring fewer parameters than standard Bayesian conditional probability tables (CPTs). In CAST:

- A *base probability* is specified for each node, indicating the independent, unconditional likelihood of an event.
- Each edge contains two influence values indicating the strength of the influence when the parent is true or false

– Using these values, the CAST algorithm computes the remaining CPT parameters necessary for a Bayesian network or influence diagram.

Inference is performed by computing posterior marginal probabilities for nodes in the network—including the intended outcome node representing the likelihood of famine—using a Bayesian network algorithm, such as variable elimination or belief propagation. In cases of very large models, an approximate inference can be performed using Monte Carlo sampling approaches, but for this case study we only examine exact inference.

The graphical structure of the model, shown in Fig. 1 as rendered using the Pythia modeling environment developed by System Architectures Laboratory at George Mason University [7,10,16], was developed through qualitative assessment of data [3] and domain expertise on agricultural systems and food security. Information from the temporal dynamics of the agricultural cycle—which months are for sowing versus growing versus harvesting, etc.— were used to derive the model parameters and the state of each event for a specific month or period of the year. A textbook Solow-Swan model [14,15] was used to construct parts of the scenario and derive the interaction parameters, modeling human migration dynamics as an exogenous factor in the agroeconomics of the region. Model inference can then be used to estimate the impact of various COAs and compute the crop yield increase, crop production increase, adequate quantities of urban food, adequate quantities of rural food, and ultimately the likelihood of famine conditions.

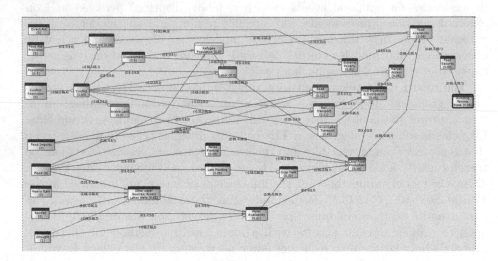

Fig. 1. TIN representation of the Gambella food security model in Pythia.

The above model was developed using the Pythia implementation of TINs in C# to compute the standard Bayesian CPT parameters via the CAST algorithm. Inference—computation of posterior marginal probabilities of network variables and maximum expected utility of possible COAs—was performed using a custom Python implementation of Bayesian inference through variable elimination.

3 Baseline Decision Analysis

In this section, we will use the above case study model to examine a baseline decision-making approach for utilizing computational models. The goal of a policy maker using this model is to maximize the probability that a famine can be avoided in Gambella, Ethiopia during a particular calendar year. This model can be used to explore three scenarios: (1) typical weather and growing conditions; (2) heavy rainfall and potential flooding; or (3) possible drought conditions. The impact of the latter two scenarios will depend greatly on when they occur during the agricultural cycle. Gambella typically experiences seasonal variations in rainfall; heavier than usual rain during the regular growing season will not impact crop yield, as the crops will be able to adapt. However, heavy rain may have severe impacts on food production if it disrupts the sowing and harvesting activities. A drought throughout any time of the season may have a negative impact, but may also be particularly impactful during sowing and harvesting. Therefore, the model parameters—i.e., the conditional influences in the network—will be different for each month and can be derived from the timelines in Sect. 2.

3.1 Optimal Course of Action

For each scenario, the decision maker can use the probabilistic model to compute the COA (or top k COAs) that maximizes the probability of famine avoidance. This analysis enables both an examination of the best policy given known conditions , as well as a hypothetical "what if?" analysis . To enable meaningful comparisons where the effects of scenario variations will be present, we parameterized the model to represent the sowing phase in April.

In Scenario 1, the probability of drought is set to 1.0, simulating climate conditions where there is no or very low rainfall leading to drought conditions. Using Bayesian inference via variable elimination [17], the Gambella model was used to predict the probability of the Avoid Famine variable under each possible course of action. Table 1 contains the results of this simulation; the highlighted rows indicate courses of action that led to $p(AvoidFamine) \geq 0.5$. In this scenario, there are several COAs that can achieve $p(AvoidFamine) \geq 0.5$; the optimal course of action that leads to the highest chance of famine avoidance is COA 15.

In Scenario 2, the probability of heavy rainfall is set to 1.0, simulating the condition under which there will be likely flooding. Again, the model was used to predict the probability of Avoid Famine for each possible course of action (see Table 2). In this scenario COAs 8 – 15 are all policies where conflict resolution is included as part of the intervention; the analogous policies without conflict resolution did not have similarly positive impacts on famine avoidance. COA 7 does not require conflict resolution, but instead assumes that all other

Table 1. Impact of each possible COA on famine avoidance during drought conditions.

COA	Conflict Resolution	Food Aid Provision	Direct Aid	Food Imports	Avoiding Famine State
0	0	0	0	0	0.06
1	0	0	0	1	0.10
2	0	0	1	0	0.17
3	0	0	1	1	0.27
4	0	1	0	0	0.15
5	0	1	0	1	0.20
6	0	1	1	0	0.40
7	0	1	1	1	0.46
8	1	0	0	0	0.42
9	1	0	0	1	0.53
10	1	0	1	0	0.53
11	1	0	1	1	0.66
12	1	1	0	0	0.59
13	1	1	0	1	0.66
14	1	1	1	0	0.71
15	1	1	1	1	0.76

Table 2. Impact of each possible COA on famine avoidance during flood conditions.

COA	Conflict Resolution	Food Aid Provision	Direct Aid	Food Imports	Avoiding Famine State
0	0	0	0	0	0.16
1	0	0	0	1	0.21
2	0	0	1	0	0.31
3	0	0	1	1	0.41
4	0	1	0	0	0.25
5	0	1	0	1	0.34
6	0	1	1	0	0.47
7	0	1	1	1	0.54
8	1	0	0	0	0.50
9	1	0	0	1	0.57
10	1	0	1	0	0.60
11	1	0	1	1	0.68
12	1	1	0	0	0.62
13	1	1	0	1	0.68
14	1	1	1	0	0.72
15	1	1	1	1	0.75

possible policy options—food imports, food aid provision, and direct financial aid provision—are implemented.

Based on these two analyses, it appears that COA 15, which is a policy that requires conflict resolution, increased food imports, food aid, and direct aid, is the "optimal" choice for avoiding famine in Gambella, giving the highest probability of success. While these results can provide some general guidance for a policy maker, there is nothing to indicates what the limitations of this model might be. In order to better understand how to apply the outputs of this model, it is necessary to know to what extent uncertainty in the parameter settings (i.e., what are the characteristics of time period we are modeling? what is the influence among model variables?) or assumptions about the scenario (i.e., does flooding occur? is there an ongoing conflict?) can alter the validity of the inference results. In the next section, we examine advanced sensitivity analysis techniques that rigorously quantify the uncertainty in the model dynamics.

4   Sensitivity Analysis for Uncertainty Quantification

In many decision support applications, decision makers will simply use the optimal COA analysis above to support their choice of policy interventions. However, isolated predictions and blackbox optimizations alone cannot provide decision makers with details of **HOW** and **WHEN** these predictive inference results should be applied and what sources of uncertainty may be impacting their validity. Standard decision-analysis procedures suggest leveraging sensitivity analysis to better understand the model behaviors by quantifying which model variables are most influential in determining the outcome of the predictive inference. We will refer to this approach as *standard tornado analysis*, since the results are often visualized in a graph called a tornado diagram (see Sect. 4.1).

We propose a technical approach that would enhance basic decision analysis by applying several variations of sensitivity analysis techniques to rigorously quantify and assess not just influence within the model, but other possible aspects related to model uncertainty: (1) factor-prioritization; (2) trend determination; (3) understanding interactions; and (4) understanding stability. *Factor-prioritization* allows the analyst to understand which features drive the results of a model prediction, and whether it is worth it to collect further data/information (and if so, where to collect). It helps identify factors that may cause the optimal policy interventions or COA to change or those that may deserve attention during policy implementation to assess the effects of these policy interventions. *Trend determination* helps the analyst understand whether the model responds positively or negatively to input variations and can enable comparision of model results under different starting conditions. Probabilistic graphical models, such as the influence diagrams used in our Gambella case study, are capable of representing complex interactions between multiple variables at a time. Yet, standard sensitivity analysis approaches only examine a single variable at a time isolation, *ceteris parabus*. To *understand interactions*, a system must address the question: if two or more variables vary simultaneously, does their joint variation produce a synergistic (positive interaction) or an antagonistic effect? Similarly, the basic optimal COA analysis only indicates the "optimal" policy option under a single set of conditions. To understand *stability* of this COA, decision maker if a preferred strategy is robust to variations in external factors.

To quantify model uncertainty along the above dimensions, we propose additional sensitivity measures beyond a standard single-factor examination, both of which can help address all four types of uncertainty identified above:

- *Generalized Tornado Analysis* [2], which is a novel approach to assessing the interaction effects of combinations of model variables
- *Frequency Analysis* [2], which assesses frequency with which each COA is selected as the optimal strategy
- *Expected Value of Perfect Information (EVPI)* [2,4], which can quantify the degree to which the model results may be improved by gathering additional targeted data on specific variables or parameters

In most models, especially those of complex sociocultural systems, it will be computationally intractable to compute the full generalized tornado analysis, which would look at all possible combinations of all possible variable values and then run standard model inference with these settings. Instead, a Monte Carlo sampling technique can be applied to approximate this behavior by examining a random subset of these possible combinations.

Leveraging the above sensitivity analysis tools, our proposed technical approach for uncertainty analysis consists of the following steps:

Step 1. Perform basic static model inference and optimal COA generation
Step 2. Compute standard single-variable tornado analysis for each possible COA and scenario
Step 3. Use Monte Carlo analysis to compute the generalized tornado for interactions of variables for each possible COA and scenario.
Step 4. Using the Monte Carlo results, estimate the frequency that each COA is selected as optimal
Step 5. Compute the EVPI for each variable and possible scenario

Step 1 of this procedure was already discussed in Sect. 3. In the remainder of this section, we discuss the details of our technical approach, illustrating the application of these techniques using the Gambella food security test case.

4.1 Standard Tornado Analysis

In this section we describe a standard decision analysis technique [2] using sensitivity to quantify the impact of individual model variables, indicating the robustness of an intervention or course of action, as well as the validity of model behavior. The algorithm for computing this "tornado" analysis, so called due to the shap of the graphed results, is shown in Fig. 2.

#	Step	Quantity obtained
1	Evaluate the model given the base case	$y^0 = g(\mathbf{x}^0)$
2	Move one model input at a time from x_i^0 to x_i^+ and evaluate the model	$y_i^+ = g(x_i^+, \mathbf{x}_{\sim i}^0)$
3	Compute the difference $y_i^+ - y^0$	Δy_i^+
4	Move one model input at a time from x_i^0 to x_i^- and evaluate the model	$y_i^- = g(x_i^-, \mathbf{x}_{\sim i}^0)$
5	Compute the difference $y_i^- - y^0$	Δy_i^-
6	Draw the sensitivity measures as horizontal bars	

Fig. 2. Steps for computing the standard "tornado" sensitivity analysis to quantify the impact of each model variable on the inferred outcome.

First, we compute the *baseline* outcome of the model, y^0, which represents the value of the outcome variable (i.e., *AvoidFamine*) when all variables in the model operate according to their stochastic relationships. Next, for each variable i, the tornado analysis looks at the minimum and maximum possible values, allowing all the other variables to remain stochastic. The difference is computed between the baseline outcome and the outcome variable at the min and max points, φ_i^- and φ_i^+, respectively. In a stochastic model, these φ values represent a potential reduction in uncertainty, indicating how the model predictions might change if we know the exact value of the variable. For example, comparing the probability of $AvoidFamine$ when $Conflict = 0$ and when $Conflict = 1$ under different policy interventions can help a decision maker determine if the optimal COA is robust to potential conflict in the region or not—if the sensitivity analysis indicates large φ range, this indicates the intervention may not be stable depending on the conflict situation.

Using the Gambella model, we computed a standard tornado analysis for both scenarios (Scenario1 models drought conditions during the sowing season, Scenario2 contains heavy rainfall and flooding during the sowing season), looking at the impact of each of the model variables on the probability of famine avoidance. This analysis was performed separately for each possible policy intervention, enabling us to explore the impact of model variation under each possible COA. Figures 3 and 4 show two examples of these results. The graph is organized with the most influential variables at the top and least influential at the bottom, giving it a "tornado" shape. The vertical line indicates the baseline value of $AvoidFamine$, and the bars indicate the φ_i^- and φ_i^+ variations from baseline.

In both scenarios, our variables capturing the degree of famine conditions and food availability are the most influential, which is expected as they are direct parents of the probability of famine avoidance. However, as hinted at by our COA analysis, the existence of civil conflict is also extremely impactful, and can cause massive shifts in the likelihood of famine avoidance. This analysis indicates several things. First, it can provide transparency into the model dynamics and indicate whether the model comports with intuitive or theoretical understandings of the world. Second, more certainty about the possibility of conflict can potentially change a decision maker's chosen COA.

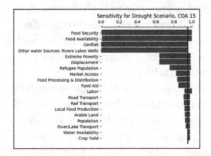

Fig. 3. Tornado diagram for Drought.

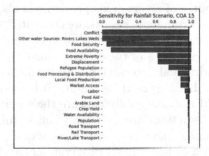

Fig. 4. Tornado diagram for Rainfall.

While the basic tornado analysis certainly provides a better understanding of the model behaviors and some view into the necessity of additional data collection, as a stochastic system, our model still remains quite complex and opaque. Using our sensitivity measures described above, a decision maker can more accurately quantify the uncertainty that might be present due to lack of data or information, as well as increase their confidence (or lack thereof) in the predicted optimal course of action.

4.2   Generalized Tornado Analysis

A standard tornado analysis provides information about the influence of each model variable in isolation. However, in a complex stochastic model, interaction effects are critical for understanding the actual behavior of the model and contextualizing the inferred outcomes. We use a novel generalized tornado analysis to compute the value of the outcome variable when a combination of variables takes their min/max values simultaneously. This *interaction effect*, $\varphi^{\pm}_{x \in P(vars)}$ quantifies the degree to which uncertainty could be reduced by further data about combinations of variables, even if individually the variables may not be very influential. The *total* impact of a variable is the sum of its individual and interaction effects: $\varphi^{T}_{i} = \varphi^{\pm}_{i} + \sum_{x|i \in x} \varphi^{\pm}_{x \in P(vars)}$.

To assess the interaction effects of each variable, the generalized tornado computes the effect for the full powerset $P(var)$ of variable combinations taking their min/max values. This complete analysis is often computationally intractable. For example, the Gambella model has 2^{25} possible variable combinations. Therefore, a Monte Carlo sampling approach can be used to estimate the interaction effects by running predictive inference on the model with a randomly selected set of variables set to their min/max. The overall interaction effect for any variable would now be the sum over the effect of each *sample* in which it was set to min/max rather than being left as a stochastic random variable.

Using the Gambella model, we performed a Monte Carlo analysis over 10,000 samples setting random combinations of variables to their max and min value. For each sample, we ran the probabilistic inference algorithm for each possible course of action. This process was also completed across our two climate scenarios. From these results, we computed the individual, interaction, and total effects of each model variable for each scenario and course of action. Figure 5 shows an examples of these results for COA 15 for the drought scenario, including the individual, interaction, and total effects for each variable.

Looking at these results, it is immediately evident that our baseline analysis did not adequately capture the sensitivity and uncertainty effects of the model. Interactions are extremely impactful, which may cause the decision maker to change the chosen course of action. If we can gain more certainty (i.e., gather data or make direct observations) about combinations of variables, such as those relating to available water, then the outputs of the model can be drastically different—e.g., if there is sufficient water sources despite a drought, the need for famine avoidance policies may be obviated.

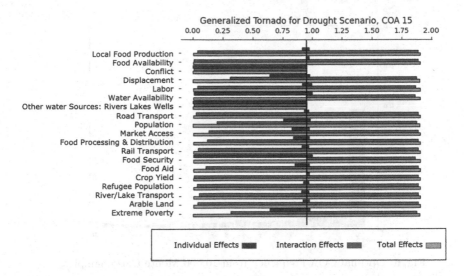

Fig. 5. Generalized Tornado Analysis from 10,000 Monte Carlo samples.

From this Monte Carlo sampling, we were also able to estimate the frequency with which each COA was selected as optimal. These results are shown in Fig. 6. In contrast to our COA analysis in Sect. 3 which indicated that under conditions of uncertainty COA 15—implementation of all policies—would be the optimal course of action, the frequency analysis under varying situations (i.e., collecting data regarding our variables rather than leaving them as random variables in the model), indicates the exact opposite—COA 0, which involves abstaining from any of the proposed policy interventions, leads to the maximum likelihood of avoiding famine. This highlights the importance of uncertainty reduction in the policy implementation. With no additional information, the best we can do is everything; with strategic acquisition of highly influential information, the optimal COA becomes to simply maintain the status quo and enable the existing agro economicprocesses to play out.

4.3 Expected Value of Perfect Information

Finally, in addition to assessing the influence of variables individually and in combination, our technical approach uses expected value of perfect information (EVPI). EVPI is a sensitivity measure that looks at the (in)stability or uncertainty introduced by certain variable values or model parameters. EVPI is responding to the question: is there a variable or model parameter that, if known with greater certainty, will make the decision maker change their mind? And what is the associated expected utility/probability increase in the optimal course of action? Similar to the generalized tornado analysis, the EVPI can be computed from a Monte Carlo sample over the model.

The results of EVPI indicate the change in utility that can be achieved by knowing each of the values with absolute certainty. Given this information, a

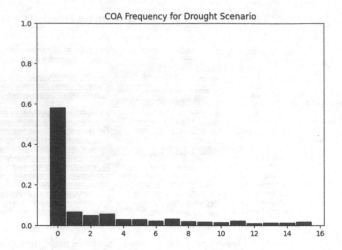

Fig. 6. Optimal COA Frequency from 10,000 Monte Carlo Samples.

decision maker would be able to gauge the stability of a recommended course of action. If critical information was missing which has a very high EVPI, then the current results may not be reliable and additional data gathering or model tuning may be required.

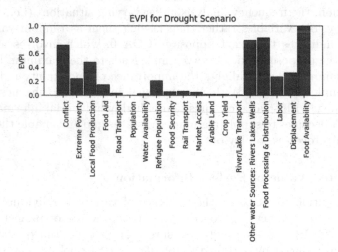

Fig. 7. Expected Value of Perfect Information for each variable.

For the Gambella model, we computed the EVPI for each variable. The results for this analysis are shown in Fig. 7 for the drought scenario. The variables with the highest bars are those that, if known with certainty, would have the largest impact on reducing the stochastic nature of the outcome variable.

For the drought scenario, the EVPI analysis shows the high value of gathering more information about the availability of alternative water sources, the conflict situation, and availability of food stores before making a decision about what is needed for famine avoidance. If these variables can be directly observed, decision makers would know with much higher certainty the liklihood of famine.

5 Conclusions and Future Work

In this paper, we have presented a methodology for using sensitivity analysis to quantify uncertainty for decision making with uncertain, complex probabilistic models. By exploring and quantifying the impact of variable interactions, the optimal decision for a given situation may be drastically different from a purly stochastic model with no observations. We demonstrated the effectiveness of this enhanced decision analysis approach using a case study in policy assessment for famine avoidance in Gambella, Ethiopia.

We have several planned directions for future work. First, there are additional approaches to quantifying model stability and reliability that we intend to implement and assess, including (1) a generalized version of expected value of information that looks at different combinations of variable values to understand the impact of increased certainty in multiple variables simultaneously; (2) partial dependence . There are also several ways in which we can increase the complexity of the modeling representation that would require additional sensitivity measures to quantify the uncertainty.

Finally, we would like to explore a similar scenario in a region without ongoing civil conflict. As discussed in Sect. 3, the presence of conflict dominated much of the model outcome, regardless of meteorological or humanitarian aid conditions. For making decisions in Gambella, Ethiopia, this is essential to consider, as other policy interventions besides conflict resolution may be largely ineffectual until the violent conflict has been tempered. However, it would be interesting to apply the same methodology to agricultural processes and dynamics in other famine-prone regions which are not necessarily also experiencing civil conflict.

Acknowledgements. Part of this work was performed under DARPA contract number W911NF-18-C-0015 to Charles River Analytics, Inc.

References

1. Begoli, E., Bhattacharya, T., Kusnezov, D.: The need for uncertainty quantification in machine-assisted medical decision making. Nature Mach. Intell. **1**(1), 20–23 (2019)
2. Borgonovo, E., Plischke, E.: Sensitivity analysis: a review of recent advances. Eur. J. Oper. Res. **248**(3), 869–887 (2016)
3. Dalal, M., et al.: Sensitivity analysis of uncertainty in causal environments. Report R1708764, Charles River Analytics Inc, Cambridge, MA, USA (2021)
4. Felli, J.C., Hazen, G.B.: Sensitivity analysis and the expected value of perfect information. Med. Decis. Making **18**(1), 95–109 (1998)

5. Getoor, L., Friedman, N., Koller, D., Pfeffer, A., Taskar, B.: Probabilistic Relational Models. Introduction to Statistical Relational Learning, vol. 8 (2007)
6. Guidotti, R., Monreale, A., Ruggieri, S., Turini, F., Giannotti, F., Pedreschi, D.: A survey of methods for explaining black box models. ACM Comput. Surv. (CSUR) 51(5), 1–42 (2018)
7. Haider, S., Levis, A.H.: Modeling time-varying uncertain situations using dynamic influence nets. Int. J. Approx. Reason. 49(2), 488–502 (2008)
8. Howard, R.A., Matheson, J.E.: Influence diagrams. Decis. Anal. 2(3), 127–143 (2005)
9. Kimmig, A., Bach, S., Broecheler, M., Huang, B., Getoor, L.: A short introduction to probabilistic soft logic. In: Proceedings of the NIPS Workshop on Probabilistic Programming: Foundations and Applications, pp. 1–4 (2012)
10. Levis, A.: Pythia 1.8 User Manu-al, v. 1.03. System Architectures Laboratory, George Mason University, Fairfax, VA (2014)
11. Pearl, J.: Probabilistic reasoning in intelligent systems: networks of plausible inference. Morgan Kaufmann, Burlington (1988)
12. Richardson, M., Domingos, P.: Markov logic networks. Mach. Learn. 62, 107–136 (2006)
13. Saltelli, A.: A short comment on statistical versus mathematical modelling. Nat. Commun. 10(1), 3870 (2019)
14. Solow, R.M.: A contribution to the theory of economic growth. Q. J. Econ. 70(1), 65–94 (1956)
15. Swan, P.L.: TW swan economic growth and capital accumulation. In: Trevor Winchester Swan, Volume II: Contributions to Economic Theory and Policy, pp. 165–197. Springer, Cham (2023)
16. Wagenhals, L.W., Levis, A.H.: Course of action analysis in a cultural landscape using influence nets. In: 2007 IEEE Symposium on Computational Intelligence in Security and Defense Applications, pp. 116–123. IEEE (2007)
17. Zhang, N.L., Poole, D.: A simple approach to Bayesian network computations. In: Proceedings of the Tenth Canadian Conference on Artificial Intelligence (1994)

EPEW 2023

Evaluating the Performance Impact of No-Wait Approach to Resolving Write Conflicts in Databases

Paul Ezhilchelvan[1]([✉]), Isi Mitrani[1], Jim Webber[2], and Yingming Wang[1]

[1] School of Computing, Newcastle University, Newcastle upon Tyne NE4 5TG, UK
{paul.ezhilchelvan,isi.mitrani,Y.Wang303}@ncl.ac.uk
[2] Neo4j UK, Union House, 182-194 Union Street, London SE1 0LH, UK
Jim.Webber@neo4j.com

Abstract. A model of a large distributed database where write conflicts are handled by wait-free aborting of transactions, is introduced and analysed in the steady state. Performance measures, including the rate of aborts, are evaluated. Numerical results are presented, illustrating the behaviour of the system when various parameters are varied. The accuracy of certain approximations is assessed by means of simulations.

1 Introduction

Transactions execute sequences of read and write operations on database objects. These operations must be carried out in a manner that satisfies the criteria of 'atomicity' and 'isolation'. The former requires that all write operations, which are done only provisionally in a volatile store during execution, are either committed (i.e., made permanent) in the database, or aborted (i.e., erased). This is commonly achieved in distributed databases by executing the 'two-phase commit' (2PC) protocol after a transaction completes its execution, whereby all servers accessed by that transaction would be exchanging messages.

Here we focus on the isolation criterion, which stipulates how concurrent transactions should behave when their executions need to access the same data object. Of the numerous isolation levels that have been identified over the years, the most desirable, but also the most expensive to implement, is 'serializable isolation'. It requires that a concurrent execution of potentially conflicting transactions should produce a database state equivalent to some serial execution of those transactions (see Bernstein et al., [5]).

In large-scale distributed databases, a less stringent isolation level called 'Read Committed' (RC) is usually adopted in order to improve performance. Under RC, transactions read only the latest committed value of an object and ignore any provisional value. Consequently, conflicts where one transaction reads an object and another writes it, regardless of the order, do not affect the integrity of the database or the ability of transactions to complete. The only problematic conflicts are 'write-write' (referred to simply as write conflicts), which occur when a transaction wishes to write an object that is already written by another.

M. Iacono et al. (Eds.): EPEW/ASMTA 2023, LNCS 14231, pp. 171–185, 2023.
https://doi.org/10.1007/978-3-031-43185-2_12

Among the possible ways of handling write conflicts are policies which cause the transaction encountering a conflict to wait (a bounded or an unbounded interval) for the other transaction to commit. Such policies reduce the need to abort transactions, but do not eliminate it. The price paid for saving some aborts is poorer performance due to overheads involved in monitoring transactions and avoiding deadlocks in case of unbounded waiting (see Li et al. [10]).

An alternative policy, referred to as 'no-wait', is to abort the transactions encountering a write conflict. It is virtually overhead-free and therefore trades good performance against an increase in the number of aborted transactions.

The contribution of this paper is to quantify the above trade-off by analysing a model of a distributed database subject to write conflicts under a no wait policy. Various performance measures, including the rate of aborts, are evaluated. Such an analysis has not been attempted before. In fact, none of the alternative policies for dealing with write conflicts has been evaluated through probabilistic modelling. We make some approximations in estimating the probabilities of conflicts, but the resulting queueing model is solved exactly.

1.1 Related Literature

A summary and discussion of isolation levels, including serializable and read committed, can be found in Adya et al. [1]. The existing policies for handling write conflicts are described in Bernstein et al. [5], and in Barthels et al. [4].

The topic of improving 2PC efficiency has been the focus of an extensive research, e.g., by batching completed transactions (Waudby et al. [12]), employing multiple coordinators (Gray and Lamport, [8]) or exploiting cloud features (Guo et al. [9]).

More closely related to the present work are papers dealing with different aspects of conflicts in graph databases. Ezhilchelvan et al. [6] highlighted the danger of corrupting large portions of the database when the 'reciprocal consistency' of nodes connected by edges is violated. Protocols for achieving reciprocal consistency were proposed and evaluated in Ezhilchelvan et al. [7] and Waudby et al. [13]. The present study is more general, as it is concerned with read committed isolation which subsumes reciprocal consistency in graph databases.

The no wait policy always aborts the transaction that observes the write conflict. We generalise it a little, by allowing a random choice of which of the two conflicting transactions is to abort.

Our conflict model is described in Sect. 2, while Sect. 3 and the Appendix contain the queueing model and its solution. Section 4 presents a number of numerical and simulation experiments.

2 Conflict Model

The database contains a relatively small subset, S, of objects that are quite frequently accessed. The remaining subset is much larger and each particular object in it is accessed rarely. Write operations whose targets are in S are referred to as being 'popular'; there is a risk of a conflict between two such writes. Those aimed outside S are 'safe', in the sense that their likelihood of conflict is negligible. Only the popular writes are relevant for the purposes of examining the effects of conflicts; the read operations and the safe writes can be ignored.

Let s be the size of S. Assume that each of its objects is equally likely to be the target of a write operation, so that any given address is accessed with probability $1/s$.

The number of writes in a transaction, W, is a generally distributed random variable with probabilities $v_i = P(W = i)$; $i = 0, 1, \ldots$. That number tends to be quite small on the average. Assuming that each write is popular with probability α, independently of the others, we can compute the probabilities, w_k, that a transaction performs k popular writes:

$$w_k = \sum_{i=k}^{\infty} v_i \binom{i}{k} \alpha^k (1 - \alpha)^{i-k} \; ; \; k = 0, 1, \ldots, s \, . \tag{1}$$

Consider two transactions running in parallel, T_1 and T_2, and suppose that the numbers of popular writes they perform are j and k, respectively. Denote by $\beta_{j,k}$ the probability that at least one conflict will occur, i.e. at least one of T_1's popular writes has the same target in S as one of T_2's popular writes. The complementary probability, $1 - \beta_{j,k}$, that all j of T_1's popular writes have different targets from T_2's popular writes, is given by

$$1 - \beta_{j,k} = \frac{s-k}{s} \cdot \frac{s-k-1}{s-1} \cdots \frac{s-k-j+1}{s-j+1} = \binom{s-k}{j} \binom{s}{j}^{-1} \, . \tag{2}$$

The last expression in the right-hand side represents the ratio between the number of ways j objects can be chosen out of $s - k$ objects, and the number of ways j objects can be chosen out of s objects. That ratio is 0 if $s < j + k$, and 1 if either $j = 0$ or $k = 0$.

Rather than tracking the individual write operations of each transaction as they unfold, which would be intractable, we assume that conflicts are detected at moments when a transaction is about to complete. Suppose that T_1 is ready to commit at time t, having issued j popular writes, while another transaction, T_2, is still running. The probability that T_2 will perform k popular writes is w_k, but perhaps not all of them will have been carried out before time t. We propose two estimates for the probability, β_j, that at least one of T_1's writes collides with a write in T_2, given j. One approach is to assume that all of T_2's writes occur before t. Then we could write

$$\beta_j = \sum_{k=0}^{\infty} w_k \beta_{j,k} \, . \tag{3}$$

This estimate is pessimistic and is likely to over-estimate the rate of collisions.

Another reasonable assumption is that, for a given k, the $k + 1$ possible partitions, where i of the k writes occur before t and $k - i$ writes occurs after

t, $i = 0, 1, \ldots, k$, are equally likely. In other words, the k writes are uniformly spaced within the run time of T_2, and the completion instant of T_1 is a random observation point of that run time. The probability β_j would then be

$$\beta_j = \sum_{k=0}^{\infty} w_k \frac{1}{k+1} \sum_{i=0}^{k} \beta_{j,i} . \tag{4}$$

Whether this expression is an under-estimate or an over-estimate depends on the actual spread of writes during the processing of T_2.

Now consider n transactions running in parallel and one of them, T_1, is about to complete, having performed j popular writes. It embarks on a sequence of $n-1$ conflict detection tests, $c_1, c_2, \ldots, c_{n-1}$, against each of the other transactions in turn. If a conflict is detected, then the other transaction is aborted with probability γ, and T_1 is aborted with probability $1 - \gamma$. In the latter case the sequence terminates. Denote by $q_{i,k}(j)$ the probability that the first $k-1$ tests result in the abort of $i-1$ transactions *other than* T_1, and T_1 is aborted at test c_k, given j (n is not relevant to this probability). Let also $q_{i,0}(n,j)$ be the probability that T_1 survives all the tests and completes successfully, but i other transactions are aborted, given n and j. These probabilities can be expressed as follows.

$$q_{i,k}(j) = \binom{k-1}{i-1} \beta_j^i (1-\beta_j)^{k-i} \gamma^{i-1} (1-\gamma) ; \ i = 1, 2, \ldots, k , \tag{5}$$

where β_j is given by either (3) or (4). Similarly,

$$q_{i,0}(n,j) = \binom{n-1}{i} \beta_j^i (1-\beta_j)^{n-1-i} \gamma^i ; \ i = 0, 1, \ldots, n-1 . \tag{6}$$

Note. If $\gamma = 0$, the sequence of tests terminates with the abort of T_1 as soon as a conflict is detected. In general, the larger the value of γ, the more tests tend to be carried out and hence more aborts tend to take place, for a given n. One might therefore conclude that a policy that fixes $\gamma = 0$ would reduce the overall average number of aborts. We shall see in Sect. 4 that this is not necessarily the case.

The probability, $q_i(n,j)$, that at the end of the sequence of tests, there will be a total of i aborted transactions, given n and j, is obtained as

$$q_i(n,j) = q_{i,0}(n,j) + \sum_{k=i}^{n-1} q_{i,k}(j) ; \ i = 0, 1, \ldots, n-1 , \tag{7}$$

where the sum in the right-hand side is 0 by definition when $i = 0$, since $q_{i,k}(j)$ is defined only when $i > 0$.

A slightly different expression, bearing in mind that transaction T_1 will depart regardless of whether it is aborted or not, yields the probability, $r_i(n,j)$, that at the end of the sequence of tests, a total of i transactions will leave the system, given n and j:

$$r_i(n,j) = q_{i-1,0}(n,j) + \sum_{k=i}^{n-1} q_{i,k}(j) \; ; \; i = 1, 2, \ldots, n \; . \tag{8}$$

Note that the sum in the right-hand side is empty when $i = n$. The only way that n departures will occur is for T_1 to abort all other transactions and then complete successfully.

Finally, we can remove the conditioning on j, to obtain the distributions of the number of aborts, $q_i(n)$, and the number of departures, $r_i(n)$, given that there are n transactions running in parallel when one of them is about to complete.

$$q_i(n) = \sum_{j=0}^{\infty} q_i(n,j) w_j \; ; \; i = 0, 1, \ldots, n-1 \; , \tag{9}$$

where w_j is given by (1). Similarly,

$$r_i(n) = \sum_{j=0}^{\infty} r_i(n,j) w_j \; ; \; i = 1, 2, \ldots, n \; . \tag{10}$$

These probabilities will be used in the analysis of the queueing process.

3 Queueing Model

Transactions arrive into the system in a Poisson stream with rate λ. Their required processing times are i.i.d. random variables distributed exponentially with parameter μ. The system hardware and scheduling policy are modelled by assuming that service is provided by N independent servers, each processing one transaction at a time. These will be referred to as the 'notional servers'. Typically, N is some multiple of the total number of hardware cores available.

When the number of transactions present, n, is less than N, they are all processed in parallel and the completion rate is $n\mu$. When $n \geq N$, N transactions are being processed and the rest wait in a FIFO queue. The completion rate is then $N\mu$. Not all completions are successful; a transaction about to complete may in fact be aborted. The actual number of departures, i, that take place at a completion instant depends on n and on the number of conflicts that are detected at that point, as discussed in the previous section.

If the set of popular objects is small, the outcome at one completion instant may influence the targets of subsequent writes. In practice however, that set is quite large, typically containing thousands of objects. It is therefore reasonable to assume that, after a completion instant, the surviving transactions behave as if the objects they access in the future are independent of those accessed in the past.

The above assumptions imply that the system state, defined as the number of transactions currently present, evolves as a Markov process whose instantaneous transitions are from state n to state $n + 1$ with rate λ, and from state n to state $n - i$ with a state-dependent rate $\xi_{n,i}$, given by

$$\xi_{n,i} = \begin{cases} n\mu r_i(n) \,; \ i = 1, 2, \ldots, n & \text{if } n < N \\ N\mu r_i(N) \,; \ i = 1, 2, \ldots, N & \text{if } n \geq N \end{cases}, \tag{11}$$

where $r_i(n)$ is computed according to (10) and the preceding expressions. Note that the sum of $\xi_{n,i}$ over all i is equal to $n\mu$ when $n < N$, and to $N\mu$ when $n \geq N$.

From now on, when dealing with the state-independent case $n \geq N$, we shall drop the index n and denote the transition rate from n to $n - i$ by ξ_i, $i = 1, 2, \ldots, N$.

An important quantity in this context is the 'downward drift' when the queue is large, i.e. the average number of departures per unit time when $n \geq N$. The necessary and sufficient condition for stability of the queue is that the arrival rate should be strictly less than that average:

$$\lambda < \sum_{i=1}^{N} i\xi_i = N\mu \sum_{i=1}^{N} i r_i(N) \,. \tag{12}$$

That condition follows from Pake's lemma (see [11]).

Denote by π_n the steady-state probability that there are n transactions present, $n = 0, 1, \ldots$. These probabilities satisfy the following state-dependent balance equations.

$$(\lambda + n\mu)\pi_n = \lambda\pi_{n-1} + \sum_{i=1}^{N} \xi_{n+i,i}\pi_{n+i} \,; \ n \leq N \,, \tag{13}$$

where $\pi_{-1} = 0$ by definition and $\xi_{n+i,i}$ becomes ξ_i when $n + i \geq N$.

When $n > N$, the coefficients cease to depend on n and the balance equations become

$$(\lambda + N\mu)\pi_n = \lambda\pi_{n-1} + \sum_{i=1}^{N} \xi_i\pi_{n+i} \,; \ n > N \,. \tag{14}$$

Queueing systems where several customers depart simultaneously have appeared in the literature in connection with different, unrelated applications. When service is provided by a single server, it has been observed on several occasions that the queue size distribution is geometric (e.g., see [2,3]). Here we have a multi-server system, so we do not expect to find the same result. However, we shall look for a distribution with a geometric tail. More precisely, we shall look for a solution to the state-independent Eq. (14) in the form

$$\pi_{N+k} = \pi_N z_0^k \,; \ k = 0, 1, \ldots \,, \tag{15}$$

where z_0 is some real number in the interval (0,1).

Substituting (15) into (14), with $n = N + k + 1$, and dividing both sides by $\pi_N z_0^{N+k}$, the equations become

$$(\lambda + N\mu)z_0 = \lambda + \sum_{i=1}^{N} \xi_i z_0^{i+1} . \tag{16}$$

Hence, all balance Eq. (14) would be satisfied by (15), if z_0 is a root of the polynomial

$$P(z) = \lambda - (\lambda + N\mu)z + \sum_{i=1}^{N} \xi_i z^{i+1} . \tag{17}$$

The existence and uniqueness of the solution is established by the following result.

Lemma. *When the inequality (12) is satisfied, the polynomial $P(z)$ has exactly one root in the interior of the unit disc. That root, z_0, is real and positive.*

The proof of this Lemma is in the Appendix.

According to (15), the sum of the steady-state probabilities π_n for $n \geq N$ is given by

$$\sum_{n=N}^{\infty} \pi_n = \frac{\pi_N}{1 - z_0} . \tag{18}$$

We can therefore write the normalizing condition (all probabilities sum up to 1) in the form

$$\sum_{n=0}^{N-1} \pi_n + \frac{\pi_N}{1 - z_0} = 1 . \tag{19}$$

Having determined z_0, the probabilities π_n for $n \leq N$ can be computed by using the balance Eq. (13) as backward recurrences. Start by setting π_N to an arbitrary positive value, e.g. $\pi_N = 1$. Then (13), for $n = N$, yields

$$\pi_{N-1} = \frac{1}{\lambda} \left[(\lambda + N\mu)\pi_N - \sum_{i=1}^{N} \xi_i \pi_N z_0^i \right] . \tag{20}$$

At step 2, π_{N-2} is obtained from (13) for $n = N - 1$:

$$\pi_{N-2} = \frac{1}{\lambda} \left[(\lambda + (N-1)\mu)\pi_{N-1} - \sum_{i=1}^{N} \xi_i \pi_N z_0^{i-1} \right] . \tag{21}$$

The state-dependent transition rates $\xi_{n,i}$ start appearing from step 3:

$$\pi_{N-3} = \frac{1}{\lambda} \left[(\lambda + (N-2)\mu)\pi_{N-2} - \xi_{n-1,1}\pi_{N-1} - \sum_{i=2}^{N} \xi_i \pi_N z_0^{i-2} \right] . \tag{22}$$

After N backward steps, the probabilities $\pi_0, \pi_1, \ldots, \pi_{N-1}$ are determined in terms of π_N. All balance equations are satisfied, but not the normalizing equation. The final step is to normalize all computed probabilities, by dividing them by the left-hand side of (19). This completes the solution algorithm.

The performance measures of interest are

1. The steady-state average number of transactions present, L.
2. The average number of aborted transactions per unit time, A, or the percentage of incoming transactions that are aborted, $a = 100A/\lambda$.
3. The useful throughput, $\lambda - A$.
4. The average response time of a successful transaction (i.e. one that commits, rather than being aborted), $\tilde{W}$.

The first three of these metrics are obtained from

$$L = \sum_{n=1}^{N-1} n\pi_n + \frac{\pi_N}{1-z_0}\left[N + \frac{z_0}{(1-z_0)}\right] , \tag{23}$$

and

$$A = \sum_{n=1}^{N-1} n\mu\pi_n \sum_{i=1}^{n} i q_i(n) + \frac{N\mu\pi_N}{1-z_0}\sum_{i=1}^{N} i q_i(N) , \tag{24}$$

where $q_i(n)$ is given by (9).

To evaluate the average response time of a committed transaction, $\tilde{W}$, note that both committed and aborted transactions spend the same average period waiting in the queue, W_Q, By Little's result, $W_Q = L_Q/\lambda$, where L_Q is the average number of waiting transactions. However, the active periods of committed and aborted transactions are different. Moreover, the conditional distribution of a transaction processing time, given that it is not aborted, is not the same as its unconditional distribution. We have been unable to determine the conditional average processing time directly and therefore adopt an indirect approach.

Consider a state with n active transactions. Let $a(n)$ be the average number of aborts that would occur if one of those n transactions completes. Similarly, let $d(n)$ be the total average number of departures that would occur if one of them completes. Those averages are obtained from Eqs. (9) and (10). One can then argue that, on the average, a fraction $a(n)/d(n)$ of the n transactions will be aborted and a fraction $1 - a(n)/d(n)$ will commit. Hence, the average number of committing transactions in state n can be approximated as $c(n) = n[1 - a(n)/d(n)]$. We can then write an expression for the average number, L_C, of *active* transactions that will commit:

$$L_C = \sum_{n=1}^{N-1} \pi_n c(n) + \pi_N c(N)\frac{1}{1-z_0} . \tag{25}$$

Bearing in mind that committing transactions arrive at rate $\lambda - A$, applying Little's result to the set of active committing transactions, and adding the average waiting time $W_Q = L_Q/\lambda$, we arrive at the following expression for $\tilde{W}$:

$$\tilde{W} = \frac{L_C}{\lambda - A} + \frac{\pi_N z_0}{\lambda(1-z_0)^2} . \tag{26}$$

The behaviour of these performance measures for different parameter settings is examined empirically in the next section.

4 Numerical and Simulation Experiments

We have applied the analytical results that have been obtained to some reasonably realistic examples of distributed databases. The objectives of the experiments are to quantify the effects of conflicts on different aspects of system performance, and at the same time to assess the accuracy of the approximations that have been made in the analysis. Of particular concern is the assumption that conflicts are detected at completion instants, rather than dynamically as soon as a write operation is performed.

The base system contains $N = 640$ notional servers, representing a typical distributed database cluster. Each transaction performs 0, 1, 2 or 3 write operations, with probabilities fixed at $v_0 = 0.12$, $v_1 = 0.45$, $v_2 = 0.3$, $v_3 = 0.13$, representing typical OLTP activity. The size of the popular subset is $s = 10^4$. The parameters that are varied are λ (increasing the offered load), α (producing different distributions w_k) and γ.

To carry out the accuracy assessment, a system with dynamic detection of conflicts was simulated. A simulation run generates a Poisson stream of transactions, and for each incoming transaction, the probabilities v_k are used in order to generate a random number of writes; then the fraction α determines which, if any of those writes are popular. Each popular write is associated with one of the objects in the popular subset, chosen with equal probability. Within a transaction those objects are distinct, but two or more active transactions may write the same object.

To simulate the processing of a transaction, an exponentially distributed interval (with mean 1, without loss of generality) is generated and inserted before the first write, then another one is inserted between the first and the second write, ..., and one after the last write. Thus the distribution of the simulated required processing times in the examples is a random mixture of Erlang distributions, with an average given by

$$\frac{1}{\mu} = 1 + \sum_{j=0}^{3} jv_j \ . \tag{27}$$

This mechanism was suggested by the practical observation that transactions performing more writes tend to have longer processing times. At each write event, the currently written object is compared with those already written by other active transactions and if a conflict is detected, one of the two transactions involved is aborted according to the probability γ.

Performance measures obtained from the simulation are compared with those provided by the model, where the processing times are assumed to be exponentially distributed with parameter μ given by (27). Both proposed expressions for β_j, (3) and (4), are used in the evaluations.

In the first example, the steady-state average percentage of aborted transactions, $a = 100A/\lambda$, is plotted against the arrival rate. All writes are assumed to be popular (i.e. $\alpha = 1$), so that the distribution of popular writes is $(w_0, w_1, w_2, w_3) = (0.12, 0.45, 0.30, 0.13)$. This implies that a transaction performs about 1.5 popular writes, and its required processing time is about 2.4 on the average. For the range of λ values chosen, the system varies from moderately to heavily loaded. In fact, when $\lambda = 270$, the offered load exceeds 640, so without the aborts the system would be saturated.

The top and bottom plots in Fig. 1 correspond to values of a computed using expressions (3) and (4), respectively, while the middle plot shows the simulated values.

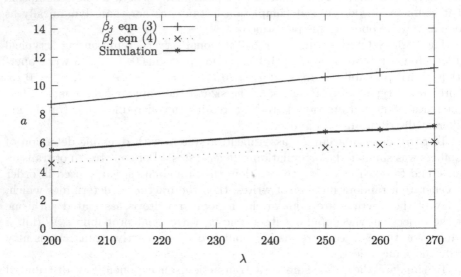

Fig. 1. Percentage of aborted transactions, $a = 100A/\lambda$; $\alpha = 1$, $\gamma = 0.5$

As expected, Eq. (3) leads to an over-estimate of the fraction of aborts because it includes more writes in the collision detection process. Equation (4) does not have that defect and so its approximation is better. What is less obvious is why that approximation consistently under-estimates a. A likely explanation is that making decisions at completion instants, rather than at write instants, tends to lengthen the intervals between aborts, thus decreasing the rate at which they occur.

If λ is increased further, so that the system approaches saturation, all three plots would be observed to flatten. The abort rate depends on the average number of active transactions, not the total number present. That average almost ceases to increase as it approaches N from below.

It should be pointed out that the model computations are vastly faster than the simulations. Each simulated point in Fig. 1 (and in the subsequent figures) is the result of a run where half a million transactions commit. Such a run takes about two hours of computer time. By contrast, both model plots together take less than a minute to compute.

For the same set-up and parameters, Fig. 2 shows two computed and one simulated plots of the average number of jobs in the system, L, as function of the arrival rate.

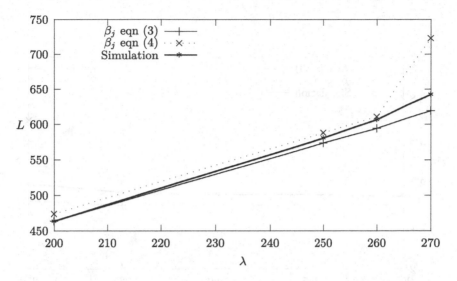

Fig. 2. Average number of transactions present, L; $\alpha = 1$, $\gamma = 0.5$

There is a close agreement between the three for most of the range. Only when the system is very heavily loaded do the estimates begin to diverge. This time Eq. (3) under-estimates the performance measure, and (4) over-estimates it. This is not surprising, because aborts actually decrease the load by making transactions depart faster. Thus, over-estimating the abort rate is associated with under-estimating the number present, and vice versa.

The average response time of committed transactions, $\tilde{W}$, was evaluated according to expression (26). The three plots in Fig. 3 follow the pattern of Fig. 2. The two estimates are quite close to the simulation for most of the range, with (3) under-estimating the response time and (4) over-estimating it.

A notable feature of Fig. 3 is that the observed values of $\tilde{W}$ are, for much of the range, lower than the average required processing time, which is about 2.4 for these parameters.

·This is a clear illustration of the fact that the conditional average processing time, given that the transaction is not aborted, is shorter than the unconditional one. The approximation that over-estimates the aborts, (3), also under-estimates the average response time.

In the next experiment, the percentage of aborted transaction, a, is evaluated and simulated as a function of the fraction of popular writes, α. The distribution of writes per transaction is as before, and the arrival rate is kept fixed at $\lambda = 250$. The results are plotted in Fig. 4.

Since the likelihood of conflicts increases with α, we expect the percentage of aborted transactions to do so. Indeed, that is what is observed. As in Fig. 1, the under-estimate computed with expression (4) is a better approximation to the simulated values of a than the over-estimate provided by (3).

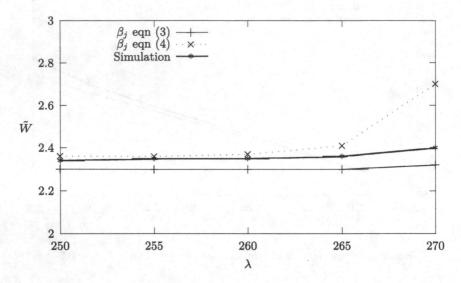

Fig. 3. Average response time of committed transactions, $\tilde{W}$; $\alpha = 1$, $\gamma = 0.5$

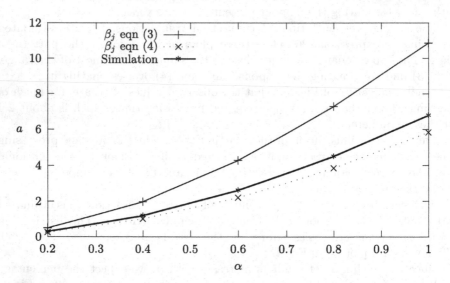

Fig. 4. Percentage of aborted transactions as function of α $\lambda = 250$, $\gamma = 0.5$

The last experiment examines the effect of the parameter γ, i.e. the probability that when a conflict occurs, the detected transaction is aborted, rather than the detecting one. The other parameters are as before, with fixed $\lambda = 250$ and $\alpha = 0.5$.

In the simulated system, conflicts are detected, and acted upon, at write instants. Hence, only one other transaction is ever involved in a conflict. Since these events occur at random, it should not matter which of the two transactions is aborted. On the other hand, the model assumes that detections take place at completion instants, which means that the completing transaction, while surviving conflicts, may cause more than one other transaction to be aborted. In those circumstances, it may be supposed that the percentage of aborts would increase with γ.

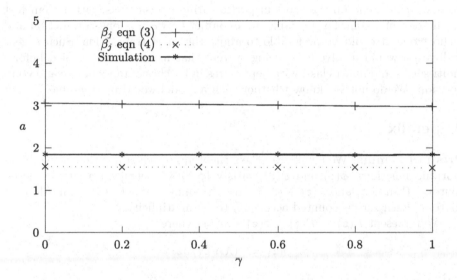

Fig. 5. Percentage of aborted transactions as a function of γ; $\lambda = 250$, $\alpha = 0.5$

In Fig. 5, the computed and simulated values of a are plotted against γ. While the simulated plot confirms the intuition that the performance measure does not depend on γ, both computed plots show a very slight, but noticeable decrease in a. That seems to be counter-intuitive.

In fact, there are two trends that act in opposite directions. For a given number of transactions present, n, the average number of aborts increases with the probability γ. Consequently, the downward transition rates, from n to $n - i$, increase with γ, which in turn increases the probabilities of states with lower n. In those states, the probability of detecting a conflict is lower and the aborts are fewer. It appears that, after averaging over all states, the tendency towards fewer aborts slightly dominates the tendency towards more aborts.

5   Conclusions

The analysis presented here enables us to compute performance estimates for systems where write conflicts lead to aborts of transactions. Those computations are several orders of magnitude faster than simulations. Expression (3) for the conditional probability of conflict is pessimistic and can be expected to yield an upper bound for the rate of aborts, regardless of the spread of write operations during the processing of transactions. The second expression, (4), has been observed to provide a lower bound when the writes are spread uniformly during the processing. Moreover, the distance between the upper and lower bounds is quite small over a wide range of parameter values.

It would be desirable to extend the model by including different types of transactions, with different characteristics. Then one could handle abort decisions which depend on the types of participating transactions, rather than just on the toss of a coin. For example, in a conflict between a long transaction and a short one, it could be preferable to abort the short transaction. Such a generalization would involve a queueing system that has not been studied before: multi-server and multi-class, with departures in batches of random size and composition. We do not yet know whether such a model would be tractable.

Appendix

Proof of Lemma. We invoke Rouché's theorem, which states that if two holomorphic functions, $\phi(z)$ and $\psi(z)$, satisfy $|\phi(z)| > |\psi(z)|$ on a simple closed contour, then $\phi(z)$ and $\phi(z) + \psi(z)$ have the same number of zeros inside that contour. Each zero is counted according to its multiplicity.

We represent $P(z)$ as $P(z) = \phi(z) + \psi(z)$, where

$$\phi(z) = -(\lambda + N\mu)z$$

and

$$\psi(z) = \lambda + \sum_{i=1}^{N} \xi_i z^{i+1}$$

The closed contour is the unit circle. When $|z| = 1$, $|\phi(z)| = \lambda + N\mu$. Applying the triangle inequality to $\psi(z)$, we find

$$|\psi(z)| \leq \lambda + \sum_{i=1}^{N} \xi_i = \lambda + N\mu \ .$$

Moreover, the inequality is strict everywhere on the contour, except at $z = 1$, where it is an equality.

Note that the derivative of $P(z)$ is positive at $z = 1$. This is because

$$P'(1) = -\lambda + \sum_{i=1}^{N} i\xi_i > 0 \ ,$$

according to (12). Hence, we can choose a sufficiently small number, ϵ, such that $P(1-\epsilon) < 0$. Modifying the contour slightly in the vicinity of $z = 1$, by making it pass through the point $z = 1-\epsilon$, would ensure that the inequality $|\phi(z)| > |\psi(z)|$ is strict on the entire modified contour.

The function $\phi(z)$ is linear and has a single zero, $z = 0$, inside the contour. Therefore, by Rouché's theorem, $P(z)$ also has a single zero inside the contour. That zero, z_0, must be in the interval $(0, 1 - \epsilon)$, because $P(0) = \lambda > 0$ and $P(1 - \epsilon) < 0$. This completes the proof.

References

1. Adya, A., Liskov, B., O'Neil, P.: Generalized isolation level definitions. In: Proceedings of 16th International Conference on Data Engineering, pp. 67–78 (2000)
2. Bailey, N.T.J.: On queueing processes with bulk service. J. Roy. Stat. Soc. B **16**(1), 80–87 (1954)
3. Balsamo, S., Malakhov, I., Marin, A., Mitrani, I.: Transaction confirmation in proof-of-work blockchains: auctions, delays and droppings. In: 20th Mediterranean Communication and Computer Networking Conference (MedComNet), pp. 140–149 (2022)
4. Barthels, C., Müller, I., Taranov, K., Alonso, G., Hoefler, T.: Strong consistency is not hard to get: two-phase locking and two-phase commit on thousands of cores. Proc. VLDB Endowment **12**(13), 2325–2338 (2019)
5. Bernstein, P.A., et al.: Concurrency Control and Recovery in Database Systems (1987)
6. Ezhilchelvan, P., Mitrani, I., Webber, J.: On the degradation of distributed graph databases with eventual consistency. Queueing Models Serv. Manag. **3**(2), 235–253 (2020)
7. Ezhilchelvan, P., Mitrani, I., Waudby, J., Webber, J.: Design and evaluation of an edge concurrency control protocol for distributed graph databases. In: Gribaudo, M., Iacono, M., Phung-Duc, T., Razumchik, R. (eds.) EPEW 2019. LNCS, vol. 12039, pp. 50–64. Springer, Cham (2020). https://doi.org/10.1007/978-3-030-44411-2_4
8. Gray, J., Lamport, L.: Consensus on transaction commit. ACM Trans. Database Syst. (TODS) **31**(1), 133–160 (2006)
9. Guo, Z., et al.: Cornus: atomic commit for a cloud DBMS with storage disaggregation. Proc. VLDB Endowment **16**(2), 379–392 (2022)
10. Li, C., et al.: ByteGraph: a high-performance distributed graph database in ByteDance. Proc. VLDB Endowment **15**(12), 3306–3318 (2022)
11. Pakes, A.G.: Some conditions for ergodicity and recurrence of Markov chains. Oper. Res. **17**(6), 1058–1061 (1969)
12. Waudby, J., Ezhilchelvan, P., Mitrani, I., Webber, J.: A performance study of epoch-based commit protocols in distributed OLTP databases. In: 41st International Symposium on Reliable Distributed Systems (SRDS) (2022)
13. Waudby, J., Ezhilchelvan, P., Webber, J., Mitrani, I.: Preserving reciprocal consistency in distributed graph databases. In: 7th Workshop on Principles and Practice of Consistency for Distributed Data, pp. 1–7 (2020)

Stochastic Matching Model
with Returning Items

A. Busic[1] and J. M. Fourneau[2(✉)]

[1] INRIA and DIENS, École Normale Supérieure,
PSL University, CNRS, Paris, France
`ana.busic@inria.fr`
[2] DAVID, Univ. Paris-Saclay, UVSQ, Versailles, France
`Jean-Michel.Fourneau@uvsq.fr`

Abstract. We consider a stochastic matching model with general compatibility graph and returning items. The return time is modeled by the sojourn time in a Jackson network. After being matched at their arrival, the items move to a Jackson network where they spend some time before eventually returning to the matching system. This general setting includes phase-type return times as a special case. We prove that under some assumptions about the routing and service in the Jackson network, the steady-state distribution of the Markov chain has a product form solution.

1 Introduction

A matching model describes items waiting in a system until they move upon arrival of a compatible item. Following [1] a Matching model is a triple (G, Φ, μ) formed by

1. an undirected graph $G = (V, E)$ whose vertices in V are classes of items and whose edges in E model the allowed matching of items. G is called the matching graph.
2. Φ is a matching policy. When an arriving item matches several types of items already waiting, it gives the couple of items which is selected.
3. a distribution of probability μ to model the arrivals of items. Alternatively one can consider a collection of Poisson processes for a continuous-time model.

The matching graph represents the classes of items and the compatibility among classes of items. Upon arrival, an item is queued if there are not compatible items present in the system. A matching occurs when two compatible items are present and it is performed according to the matching discipline. Typical matching disciplines are First Come First Match (FCFM), Random, or Match the Longest Queue. Once they are matched, in the initial general matching model, both items leave the matching system immediately. Here, we study systems where some of the items, after they have been matched, enter a network of queues to receive ser-

© The Author(s), under exclusive license to Springer Nature Switzerland AG 2023
M. Iacono et al. (Eds.): EPEW/ASMTA 2023, LNCS 14231, pp. 186–200, 2023.
https://doi.org/10.1007/978-3-031-43185-2_13

vice. Our model is an hybridization of the general matching model and Jackson network of queues and such an hybridization was, to the best of our knowledge, not considered in the literature.

Assuming independent Poisson arrivals of items, and FCFM discipline the general matching model is associated with a Markov chain with an infinite state space. Under these assumptions, a necessary and sufficient condition for stability and a product form solution were proved in [2] and [1]. We also recently established that there exists some performance paradox for FCFM matching models [3]. When one add new edges in the matching graph, one may expect that the expectation of the total number of customers decreases. We have given some examples which show that it is not always true and we prove a sufficient condition for such a performance paradox to exist. Thus adding flexibility on the matching does not always result in a performance improvement.

The general matching model proposed in [2] and [1] was considering a general undirected matching graph G and it is assumed that the arrivals of items occur one at a time. It is important to avoid the confusion with Bipartite Matching Model (see for instance [4] and references therein) where the matching graph is bipartite and two items of distinct classes arrive at the same time. Bipartite Matching Models were motivated by analysis of the public housing [5] and the kidney exchanges [6,7]. The kidney exchange arises when a healthy person who wishes to donate a kidney is not compatible (blood types or tissue types) with the receiver. Two incompatible pairs (or maybe more) can form a cyclic exchange, so that each patient can receive a kidney from a compatible donor (see [8] and reference therein for a presentation of the problem and the modeling and algorithmic issues).

Here, we investigate how we can hybrid matching models with Jackson networks of queues. The main idea was to send the items leaving the matching system into a Jackson network where they can spend some time in service before eventually coming back to the matching system. Here, we restrict ourselves to the case where the matching graph is complete graph or a graph which contains a star.

The technical part of the paper is as follows. We begin in the next section with the notations. Section 3 is devoted to the description of the Stop Protocol. In Sect. 4, we prove the existence of the steady-state distribution and checked that a multiplicative solution holds when the matching graph is a complete graph. Section 5 is devoted to a generalization with several items and several Jackson networks connected to a matching system. Finally in Sect. 6, we study a more complex matching graph (i.e. a connected graph with a star) associated with the Stop protocol and we also prove under some technical assumptions that a product form stationary distribution exists.

2   Notation and Assumptions

Let $G = (V, E)$ be the matching graph. Nodes in V are also denoted as letters. An ordered list of letters is called a word. Assume that x is a letter from V,

$\Gamma(x)$ is the set of neighbors of x in G. The cardinality of V is N: i.e. we have N distinct types of item. Items of type 1 have a particular behavior when we deal with the hybridization with the Jackson network. In the matching model, they behave like the other items.

We consider a continuous time model for the matching model. We consider the First Come First Match policy or FCFM. An arriving letter will be added at the end of the word if it does not match any letter in the current word. If the arriving letter matches one or several letters of the current word, both the oldest matching letter and the arriving letter move immediately. We assume Poisson arrivals for the items and λ_i will denote the arrival rate of type i item in the matching model.

After matching of letter of type 1 by an arriving letter, the letter 1 joins a Jackson network as a usual customer. It joins queue j with probability m_j. We restrict ourselves to movement of type 1 customers. Furthermore the customers which join the network are the type 1 items which were waiting in the matching system and which are matched by an arriving item. Item of type 1 which arrives in the matching system and matches a waiting item are not sent to the Jackson network. We assume that there are no arrivals of fresh customers from the outside to the Jackson network. In the Jackson network, the service durations follow exponential distributions with rate μ_i. The routing matrix inside Jackson network is denoted as matrix R. At queue i there is a departure probability d_i to leave for the outside and a return probability r_i to return to the matching system as type 1 item (see the left part of Fig. 1). Of course, for all queue index i, we have:

$$r_i + d_i + \sum_{j=1}^{m} R(i,j) = 1 \qquad (1)$$

Definition 1 *(Independent Set). Let $G = (V,E)$ be a graph, an independent set of G is a set of nodes IS such that $IS \subset V$ and for all $i,j \in IS$, $(i,j) \notin E$.*

As usual with FCFM discipline, the state of the matching system at time t is a word, $\{W_t\}$, with $t \in \mathbb{R}+$. By construction, all the states of the matching system are independent sets of the matching graph. Indeed, if two letters are neighbors in the matching graph, they cannot be both waiting at the same time in the matching system. In this paper, we assume that the matching graph is a complete graph with p nodes: K_p. This implies a strong simplification of the state space as stated in the following.

Lemma 1. *The state space of a matching system based on graph K_N and FCFM discipline is a collection of N bidirectional paths connected by the empty state (see the right part of Fig. 1). Therefore the states $\{W_t\}$ can be replaces by a vector $\vec{X} = (X_1, .., X_N)$ where X_i is the number of type i items and such that $X_i \times X_j = 0$ for all $i \neq j$.*

Similarly we denote by $\vec{Y} = (Y_1, ..., Y_m)$ the state of the Jackson network (with M queues). As usual Y_j is the number of customers at queue j. Under these assumptions, the process $(\vec{X}, \vec{Y})_t$ is a Continuous Time Markov chain.

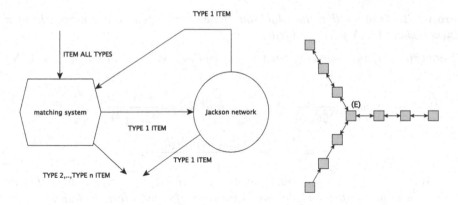

TYPE 1 ITEM

ITEM ALL TYPES

matching system

Jackson network

TYPE 1 ITEM

(E)

TYPE 1 ITEM

TYPE 2,..,TYPE n ITEM

Fig. 1. Hybridization first model (left), Markov chain of K3 matching graph (right).

3 The Stop Protocol

We call $J1$ the Jackson network where the items of type 1 arrive after they have been matched. We describe the Stop Protocol for Jackson network $J1$ as follows:

- When the matching system contains items which are not type 1 items, the servers stop their activity. The services are resumed when all the customers of type 2 to N have leaved the matching system.
- As there is no arrivals from the outside and only type 1 items are entering the Jackson network from the matching system, there are no arrivals in the Jackson network when the matching system does not contain type 1 items.

Thus when the matching system contains items of type 2 to N, Jackson network $J1$ does not evolve. Please remark that when the matching system is empty, the servers are still in operation in Jackson network $J1$.

Definition 2. *A Jackson network is open and connected if:*

- *All queues may receive customer from the matching system through the routing. More precisely, for all queue i, there exist a queue j such that $m_j > 0$ and there exists a directed path from j to i.*
- *From all queues, a customer can leave the network after some routing, either to join the matching system or to reach the outside: i.e. for all queue i, there exist a queue j such that $r_j + d_j > 0$ and there exists a directed path from i to j.*

4 Product Form for the Steady State Distribution

We assume that the Jackson network operates under the Stop Protocol we have just defined. We first state the irreducibility of the Markov chain. The proof is in an appendix in the full paper.

Lemma 2. *Assume that the Jackson network is open and connected, then Markov chain* $(\vec{X}, \vec{Y})_t$ *is irreducible.*

Theorem 1. *If the following fixed point equation has a solution* ω_i *($i = 1..N$) and* ρ_j *($j = 1..M$),*

$$\omega_i = \frac{\lambda_i}{\sum_{k \neq i} \lambda_k}, \forall i > 1, \quad and \quad \omega_1 = \frac{\Lambda_1}{\sum_{k \neq 1} \lambda_k},$$

$$\rho_j = \frac{\Lambda_1 m_j + \sum_{k=1}^m \mu_k \rho_k R(k, j)}{\mu_j},$$

with $\Lambda_1 = \lambda_1 + \sum_j \mu_j \rho_j r_j$, *such that* $\rho_j < 1$ *and* $\omega_i < 1$, *then the following product form probability distribution is invariant for the Markov chain:*

$$\pi(\vec{X}, \vec{Y}) = C \prod_{i=1}^n \omega_j^{X_j} \prod_{j=1}^m \rho_j^{Y_j}.$$

Proof. First it is interesting to remark that ω_i is equal to $\frac{\lambda_i}{\sum_{k \neq i} \lambda_k}$ rather than $\frac{\lambda_i}{\Lambda_1 + \sum_{k \neq i, k \neq 1} \lambda_k}$. Indeed the arrival from the Jackson network only occurs when the matching systems does not contain type i item. Therefore we only have to consider arrivals of type 1 items from the outside with rate λ_1.

We write the global balance equations and we check the solution. We have to write 3 balance equations according to the value of $\vec{X}$: the equation when $X_1 > 0$ because of the particular behavior of type 1 item, the equation when $\vec{X} = \vec{0}$, and the balance when $X_i > 0$ for any $i > 1$. Let e_i be a null vector except component i which is 1. Let us begin with the equation when the matching system only contains type 1 items (i.e. $X_1 > 0$).

$$\begin{aligned}
\pi(\vec{X}, \vec{Y})[\sum_i \lambda_i + \sum_j \mu_j 1_{Y_j > 0}] = {} & \pi(\vec{X} - e_1, \vec{Y})\lambda_1 \\
& + \sum_j \pi(\vec{X} - e_1, \vec{Y} + e_j)\mu_j r_j \\
& + \sum_j \pi(\vec{X}, \vec{Y} + e_j)\mu_j d_j \\
& + \sum_j \sum_k \pi(\vec{X}, \vec{Y} + e_j - e_k)\mu_j R(j, k) 1_{Y_k > 0} \\
& + \sum_j \pi(\vec{X} + e_1, \vec{Y} - e_j)m_j \sum_{k \neq 1} \lambda_k 1_{Y_j > 0}
\end{aligned} \quad (2)$$

When the matching system is empty, the balance equation is:

$$\begin{aligned}
\pi(\vec{0}, \vec{Y})[\sum_i \lambda_i + \sum_j \mu_j 1_{Y_j > 0}] = {} & \sum_j \pi(\vec{0}, \vec{Y} + e_j)\mu_j d_j \\
& + \sum_j \sum_k \pi(\vec{0}, \vec{Y} + e_j - e_k)\mu_j R(j, k) 1_{Y_k > 0} \\
& + \sum_j \pi(e_1, \vec{Y} - e_j)m_j \sum_{k \neq 1} \lambda_k 1_{Y_j > 0} \\
& + \sum_{i>1} \pi(e_i, \vec{Y}) \sum_{k \neq i} \lambda_k.
\end{aligned} \quad (3)$$

Now consider the case where $X_1 = 0$ and $X_j > 0$ for some item index j. Due to the Stop Protocol, component $\vec{Y}$ does not evolve and the balance equation is much simpler:

$$\pi(\vec{X}, \vec{Y})[\sum_i \lambda_i] = \pi(\vec{X} - e_j, \vec{Y})\lambda_j + \pi(\vec{X} + e_j, \vec{Y}) \sum_{k \neq j} \lambda_k. \quad (4)$$

Let us begin with Eq. 2. Divide both sides by $\pi(\vec{X}, \vec{Y})$, assume the multiplicative solution proposed in the Theorem and simplify the expressions to get:

$$
\begin{aligned}
\sum_i \lambda_i + \sum_j \mu_j 1_{Y_j>0} &= \frac{\lambda_1}{\omega_1} \\
&+ \sum_j \frac{1}{\omega_1} \rho_j \mu_j r_j \\
&+ \sum_j \rho_j \mu_j d_j \\
&+ \sum_j \sum_k \frac{\rho_j}{\rho_k} \mu_j R(j,k) 1_{Y_k>0} \\
&+ \sum_j \omega_1 \frac{1}{\rho_j} m_j \sum_{k\neq1} \lambda_k 1_{Y_j>0}.
\end{aligned} \tag{5}
$$

After exchanging index j and k in the fourth term of the r.h.s. and factorization we obtain:

$$
\begin{aligned}
\sum_i \lambda_i + \sum_j \mu_j 1_{Y_j>0} &= \frac{1}{\omega_1}[\lambda_1 + \sum_j \rho_j \mu_j r_j] + \sum_j \rho_j \mu_j d_j \\
&+ \sum_j \frac{1}{\rho_j} 1_{Y_j>0}[\omega_1 m_j \sum_{k\neq1} \lambda_k + \sum_k \rho_k \mu_k R(k,j)].
\end{aligned} \tag{6}
$$

After substitution, we get:

$$
\begin{aligned}
\sum_i \lambda_i + \sum_j \mu_j 1_{Y_j>0} &= \frac{\Lambda_1}{\omega_1} + \sum_j \rho_j \mu_j d_j \\
&+ \sum_j \frac{1}{\rho_j} 1_{Y_j>0}[\Lambda_1 m_j + \sum_k \rho_k \mu_k R(k,j)].
\end{aligned} \tag{7}
$$

Let us first consider the equality between the step functions. We must have for all j:

$$
\mu_j = \frac{1}{\rho_j}[\Lambda_1 m_j + \sum_k \rho_k \mu_k R(k,j)],
$$

and this relation clearly holds due to the definition of rate ρ_j. It remains to check that:

$$
\sum_i \lambda_i = \frac{\Lambda_1}{\omega_1} + \sum_j \rho_j \mu_j d_j.
$$

From the definition of ω_1, we get: $\frac{\Lambda_1}{\omega_1} = \sum_{i\neq1} \lambda_i$. Thus, after cancellation of some λ_i on both sides of the equation we finally get:

$$
\lambda_1 = \sum_j \rho_j \mu_j d_j.
$$

To prove this last equality, we consider the flow equation on the queues of the Jackson network, we multiply by μ_j and we sum for all queue index i:

$$
\sum_{j=1}^m \mu_j \rho_j = \sum_{j=1}^m \Lambda_1 m_j + \sum_{j=1}^m \sum_{k=1}^m \mu_k \rho_k R(k,j).
$$

As $\sum_j m_j = 1$ and for all k, $\sum_{j=1}^m R(k,j) = 1 - r_k - d_k$, we get after simplification:

$$
\sum_{j=1}^m \mu_j \rho_j = \Lambda_1 + \sum_{k=1}^m \mu_k \rho_k (1 - r_k - d_k).
$$

As $\Lambda_1 = \lambda_1 + \sum_j \mu_j \rho_j r_j$, the last equality is proved.

Now consider Eq. 3. Divide both sides of the equation by $\pi(\vec{0}, \vec{Y})$, use the multiplicative form of the solution and make some simplification:

$$\begin{aligned}
\sum_i \lambda_i + \sum_j \mu_j 1_{Y_j > 0} &= \sum_j \rho_j \mu_j d_j \\
&+ \sum_j \sum_k \frac{\rho_j}{\rho_k} \mu_j R(j,k) 1_{Y_k > 0} \\
&+ \sum_j \omega_1 \frac{1}{\rho_j} m_j \sum_{k \neq 1} \lambda_k 1_{Y_j > 0} \\
&+ \sum_{i > 1} \omega_i \sum_{k \neq i} \lambda_k.
\end{aligned} \tag{8}$$

We already know that $\lambda_1 = \sum_j \rho_j \mu_j d_j$, $\omega_1 = \frac{\Lambda_1}{\sum_{k \neq 1} \lambda_k}$ and that $\omega_i = \frac{\lambda_i}{\sum_{k \neq i} \lambda_k}$. Therefore:

$$\begin{aligned}
\sum_i \lambda_i + \sum_j \mu_j 1_{Y_j > 0} &= \lambda_1 \\
&+ \sum_j \sum_k \frac{\rho_j}{\rho_k} \mu_j R(j,k) 1_{Y_k > 0} \\
&+ \sum_j \Lambda_1 \frac{1}{\rho_j} m_j 1_{Y_j > 0} \\
&+ \sum_{i > 1} \lambda_i.
\end{aligned} \tag{9}$$

We have established previously that

$$\sum_j \mu_j 1_{Y_j > 0} = \sum_j \sum_k \frac{\rho_j}{\rho_k} \mu_j R(j,k) 1_{Y_k > 0} + \sum_j \Lambda_1 \frac{1}{\rho_j} m_j 1_{Y_j > 0}.$$

Thus the global balance equation for these states is satisfied.

Finally we consider Eq. 4 with a similar approach and we obtain:

$$\sum_i \lambda_i = \frac{1}{\omega_j} \lambda_j + \omega_j \sum_{k \neq j} \lambda_k. \tag{10}$$

And this clearly holds as $\omega_j = \frac{\lambda_j}{\sum_{k \neq j} \lambda_k}$.

The proof is complete as we have checked all the global balance equations.

We still have to give simple formula for C. For arbitrary matching graph, it is not possible to give a closed form expression form the normalization constant (see [9] for a recurrence formula). However as the graph is a complete graph, one can derive by an easy enumeration a simple expression for the normalization constant.

Property 1. *One can obtain (the proof in an appendix of the full paper):*

$$C = [1 + \sum_{k=1}^{n} \frac{\omega_k}{1 - \omega_k}]^{-1} \prod_{j=1}^{m} (1 - \rho_j).$$

We now prove the ergodicity of the chain.

Theorem 2. *As the Markov chain is irreducible and we have found an invariant probability distribution which can be renormalized, then the chain is ergodic (see Brémaud [10], Chap 8, Th 5.3).*

5   Several Types of Item and Networks

We now assume that item of 2 behaves like item of type 1 in the previous section. But after they have been matched, they join another Jackson network (say $J2$) which is not connected to $J1$. We modify the Stop Protocol to take into account items of type 2:

– When the matching systems is empty, both Jackson networks $J1$ and $J2$ operates.
– When the matching system contains type 1 items, only $J1$ is operating.
– Similarly, when the matching system contains type 2, only the servers of $J2$ are in operation.
– Finally, when the matching system contains item of type $i > 2$, both networks are stopped.

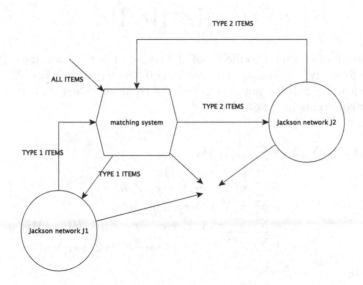

Fig. 2. Hybridization second model.

We modify the description of the state space of the Markov chain to include the states of $J2$. Let $\vec{Z} = (Z_1, .., Z_l)$ the state of $J2$. Z_j is the number of customers at queue j in $J2$. We slightly change the notation as follows:

– the service rate for queue j in $J1$ is $\mu 1_j$ while it is denoted as $\mu 2_k$ for queue k in $J2$.
– Similarly the routing matrices are called $R1$ and $R2$ for networks $J1$ and $J2$.
– The probabilities m, d and r in the initial model are now called $m1$, $d1$ and $r1$ or $m2$, $d2$ and $r2$ for networks $J1$ or $J2$.

Theorem 3. *Assume that both Jackson networks J1 and J2 are open and connected and operate under the Stop Protocol. We also assume that Markov chain $(\vec{X}, \vec{Y}, \vec{Z})_t$ is ergodic. If the following fixed point equation has a solution ω_i $(i = 1..N)$, ρ_j $(j = 1..M1)$, and ν_j $(j = 1..M2)$,*

$$\omega_i = \frac{\lambda_i}{\sum_{k \neq i} \lambda_k}, \forall i > 1, \quad and \quad \omega_1 = \frac{\Lambda_1}{\sum_{k \neq 1} \lambda_k}, \quad and \quad \omega_2 = \frac{\Lambda_2}{\sum_{k \neq 2} \lambda_k},$$

and

$$\rho 1_j = \frac{\Lambda_1 m 1_j + \sum_{k \in J1} \mu 1_k \rho 1_k R1(k,j)}{\mu 1_j}, \quad and \quad \rho 2_j = \frac{\Lambda_1 m 2_j + \sum_{k \in J2} \mu 2_k \rho 2_k R2(k,j)}{\mu 2_j},$$

with $\Lambda_1 = \lambda_1 + \sum_j \mu 1_j \rho 1_j r 1_j$, and $\Lambda_2 = \lambda_2 + \sum_j \mu 2_j \rho 2_j r 2_j$, then the steady state distribution has a multiplicative form:

$$\pi(\vec{X}, \vec{Y}) = C \prod_{i=1}^{n} \omega_j^{X_j} \prod_{j=1}^{m} \rho_j^{Y_j} \prod_{j=1}^{p} \nu_j^{Z_j}.$$

Proof. The proof is similar to the proof of Theorem 1 but we now have 4 balance equations to study: the equation when $X_1 > 0$, the equation when $X_2 > 0$, the balance when $X_i > 0$ for any $i > 2$ and the equation when $\vec{X} = \vec{0}$. We write these four equations in that order.

For $X_1 > 0$:

$$
\begin{aligned}
\pi(\vec{X}, \vec{Y}, \vec{Z}))[\textstyle\sum_i \lambda_i + \sum_{j \in J1} \mu 1_j 1_{Y_j > 0}] \\
= \lambda_1 \pi(\vec{X} - e_1, \vec{Y}, \vec{Z}) \\
+ \textstyle\sum_j \pi(\vec{X} - e_1, \vec{Y} + e_j, \vec{Z}) \mu 1_j r 1_j \\
+ \textstyle\sum_j \pi(\vec{X}, \vec{Y} + e_j, \vec{Z}) \mu 1_j d 1_j \\
+ \textstyle\sum_j \sum_k \pi(\vec{X}, \vec{Y} + e_j - e_k, \vec{Z}) \mu 1_j R1(j,k) 1_{Y_k > 0} \\
+ \textstyle\sum_j \pi(\vec{X} + e_1, \vec{Y} - e_j, \vec{Z}) m 1_j \textstyle\sum_{k \neq 1} \lambda_k 1_{Y_j > 0}
\end{aligned}
\tag{11}
$$

For $X_2 > 0$:

$$
\begin{aligned}
\pi(\vec{X}, \vec{Y}, \vec{Z}))[\textstyle\sum_i \lambda_i + \sum_{j \in J2} \mu 2_j 1_{Z_j > 0}] \\
= \lambda_1 \pi(\vec{X} - e_1, \vec{Y}, \vec{Z}) \\
+ \textstyle\sum_j \pi(\vec{X} - e_1, \vec{Y}, \vec{Z} + e_j) \mu 2_j r 2_j \\
+ \textstyle\sum_j \pi(\vec{X}, \vec{Y}, \vec{Z} + e_j) \mu 2_j d 2_j \\
+ \textstyle\sum_j \sum_k \pi(\vec{X}, \vec{Y}, \vec{Z} + e_j - e_k) \mu 2_j R2(j,k) 1_{Z_k > 0} \\
+ \textstyle\sum_j \pi(\vec{X} + e_1, \vec{Y}, \vec{Z} - e_j) m 2_j \textstyle\sum_{k \neq 2} \lambda_k 1_{Z_j > 0}
\end{aligned}
\tag{12}
$$

For $X_1 = 0$ and $X_2 = 0$ and $X_j > 0$ for some $j > 2$:

$$
\begin{aligned}
\pi(\vec{X}, \vec{Y}, \vec{Z})[\textstyle\sum_i \lambda_i] = \pi(\vec{X} - e_j, \vec{Y}, \vec{Z}) \lambda_j \\
+ \pi(\vec{X} + e_j, \vec{Y}, \vec{Z}) \textstyle\sum_{k \neq j} \lambda_k
\end{aligned}
\tag{13}
$$

For $X_i = 0$ for all i

$$\pi(\vec{0}, \vec{Y}, \vec{Z})[\sum_i \lambda_i + \sum_j \mu_j 1_{Y_j > 0} + \sum_{j \in J2} \mu 2_j 1_{Z_j > 0}]$$
$$= \sum_{j \in J1} \pi(\vec{0}, \vec{Y} + e_j, \vec{Z}) \mu 1_j d1_j$$
$$+ \sum_{j \in J2} \pi(\vec{0}, \vec{Y}, \vec{Z} + e_j) \mu 2_j d2_j$$
$$+ \sum_{j \in J1} \sum_{k \in J1} \pi(\vec{0}, \vec{Y} + e_j - e_k, \vec{Z}) \mu 1_j R1(j,k) 1_{Y_k > 0} \quad (14)$$
$$+ \sum_{j \in J2} \sum_{k \in J2} \pi(\vec{0}, \vec{Y}, \vec{Z} + e_j - e_k) \mu 2_j R2(j,k) 1_{Z_k > 0}$$
$$+ \sum_{j \in J1} \pi(e_1, \vec{Y} - e_j, \vec{Z}) m1_j \sum_{k \neq 1} \lambda_k 1_{Y_j > 0}$$
$$+ \sum_{j \in J2} \pi(e_2, \vec{Y}, \vec{Z} - e_j) m2_j \sum_{k \neq 2} \lambda_k 1_{Z_j > 0}$$
$$+ \sum_{i > 2} \pi(e_i, \vec{Y}, \vec{Z}) \sum_{k \neq i} \lambda_k$$

Clearly, we only have to examine the last equation (i.e. Eq. 14) as the first three ones have already been established in the proof of Theorem 1. As usual we divide both sides of the equation by the steady-state probability and we make the necessary cancellation of terms after having used the proposed multiplicative solution:

$$\sum_i \lambda_i + \sum_j \mu_j 1_{Y_j > 0} + \sum_{j \in J2} \mu 2_j 1_{Z_j > 0} =$$
$$\sum_{j \in J1} \rho 1_j \mu 1_j d1_j + \sum_{j \in J2} \rho 2_j \mu 2_j d2_j$$
$$+ \sum_{j \in J1} \sum_{k \in J1} \frac{\rho 1_j}{\rho 1_k} \mu 1_j R1(j,k) 1_{Y_k > 0}$$
$$+ \sum_{j \in J2} \sum_{k \in J2} \frac{\rho 2_j}{\rho 2_k} \mu 2_j R2(j,k) 1_{Z_k > 0} \quad (15)$$
$$+ \sum_{j \in J1} \frac{\omega_1}{\rho 1_j} m1_j \sum_{k \neq 1} \lambda_k 1_{Y_j > 0}$$
$$+ \sum_{j \in J2} \frac{\omega_2}{\rho 2_j} m2_j \sum_{k \neq 2} \lambda_k 1_{Z_j > 0}$$
$$+ \sum_{i > 2} \omega_i \sum_{k \neq i} \lambda_k.$$

Remember that for $i > 2$, we have $\omega_i = \frac{\lambda_i}{\sum_{k \neq i} \lambda_k}$ and $\omega_1 = \frac{\Lambda_1}{\sum_{k \neq 1} \lambda_k}$ while $\omega_2 = \frac{\Lambda_2}{\sum_{k \neq 2} \lambda_k}$. Therefore after substitution and cancellation we get:

$$\lambda_1 + \lambda_2 + \sum_j \mu_j 1_{Y_j > 0} + \sum_{j \in J2} \mu 2_j 1_{Z_j > 0} =$$
$$\sum_{j \in J1} \rho 1_j \mu 1_j d1_j + \sum_{j \in J2} \rho 2_j \mu 2_j d2_j$$
$$+ \sum_{j \in J1} \sum_{k \in J1} \frac{\rho 1_j}{\rho 1_k} \mu 1_j R1(j,k) 1_{Y_k > 0}$$
$$+ \sum_{j \in J2} \sum_{k \in J2} \frac{\rho 2_j}{\rho 2_k} \mu 2_j R2(j,k) 1_{Z_k > 0} \quad (16)$$
$$+ \sum_{j \in J1} \frac{\Lambda_1}{\rho 1_j} m1_j 1_{Y_j > 0}$$
$$+ \sum_{j \in J2} \frac{\Lambda_2}{\rho 2_j} m2_j 1_{Z_j > 0}.$$

The end of the proof mimics the treatment of Eq. 3 and is omitted for the readability of the paper.

Corollary 1. *A straight-forward generalization leads to the same type of product form solution when k types of items of the matching system leave to join k disconnected Jackson networks.*

6   Matching Graph $IS_1 \bowtie G$

We begin with the definition of an useful operation on matching graphs. In the following IS_p is an Independent Set with size p.

Definition 1 (JOIN operation). *We consider two arbitrary graphs $G1 = (V1, E1)$ and $G2 = (V2, E2)$. The JOIN of $G1$ and $G2$ is graph $G = (V, E)$ defined as follows:*

– *Nodes: $V = V1 \cup V2$,*
– *Edges: $E = E1 \cup E2 \cup \{(x, y), x \in V1, y \in V2\}$.*

Intuitively, we keep the nodes and edges of $G1$ and $G2$ and we add all the edges between $V1$ and $V2$. The JOIN operation will be noted $\bowtie$.

We depict in the next figure the JOIN of an independent set (nodes 4 to 7) and a complete graph (nodes 1 to 3).

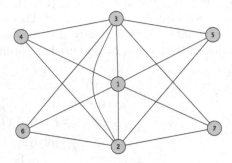

Fig. 3. Graph $IS_4 \bowtie K_3$.

In this section, we assume that the matching graph is $IS_1 \bowtie G$ with $G = (V, E)$ an arbitrary graph. Clearly graph $IS_1 \bowtie G$ contains a star and this provides a structured space space for the matching model (we do not consider the Jackson network here). Without loss of generality the type of the items associated with IS_1 is 1. We depict in Fig. 4 the Markov chain associated with $IS_1 \bowtie IS_2$ and FCFM discipline.

In this section, we consider only one Jackson network (say J). It receives items of type 1 as in Sect. 3. We use the same notation as in that section. We modify slightly the Stop Protocol as follows: the Jackson network operates when the system is in the empty state or states with only type 1 items. We now give a Markovian representation of this system.

Lemma 3. *We build an adapted representation of the chain to represent the structure of the state space. Let X_t be the number of type 1 item in the system, let $\vec{Y}_t$ be the vector representing the state of the Jackson network, and let W_t be the word representing the state of the matching system when it only contains items of V. The star in the matching graph implies that $X_t = 0$ when W_t is not empty. Similarly, when X_t is positive, then W_t is empty. Empty state E is defined by $X = 0$ and $W = \emptyset$. With the previous assumptions on the Poisson arrivals, the service rates in the Jackson network and the Stop Protocol, (X_t, W_t, Y_t) is clearly a Markov chain.*

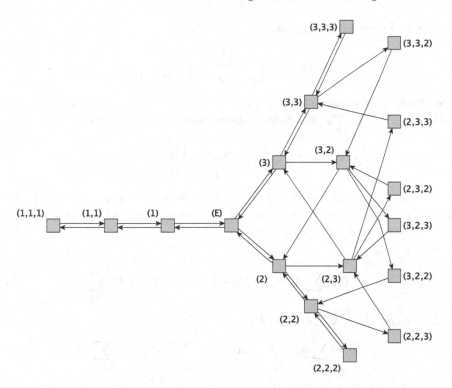

Fig. 4. The Markov chain associated with $IS_1 \bowtie IS_2$ truncated to states with less than 3 items.

We are now able to prove that, if it is ergodic, this Markov chain has a product form steady-state solution. We begin with a result established in [2] for the steady-state result of a Markov chain associated to a matching model with FCFM discipline. We first need some notation. Let W be a word on alphabet V, $|W|$ is the size of W, $W(i)$ is the type of item at position i, and (j) is the word with one letter equal to j. $Pre(W, \psi)$ is the prefix of W with ψ letters and $Ext(W, i)$ is the set of words W' with length $|W| + 1$ such that the arrivals of letter i in W' provokes a transition to state W with the FCFM discipline. Finally, $\Gamma'(i)$ will be the set of neighbors of i in the matching graph, $\Gamma(S)$ be the union for all items in S of the neighbors of these items in G, and the arrival rate of a set of item types is the sum of the arrival rates:

$$\Gamma(S) = \cup_{i \in S} \Gamma_i, \quad and \quad \lambda(S) = \sum_{i \in S} \lambda_i.$$

Theorem 4. *(see [2]) We consider the chain associated with matching graph G and FCFM discipline. The steady-state distribution (if it exists) of this Markov* *chain has a multiplicative form:*

$$\pi(W) = C' \frac{\prod_{\psi=1}^{|W|} \lambda_{W(\psi)}}{\prod_{\psi=1}^{|W|} \lambda(\Gamma(Pre(W, \psi)))}.$$

Corollary 2. *For matching graph $IS_1 \bowtie G$, the Markov chain may be represented by $(X, W)_t$. The steady state distribution (if it exists) is:*

$$\pi(X, W) = C' (\frac{\lambda_1}{\sum_{k \neq 1} \lambda_k})^X \frac{\prod_{\psi=1}^{|W|} \lambda_{W(\psi)}}{\prod_{\psi=1}^{|W|} \lambda(\Gamma(Pre(W, \psi)))}.$$

And in the second part of this formula the letters belong to set $V \setminus \{1\}$. Furthermore the global balance equation for the chain associated with this matching graph is decomposed into three cases.

– *$X > 0$, and $W = \emptyset$*

$$\pi(X, \emptyset) \sum_i \lambda_i = \lambda_1 \pi(X - 1, \emptyset) + \pi(X + 1, \emptyset) \sum_{j \neq 1} \lambda_j, \tag{17}$$

– *$X = 0$, and $W = \emptyset$ (i.e. state E):*

$$\pi(0, \emptyset) \sum_i \lambda_i = \sum_{j \neq 1} \lambda_j \pi(1, \emptyset) + \sum_{j \neq 1} \pi(0, (j)) \sum_{i \in \Gamma(j)} \lambda_i, \tag{18}$$

– *$X = 0$, and $W \neq \emptyset$. Note that $1_{\Gamma(i) \cap W \neq \emptyset}$ means that a matching occurs at the arrival of letter i when the state is W while $1_{W(|W|)=i}$ means that the last letter of W is letter i.*

$$\pi(0, W) \sum_i \lambda_i = \sum_i \lambda_i 1_{\Gamma(i) \cap W = \emptyset} 1_{W(|W|)=i} \pi(0, Pre(W, |W| - 1))$$
$$+ \sum_i \lambda_i \sum_{W' \in Ext(W, i)} 1_{\Gamma(i) \cap W' \neq \emptyset} \pi(0, W'). \tag{19}$$

Let us now consider the hybridization of the $IS_1 \bowtie G$ with a Jackson network where the type 1 items receive service.

Theorem 5. *Assume that Markov chain (X_t, W_t, Y_t) is ergodic. If the fixed point system*

$$\omega_1 = \frac{\Lambda_1}{\sum_{k \neq 1} \lambda_k}, \quad \text{with} \quad \Lambda 1 = \lambda_1 + \sum_{j \in J} \mu_j \rho_j r_j, \tag{20}$$

and

$$\rho_j = \frac{\Lambda 1 \, m_j + \sum_{k \in J} \mu_k \rho_k R(k, j)}{\mu_j}, \forall j \in J, \tag{21}$$

has a solution such that $\rho_i < 1$ and $\omega_1 < 1$, then the steady state distribution has a product form:

$$\pi(X, W, Y) = \prod_{j \in J} (1 - \rho_j) \rho_j^{Y_j} [C \, \omega_1^X \frac{\prod_{\psi=1}^{|W|} \lambda_{W(\psi)}}{\prod_{\psi=1}^{|W|} \lambda(\Gamma(Pre(W, \psi)))}].$$

Proof. We study the 3 global balance equations.

– $X > 0$, $W = \emptyset$, any Y

$$
\begin{aligned}
\pi(X,\emptyset,\vec{Y})[\textstyle\sum_i \lambda_i + \sum_j \mu_j 1_{Y_j>0}] = {} & \lambda_1 \pi(X-1,\emptyset,\vec{Y}) \\
& + \pi(X+1,\emptyset,v\mathbf{Y})\textstyle\sum_{j\neq 1}\lambda_j \\
& + \textstyle\sum_{j\in J}\sum_{k\in J}\pi(X,\emptyset,\vec{Y}+e_j-e_k)\mu_j R(j,k)1_{Y_k>0} \\
& + \textstyle\sum_{j\in J}\pi(X,\emptyset,\vec{Y}+e_j)\mu_j d_j \\
& + \textstyle\sum_{j\in J}\pi(X-1,\emptyset,\vec{Y}+e_j)\mu_j r_j.
\end{aligned}
$$
(22)

– $X = 0$, and $W = \emptyset$ (i.e. state E), any Y:

$$
\begin{aligned}
\pi(0,\emptyset,\vec{Y})[\textstyle\sum_i \lambda_i + \sum_j \mu_j 1_{Y_j>0}] = {} & \textstyle\sum_{j\in J}\pi(1,\emptyset,\vec{Y}-e_j)m_j 1_{Y_j>0}\sum_{i\neq 1}\lambda_i \\
& + \textstyle\sum_{j\neq 1}\pi(0,(j),\vec{Y})\sum_{i\in\Gamma(j)}\lambda_i \\
& + \textstyle\sum_{j\in J}\sum_{k\in J}\pi(0,\emptyset,\vec{Y}+e_j-e_k)\mu_j R(j,k)1_{Y_k>0} \\
& + \textstyle\sum_{j\in J}\pi(0,\emptyset,\vec{Y}+e_j)\mu_j d_j.
\end{aligned}
$$
(23)

– $X = 0$, and $W \neq \emptyset$, any Y:

$$
\pi(0,W,\vec{Y})\textstyle\sum_i \lambda_i = \sum_i \lambda_i 1_{\Gamma(i)\cap W=\emptyset}\; 1_{W(|W|)=i}\;\pi(0,Pre(W,|W|-1),\vec{Y})
$$

$$
+ \textstyle\sum_i \lambda_i \sum_{W'\in Ext(W,i)} 1_{\Gamma(i)\cap W'\neq\emptyset}\;\pi(0,W',\vec{Y}).
$$
(24)

We begin with Eq. 24. As $\vec{Y}_t$ does not evolve due to the Stop Protocol, as $X = 0$, and due to the multiplicative form of the steady-state solution, after dividing by $\prod_{j\in J}(1-\rho_j)\rho_j^{Y_j}$ we found again Eq. 19 which holds for this solution.

Now let us turn to Eq. 22. We proceed as usual: we divide both sides of the equation by the steady state probability and we use the multiplicative form for the solution. As component W is empty on both sides, we again found Eq. 5. The proof of Theorem 1 establish that this solution holds for the steady state solution of this part of the Markov chain.

Finally we study Eq. 23 for the last part of the Markov chain at steady-state. We divide both sides by the steady-state distribution and we simplify after using the multiplicative form for the solution.

$$
\begin{aligned}
\textstyle\sum_i \lambda_i + \sum_{j\in J}\mu_j 1_{Y_j>0} = {} & \textstyle\sum_{j\in J}\omega_1\sum_{i\neq 1}\lambda_i m_j 1_{Y_j>0} + \sum_{j\neq 1}\frac{\lambda_j}{\lambda_{\Gamma(j)}}\sum_{i\in\Gamma(j)}\lambda_i \\
& + \textstyle\sum_{j\in J}\sum_{k\in J}\frac{\rho_j}{\rho_k}\mu_j R(j,k)1_{Y_k>0} \\
& + \textstyle\sum_{j\in J}\rho_j\mu_j d_j.
\end{aligned}
$$

As in the proof of Theorem 1, $\lambda_1 = \sum_j \rho_j\mu_j d_j$, $\omega_1 = \frac{\Lambda_1}{\sum_{k\neq 1}\lambda_k}$. Thus, after exchanging index j and k in the fourth term:

$$
\begin{aligned}
\textstyle\sum_i \lambda_i + \sum_{j\in J}\mu_j 1_{Y_j>0} = {} & \textstyle\sum_{i\neq 1}\lambda_i \\
& + \textstyle\sum_{j\in J}\sum_{k\in J}\frac{\rho_k}{\rho_j}\mu_k R(k,j)1_{Y_j>0} \\
& + \textstyle\sum_{j\in J}\Lambda_1 m_j 1_{Y_j>0} + \lambda_1.
\end{aligned}
$$

And we have already proved that this equation is satisfied in the proof of Theorem 1. Thus the proof is complete.

7   Conclusions and Remarks

We hope that this first work will open avenues for new results on the hybridization of matching systems and queueing models. Currently we are investigating the effect of the matching graph topology. The existence of a star in the matching graph (i.e. the last theorem) is a first step to avoid the complete graph assumption but it is still possible to improve this result. In [11], we have developed Lumpability arguments which were proved to be useful to show the links between some operations on the matching graph and some transformations on the Markov chains. It is also possible to consider a more complex hybridization protocol than the Stop Protocol. Another idea is to explore the matching graph with loops we have recently introduced in [12]. Finally, we one can also study if it possible to combine queueing networks with bipartite matching graph model whose assumptions about the arrivals of items differs considerably from stochastic matching model we have considered here.

References

1. Mairesse, J., Moyal, P.: Stability of the stochastic matching model. J. Appl. Probab. **53**(4), 1064–1077 (2018)
2. Moyal, P., Bušić, A., Mairesse, J.: A product form for the general stochastic matching model. J. Appl. Probab. **57**(2), 449–468 (2021)
3. Cadas, A., Doncel, J., Fourneau, J.M., Busic, A.: Flexibility can hurt dynamic matching system performance. ACM SIGMETRICS Performance Evaluation Review, IFIP Performance Evaluation (short paper), vol. 49, no. 3, pp. 37–42 (2021)
4. Bušić, A., Gupta, V., Mairesse, J.: Stability of the bipartite matching model. Adv. Appl. Probab. **45**(2), 351–378 (2013)
5. Caldentey, R., Kaplan, E.H., Weiss, G.: FCFS infinite bipartite matching of servers and customers. Adv. Appl. Probab. **41**(3), 695–730 (2009)
6. United Network for Organ Sharing. https://unos.org/wp-content/uploads/unos/living_donation_kidneypaired.pdf
7. Unver, U.: Dynamic kidney exchange. Rev. Econ. Stud. **77**(1), 372–414 (2010)
8. Ashlagi, I., Jaillet, P., H. Manshadi. V.: Kidney exchange in dynamic sparse heterogenous pools. In: Proceedings of the Fourteenth ACM Conference on Electronic Commerce, EC 2013, pp. 25–26, New York, NY, USA. ACM (2013)
9. Comte, C.: Stochastic non-bipartite matching models and order-independent loss queues. Stochastic Models, 1–36 (2022)
10. Brémaud, P.: Markov Chains: Gibbs fields. Springer-Verlag, Monte Carlo Simulation and Queues (1999)
11. Cadas, A., Doncel, J., Fourneau, J.M., Busic. A.: Flexibility can hurt dynamic matching system performance, extended version on arxiv (2020)
12. Busic, A., Cadas, A., Doncel, J., Fourneau, J.M.: Product form solution for the steady-state distribution of a markov chain associated with a general matching model with self-loops. In: Gilly, K., Thomas, N. (eds.) Computer Performance Engineering - 18th European Workshop, EPEW Spain, Lecture Notes in Computer Science, vol. 13659, pp. 71–85. Springer, Cham (2022). https://doi.org/10.1007/978-3-031-25049-1_5

Construction of Phase Type Distributions by Bernstein Exponentials

A. Horváth[1(✉)] and E. Vicario[2]

[1] Dipartimento di Informatica, Università di Torino, Turin, Italy
horvath@di.unito.it
[2] Dipartimento Sistemi e Informatica, Università di Firenze, Florence, Italy
enrico.vicario@unifi.it

Abstract. Analysis of stochastic models is often hurdled by the complexity of probability density functions associated with generally distributed random variables. Among the approximation techniques that can be employed to reduce computational complexity, Bernstein polynomials (BP) exhibit some properties that make them suitable for the needs of that particular context. However, they also show some drawbacks; notably, their application is limited to bounded supports. We introduce Bernstein exponentials (BE) by transforming BP so as to enable approximation over unbounded intervals. We show that BE form a subclass of acyclic phase type (PH) distributions, thus possessing a well defined stochastic interpretation. The characteristics of this subclass allows for efficient analysis in the context of M/PH/1 queues. In particular, we develop a technique to calculate the queue length distribution in case of BE service time with linear time complexity in the number of phases of the service time distribution. Finally, we experiment BE approximations in distribution fitting and in the analysis M/G/1 queues.

1 Introduction

A typical problem in analysing systems including some form of stochasticity is to find distributions that describe well the duration of observed phenomena. An approach to tackle this problem is to choose a parametric family of distributions and apply fitting algorithms to find appropriate parameters. The choice of the parametric family has to be made in such a way that the analysis of the system can be carried then out. The family of phase type (PH) distributions [16], given by the time to absorption in Markov chains, is widely applied both because several fitting algorithms for their construction are available and because many solution techniques for models including PH distributions have been developed. Moreover, the class of PH distributions is dense in the set of probability distributions on $[0, \infty)$.

Several papers deal with PH distributions. For what concerns fitting, two typical approaches can be distinguished. The first applies the principle of maximum likelihood and, accordingly, looks for parameters that maximises the probability with which the distribution reproduces the samples. Methods in this line were

M. Iacono et al. (Eds.): EPEW/ASMTA 2023, LNCS 14231, pp. 201–215, 2023.
https://doi.org/10.1007/978-3-031-43185-2_14

presented, for example, in [1,2]. The second approach consists in constructing such PH distributions that match exactly some characteristics of the samples. The characteristics to capture can be the moments of samples or the type of tail decay of the distribution of the samples. Moment matching techniques were presented in [3,12,13,20] while a heuristic approach to approximate distributions with heavy tail can be found in [9]. It is possible to combine the two approaches and such a technique for distributions with heavy tails was presented in [10].

Many solution techniques for models with PH distributed durations have been developed in the field of queueing theory. Systems with PH distributions give rise to structured Markov chains which can often be solved by matrix analytic methods [14,18]. The inclusion of PH distributed timings in to higher level modeling languages, like Petri nets and process algebras, have also been considered [8,19].

In this paper we obtain PH approximations of a given distribution through Bernstein polynomials (BP). BP, which form a basis on the interval $[0, 1]$, can be mapped, by a change of variable, to the interval $[0, \infty]$ in such a way that the we obtain combinations of exponential functions. We will refer to the resulting class of combination of exponentials as Bernstein exponentials (BE). As we will describe in detail in the rest of the paper, using BE to construct approximations has the following advantages. The construction of the approximant requires only to sample the function which we aim to approximate and hence the cost of the approximation is very low. Positivity is guaranteed and the coefficient to normalise with, in order to obtain a proper pdf, can be calculated based on the samples without the need of integrating the approximant. The resulting approximant can be interpreted as an acyclic phase type distribution which can be advantageous from a numerical point of view. For instance, if in a model all durations are approximated by BE then the overall behaviour can be described by a Markov chain and randomisation can be used to its transient analysis.

The paper is organised as follows. In Sect. 2 we provide an introduction to Bernstein polynomials. Bernstein exponentials and their properties are discussed in Sect. 3. Section 4 provides the computations with which the queue length distribution of an M/PH/1 queue can be efficiently obtained if the service time is BE. In Sect. 5 we apply BE approximations to distribution fitting and to analyse M/G/1 queues. Conclusions are drawn in Sect. 6.

2   Bernstein Polynomials

The *Bernstein basis polynomials* [15] of degree n over the interval of reals in $[0, 1]$ are defined as

$$B_{i,n}(x) = \binom{n}{i} x^i (1-x)^{n-i} \tag{1}$$

with $i \in \mathbb{N}$ ranging in $[0, n]$. The Bernstein basis polynomials form a basis in the vector space of polynomials of degree n. In particular, a linear combination

$$B_n(x) = \sum_{i=0}^{n} \beta_i B_{i,n}(x) \tag{2}$$

is called a *Bernstein polynomial* (BP) (of degree n), and the coefficients β_i are called *Bernstein coefficients*.

Given a function $f(x)$ defined over $[0,1]$, the n samples of f taken at equidistant points in $[0,1]$ identify an approximant function $B_n f(x)$ in the class of Bernstein polynomials, obtained as

$$B_n f(x) = \sum_{i=0}^{n} f\left(\frac{i}{n}\right)\binom{n}{i} x^i (1-x)^{n-i} . \tag{3}$$

Bernstein basis polynomials and their application to approximate a given function are illustrated in Fig. 1.

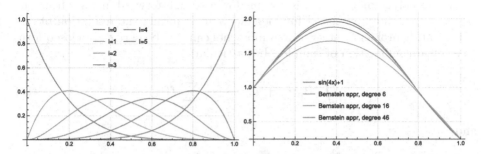

Fig. 1. Bernstein basis polynomials of degree 5 (left) and Bernsteins approximations of different degrees of $\sin(4x+1)$ over $[0,1]$ (right).

BP can be extended to approximate functions supported over an arbitrary compact interval $[a, b]$ by mapping the interval $[a, b]$ to the interval $[0,1]$ according to the function

$$\phi : [a, b] \to [0,1], \quad \phi(x) = \frac{x-a}{b-a}$$

leading to the following closed form for the approximation of function $f : [a, b] \to \mathbb{R}$:

$$B_n f(x) = \sum_{i=0}^{n} f\left(a + \frac{i}{n}(b-a)\right)\binom{n}{i}\frac{(x-a)^i(b-x)^{n-i}}{(b-a)^n} .$$

Without loss of generality, in the remainder of the article we will refer to the formulation of Bernstein polynomials on the $[0,1]$ interval.

2.1 Properties of Bernstein Polynomials

The class of Bernstein polynomials exhibits a number of fortunate properties that are relevant in the approximation of probability density functions.

Positivity Preservation. Since all Bernstein polynomials are positive over the entire $[0,1]$ interval, and since all the sample taken from pdfs are non-negative by definition, the non-negativity of the approximant $B_n f(x)$ is assured over the entire domain. Moreover, monotonicity is also preserved in the sense that $f(x) \leq g(x) \implies B_n f(x) \leq B_n g(x)$.

Globality. The approximant $B_n\ f(x)$ is global in the sense that it is given by a single analytic form over the entire support of the function, avoiding piecewise composition and partition in sub-intervals. This assumes major relevance when these functions are integrated over possibly bounded domains, leading to sub-partitioning problems that increase computational and numerical complexity. In particular, this becomes a crucial factor in the symbolic analysis approach devised in [5,6].

Simplicity of Derivation. Unlike most other approximation methods, the Bernstein polynomial $B_n\ f(x)$ is obtained in a straightforward manner from the samples of the approximated function, without requiring the resolution of any optimisation problem. Integrals and moments can also be derived in closed form as a linear combination of samples. In particular:

$$\int_0^1 B_n\ f(x)dx = \frac{1}{n+1} \sum_{i=0}^n f(i) \tag{4}$$

and

$$E[B_n\ f(x)] = \int_0^1 x \cdot B_n\ f(x)dx = \frac{1}{n^2 + 3n + 2} \sum_{i=0}^n (i+1) \cdot f(i) \ . \tag{5}$$

Partition of Unity. The Bernstein basis polynomials of degree n form a partition of unity, i.e.

$$\sum_{i=0}^n B_{i,n}(x) = \sum_{i=0}^n \binom{n}{i} x^i (1-x)^{n-i} = (x + (1-x))^n = 1 \ .$$

Unit Measure. Bernstein polynomial approximants do not preserve the measure of the approximated function. However, it can be recovered in closed form from samples according to (4) as

$$\bar{B} f_n(x) = \frac{(n+1) \cdot B_n\ f(x)}{\sum_{i=0}^n f(i)} \ .$$

In particular, if $f(x)$ has unit measure over $[0,1]$, then

$$\int_0^1 \bar{B} f_n(x) = 1.$$

Stability and Localisation. The approximation performed through Bernstein polynomials is *stable* in the sense that small deviations in the values of samples

result in smaller deviations of the approximant. In fact, assuming to independently perturb each sample adding an error $\epsilon_i \leq \epsilon$:

$$B_n f(x) - \sum_{i=0}^{n} \left(f\left(\frac{i}{n}\right) + \epsilon_i \right) \binom{n}{i} x^i (1-x)^{n-i} =$$

$$= \sum_{i=0}^{n} \epsilon_i \binom{n}{i} x^i (1-x)^{n-i} \qquad \leq$$

$$\leq \quad \epsilon \cdot \sum_{i=0}^{n} \binom{n}{i} x^i (1-x)^{n-i} = \epsilon$$

Moreover, BP approximation is *local*: even if globally defined, BP are bell-
shaped, so that the impact of changes in a sample is localised in its neighbourhood.

2.2   Drawback of Bernstein Polynomials

The usage of BP in the approximation of probability density functions also
encounters some drawbacks.

Applicable Only Over Bounded Support. As every polynomial, BP diverge
to $\pm\infty$ for $x \to \infty$, unless they are constant over their entire interval. For this
reason, they cannot be used to approximate a distribution over $[0, \infty]$.

Slow Convergence. If the approximated function is continuous, $B_n f(x)$ converges uniformly to $f(x)$ as $n \to \infty$, i.e.

$$\lim_{n \to \infty} \cdot \sup \{|f(x) - B_n f(x)| : x \in [0, 1]\} = 0 .$$

However, convergence is relatively slow. In fact, if function f satisfies the Lips-
chitz condition $|f(x) - f(y)| < L|x - y| \ \forall x \in [0, 1]$, the approximation error is
bounded as

$$|B_n f(x) - f(x)| < \frac{L}{2\sqrt{n}}$$

which runs to zero as $\frac{1}{\sqrt{n}}$ when $n \to \infty$. Moreover, the following asymptotic
formula provides a tighter error bound when $n \gg 1$:

$$B_n f(x) - f(x) = \frac{f''(x)}{2n} \cdot x(1-x) + \frac{\delta(n)}{n}$$

where $\delta(n) \to 0$ for $n \to \infty$. These results state that approximation error cannot
narrow faster than $\frac{1}{n}$. This suggests that approximation through BP may be
used to obtain a first rough estimate but over a certain limit determined by the
second derivative of f it is not convenient to increase n in order to get better
approximations.

Errors of Moments. BP approximation does not generally preserve moments
of the approximated function. In fact, BP rely on punctual values and do not

encompass any integral information, so that the error cannot be characterised unless specific assumptions can be made on the approximated function.

For instance, the expected value of the approximant given in (5) is in general not equal to the original expected value $\int_0^1 x f(x)dx$.

3   Bernstein Exponentials

Bernstein exponentials (BE) are obtained by applying Bernstein polynomial approximation after a change of variable $y = -\log(x)$ that maps the unbounded support $[0, \infty)$ into $[0, 1]$.

Specifically, given $f : [0, \infty) \to \mathbb{R}_0^+$, let $g : [0, 1] \to \mathbb{R}_0^+$ be defined as $g(y) = f(-\log(y))$, and let $\mathrm{B}_n\, g(y)$ be its BP approximant of degree n. Accordingly, the BE approximation of degree n of $f(x)$ is defined as

$$\mathrm{BE}_n\, f(x) = \mathrm{B}_n\, g(e^{-x}).$$

This results in the following closed form:

$$\mathrm{BE}_n\, f(x) = \sum_{i=0}^n f\left(-\log\left(\frac{i}{n}\right)\right) \cdot \binom{n}{i} e^{-ix}(1 - e^{-x})^{n-i} \qquad (6)$$

where we assume the notational convention: $f(\infty) = \lim_{x \to \infty} f(x)$. Note that in the summation of (6) when $i = 0$ then $f(\infty)$ is used which must be zero if $f(x)$ is a pdf. The expression given in (6) can be regarded as a combination of the exponential functions

$$BE_{i,n}(y) = \binom{n}{i} e^{-iy}(1 - e^{-y})^{n-i}$$

weighted according to the samples taken at points $-\log\left(\frac{i}{n}\right)$. The functions $BE_{i,n}$ are illustrated in Fig. 2. Examples of approximations combining these functions are given in Fig. 4.

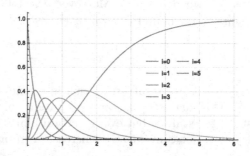

Fig. 2. The functions $BE_{i,n}$ with $n = 5$.

As for BP, the definition of Bernstein exponentials can be extended to an arbitrary interval $[a, b]$ with $a \in \mathbb{R}_0^+$ and b in $\mathbb{R}_0^+ \cup \infty$ as $\mathrm{BE}_n \, f(x) =$

$$\sum_{i=0}^{n} f\left(-\log\left(e^{-a} - \frac{i}{n}(e^{-a} - e^{-b})\right)\right) \cdot \binom{n}{i} \frac{(e^{-a} - e^{-x})^i(e^{-x} - e^{-b})^{n-i}}{(e^{-a} - e^{-b})^n}.$$

Note that as in case of BP approximations, BE approximations capture exactly the value of the approximated function at the limits of the considered interval, i.e., $\mathrm{BE}_n \, f(x)(a) = f(a)$ and $\mathrm{BE}_n \, f(x)(b) = f(b)$.

3.1   Properties Bernstein Exponentials

Bernstein exponentials preserve all the good properties of Bernstein polynomials. *Positivity, globality, simplicity* and the property we refered to as *partition of unity* are trivially maintained by BE. Other properties deserve some further discussion.

Rate of Convergence. Bounds for the approximation error of BE can be derived from those defined for BP:

$$|\tilde{g}(y) - g(y)| \le \frac{L_g}{2\sqrt{n}} \quad \forall y \in [0,1] \Rightarrow |\tilde{f}(x) - f(x)| \le \frac{L_g}{2\sqrt{n}} \quad \forall x \in [0, \infty).$$

From Lipschitz inequality, we obtain:

$$|g(y_1) - g(y_2)| = |f(-\log(y_1)) - f(-\log(y_2))| \le$$
$$\le \quad L_f| - \log(y_1) + \log(y_2)| \quad =$$
$$= \quad L_f|y_1 - y_2| \cdot \frac{\left|\log\left(\frac{y_2}{y_1}\right)\right|}{|y_1 - y_2|} \quad = L_g|y_1 - y_2|$$

In principle, this is bad news because $\frac{\left|\log\left(\frac{y_2}{y_1}\right)\right|}{|y_1 - y_2|}$, which is the Lipschitz bound that can be applied to the log function, is greater than 1 for all $y_1, y_2 \in [0,1]^2$, so that:

$$L_g = \frac{\left|\log\left(\frac{y_2}{y_1}\right)\right|}{|y_1 - y_2|} \cdot L_f \ge L_f,$$

which means that the error bound is relaxed. Note that if g takes values over $[0, b]$ (with $b \in (0,1]$), then L_g becomes unbounded, which means that uniform convergence is no more guaranteed. However, for any $a > 0$, $\frac{\left|\log\left(\frac{y_2}{y_1}\right)\right|}{|y_1 - y_2|} \le \frac{1}{a}$, which means that uniform convergence is assured for any approximation constructed over $[a, b]$.

The formula for the asymptotic bound is modified as well:

$$|\tilde{f}(x) - f(x)| = |\tilde{g}(e^{-x}) - g(e^{-x})| \simeq$$

$$\simeq \frac{1}{n} \cdot \frac{e^{-x}(1 - e^{-x})}{2} \cdot \frac{d^2}{dx^2} g(e^{-x}) = \frac{1}{n} \cdot \frac{e^{-x}(1 - e^{-x})}{2} \cdot \frac{d^2}{dx^2} f(x)$$

The above results show that the convergence properties of BE are worse than those of BP. On the other hand, from other points of view, BE have benefits with respect to BP.

Unbounded Support. BE are defined over the whole range of non-negative real numbers. For this reason they can be conveniently employed to approximate probability density functions with infinite support. Moreover, as the last sample is taken at $f(\infty) = \lim_{x \to \infty} f(x)$, the approximant catches exactly the limit of function f, enabling the approximation of transient probabilities even if the interval of interest reaches the steady state regime.

Exponential Mixture. In general, quantitative evaluation of stochastic models much more widely relies on distributions based on the exponential function rather than on polynomial distributions. Moreover, a combination of exponentials is much more suited than a polynomial in approximating transient probabilities of a CTMC, which are structurally bound to take the form of expolynomial functions of type

$$\sum_{k=0}^{K} \sum_{i=0}^{n} t^i e^{-\lambda_k t} .$$

From now on we investigate properties that are useful when the function $f(x)$ to be approximated is a pdf over the support $[0, \infty]$. We assume as well that the approximation is applied to $f(x)$ itself and not to a scaled version. The extension of the results below to a scaled version of $f(x)$ is straightforward.

Unit Measure. When approximating a pdf, the unit measure of the approximant is not guaranteed. It is easy to see however that

$$\int_{x=0}^{\infty} \binom{n}{i} e^{-ix} (1 - e^{-x})^{n-i} dx = \frac{1}{i}, \quad i \leq 1$$

and hence

$$c = \int_{x=0}^{\infty} \mathrm{BE}_n f(x) dx = \sum_{i=1}^{n} \frac{f\left(-\log\left(\frac{i}{n}\right)\right)}{i} \tag{7}$$

provides the coefficient to normalise with, in order to obtain a proper distribution. Note that different samples have different weights in the normalising coefficient. This is due to the fact that sampling points are not equidistant.

BE as Acyclic Phase Type Distributions. The Laplace transform of a given term in the approximation is

$$\int_{x=0}^{\infty} e^{-sx} f(-\log(i/n)) \binom{n}{i} e^{-ix} (1 - e^{-x})^{n-i} dx = \frac{f(-\log(i/n))}{i} \prod_{j=i}^{n} \frac{j}{s+j}, \quad i \leq 1 \tag{8}$$

and this expression provides a stochastic interpretation of the BE approximation. Namely, the term in (8) corresponds to the convolution of exponential distributions with parameters $i, i+1, ..., n$. The weight of the term is given by $f(-\log(i/n))/i$. This implies that, after normalisation, the approximant is an acyclic phase type distribution. We have depicted the phase type distribution corresponding to BE_n $f(x)$ in Fig. 3 where the absorbing phase is indicated by light-gray colour and c is as defined in (7). The structure of the PH distribution depicted in Fig. 3 is in the canonical form of acyclic PH distributions provided in [7]. It follows as well that the family of BE approximations is a subset of the acyclic phase type distributions whose pole structure is restricted and fixed.

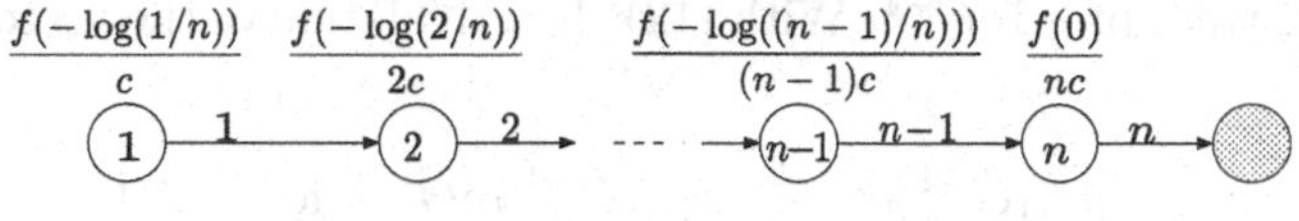

Fig. 3. Representation of BE approximations as acyclic phase type distributions.

The vector-matrix description of the PH distribution shown in Fig. 3, i.e., the initial probability vector and the matrix of the transition intensities among the transient states, are

$$
a = \begin{pmatrix} a_1 & \cdots & a_n \end{pmatrix} \text{ with } a_i = \frac{f(-log(i/n))}{ic}, \quad A = \begin{pmatrix} -1 & 1 & 0 & & \cdots & \\ 0 & -2 & 2 & 0 & \cdots & \\ & & \ddots & & & \\ & \cdots & & 0 & -(n-1) & n-1 \\ & \cdots & & & 0 & -n \end{pmatrix}
$$

$$(9)$$

Adjusting the Mean of the Approximation. From the interpretation depicted in Fig. 3, it is straightforward to obtain the mean of the approximation as

$$
m = \int_{x=0}^{\infty} x\, DE_n\, f(x)dx = \sum_{i=1}^{n} \left(\frac{f\left(-\log\left(\frac{i}{n}\right)\right)}{ic} \sum_{j=i}^{n} \frac{1}{j} \right)
$$

and this quantity can be used to adjust the mean of the approximation. If the mean given by the original function is m' then m/m' BE_n $f(x)(m/m'$ $x)$ provides a distribution that matches exactly the mean. For what concerns the PH representation of a BE, adjusting the mean is obtained simply by using $\frac{m'}{m} A$ as infinitesimal generator.

4   M/PH/1 Queue with BE Service Time

In an M/PH/1 queue with arrival intensity λ and PH service time described by initial probability vector a and infinitesimal generator A the steady state queue length probabilities, p_i, can be computed as [17]

$$
p_0 = 1 - \lambda m, \quad p_i = v_i \mathbb{1} \text{ for } i \geq 1 \text{ with} \tag{10}
$$

$$
v_0 = p_0 a, \quad v_i = v_{i-1} R \text{ for } i \geq 1 \text{ with } R = \left(I - \mathbb{1}a - \frac{A}{\lambda} \right)^{-1} \tag{11}
$$

where m is the mean service time, $\mathbb{1}$ is a column vector of ones and I is the identity matrix.

Since $\mathbb{1}a$ is of rank one, R can be calculated based on the Sherman-Morrison formula as

$$R = (G - \mathbb{1}a)^{-1} = G^{-1} + \frac{1}{1+g}G^{-1}\mathbb{1}aG^{-1} \quad \text{with} \quad G = \left(I - \frac{A}{\lambda}\right) \quad (12)$$

where g is the trace of $-\mathbb{1}aG^{-1}$. With (12) the calculation of the vectors $v_1, v_2, \dots$ becomes

$$v_i = v_{i-1}G^{-1} + \frac{1}{1+g}v_{i-1}G^{-1}\mathbb{1}aG^{-1} \quad \text{for} \quad i \geq 1 \quad (13)$$

Note that $G^{-1}\mathbb{1}$ is a column vector while aG^{-1} is a row vector. Moreover, if A describes a BE (i.e., it is as in (9)), then the entries of G^{-1} can be obtained simply as

$$\left[G^{-1}\right]_{i,j} = \lambda\frac{(j-1)!}{(i-1)!}\prod_{k=i}^{j}(\lambda+k)^{-1} \quad \text{if} \quad j \geq i \quad \text{and} \quad 0 \quad \text{otherwise}$$

from which it follows that we have a simple relation between consecutive non-zero entries in each column:

$$\left[G^{-1}\right]_{i,j} = \left[G^{-1}\right]_{i,j-1}\frac{j-1}{\lambda+j} \quad \text{for} \quad j \geq i \quad (14)$$

Based on the above, the entries of $G^{-1}\mathbb{1}$ can be calculated as

$$\left[G^{-1}\mathbb{1}\right]_n = \frac{\lambda}{\lambda+n}, \quad \left[G^{-1}\mathbb{1}\right]_i = \left[G^{-1}\mathbb{1}\right]_{i+1}\frac{i}{\lambda+i} + \frac{\lambda}{\lambda+i} \quad \text{for} \quad i \leq n-1 \quad (15)$$

while multiplication of a given vector b by G^{-1} from the right, required in (13) to obtain $v_{i-1}G^{-1}$ and aG^{-1}, can be carried out by

$$\left[bG^{-1}\right]_1 = \frac{b_1\lambda}{\lambda+1}, \quad \left[bG^{-1}\right]_i = \frac{\left[bG^{-1}\right]_{i-1}(i-1)}{\lambda+i} + \frac{b_i\lambda}{\lambda+i} \quad \text{for} \quad i \geq 2 \quad (16)$$

The coefficient g can be obtained as

$$g = -\sum_{i=1}^{n}\left[aG^{-1}\right]_i \quad (17)$$

Based on (15–17), given v_{i-1} the next vector v_i can be computed with linear time complexity in terms of the number of phases n. (In order to speed up computations, it is useful to memorize $G^{-1}\mathbb{1}$, aG^{-1} and g.) While a similar linear time approach could be developed for acyclic PH distributions in general, this is particularly interesting in the context of BE since constructing a BE approximant is also of linear time complexity as it requires only to sample a distribution at n points. Numerical illustration of approximating M/G/1 queues by M/BE/1 queues will be shown in Sect. 5.

5   Application

In this section we illustrate the application of BE to distribution fitting using two distributions of the benchmark defined in [4] for the assessment of PH approximations. The two distributions are

- Weibull with pdf

$$f(x) = \frac{\beta}{\eta} \left(\frac{x}{\eta}\right)^{\beta-1} e^{-\left(\frac{x}{\eta}\right)^\beta}$$

  and parameters $\eta = 1, \beta = 1.5$;
- uniform on $[0, 1]$.

The first, second and third moments of the Weibull distribution are 0.902745, 1.19064 and 2, respectively, and this distribution can be approximated well by even low order PH distributions because of its smooth pdf and light tail. On the other hand, the uniform distribution is notoriously difficult to fit with PH distributions due to its finite support. The first three moments of the uniform distribution are $1/2$, $1/3$ and $1/4$.

The goodness of fit achieved by the BE approximations will be evaluated by

- visual appearance of probability density function plots;
- relative error of the second and the third moment (as we adjust all the considered PH approximations to have correct mean value the error in the first moment is zero);
- application of the distributions to approximate an M/G/1 queue.

Beyond BE approximations we apply also acyclic PH distributions that capture the first three moments exactly [3] and acyclic PH distributions resulting from maximum likelihood estimation as described in [4] and implemented in [11]. Computational times are very low in case of BE approximations (it is sufficient to sample the pdf we aim to fit) and to match three moments (explicit expressions give the parameters of the PH distribution based on the moments) while maximum likelihood estimation requires much more time (in the order of 20–30 seconds) as several parameters must be optimized in an iterative manner.

The pdf of the obtained approximations are visualized in Fig. 4. For both distributions matching three moments (APH, 3 moments) results in a distribution with very different shape. In case of the Weibull distribution, the approximation obtained by maximum likelihood estimation with 10 (APH, ML, 10) and with 20 (APH, ML, 20) phases and BE approximation with 1000 phases cannot be distinguished from the original Weibull pdf. BE approximations with 10 (BE, 10) and with 50 (BE, 50) phases are worse but provide pdfs that are similar to the original. In case of the uniform distribution, maximum likelihood approximations (APH, ML, 10 and APH, ML, 20) are poor. Increasing the number of phases does not improve the fitting because of the numerical problems that arise with more parameters to optimize. With BE instead, an order 100000 approximation can easily be obtained and it gives a very accurate approximation of the original uniform pdf. In the interval [0,0.975] the difference between the uniform

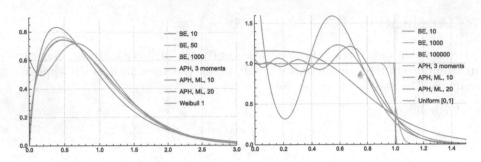

Fig. 4. Approximations of the Weibull (left) and the uniform (right) distributions.

pdf and that of the order 100000 BE approximation (BE, 100000) is less than 10^{-5}. After 1.025 the value of the pdf of the order 100000 BE approximation is less than 10^{-5} and decreases exponentially.

Table 1 reports relative error of the second and third moments of the maximum likelihood and the BE approximations denoted by e_2 and e_3. Order 1000 BE approximations, which are very fast to obtain, outperforms order 20 maximum likelihood approximations and, as mentioned before, beyond 20 phases maximum likelihood approximation suffers from numerical problems and does provide significant improvement. In case of BE approximations, using 10 times more phases results in about 10 times lower relative error.

Table 1. Relative error of the second and third moment of the approximations.

		ML		BE				
distribution/order		10	20	10	100	1000	10000	100000
Weibull	e_2	0.003806	0.003667	0.120718	0.017859	0.00176	0.000176	0.000018
	e_3	0.012042	0.012377	0.439917	0.06495	0.006124	0.000609	0.000061
uniform	e_2	0.019255	0.008539	0.139233	0.016769	0.001714	0.000172	0.000017
	e_3	0.0716	0.032688	0.460446	0.050918	0.005148	0.000515	0.000052

As described in Sect. 4 the steady state queue length distribution of an M/PH/1 queue in case of acyclic PH distributions can be calculated with linear time complexity in the number of phases of the service time PH distribution. Hereinafter we exploit this fact to approximate M/G/1 queues by M/PH/1 queues with large BE approximations of the general service time. The relative error of the queue length distribution for various PH approximations in case of Weibull service time is depicted in Fig. 5. Order 10 BE approximation (BE, 10), order 10 and 20 maximum likelihood approximations (APH, ML, 10 and APH, ML, 20) and capturing exactly three moments of the service time (APH, 3 moments) result in very similar relative errors (the corresponding curves cannot be distinguished at several points). Increasing the order of the BE approximation better and better results can be achieved. Using 10 times more phases leads to about 10 times more precise probabilities. The same is provided in Fig. 6 in case of uniform service time. Maximum likelihood approximations (APH, ML, 10 and APH, ML, 20) perform similar to the order 100 BE approximation. Capturing exactly three moments (APH, 3 moments) at some occupation levels (between 20 and 22) and at 0.9 utilization provides more precise results than the order 100000 BE approximation but its precision is less stable.

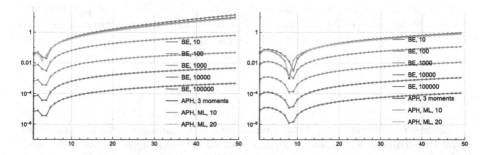

Fig. 5. Relative error of steady state queue length distribution of an M/G/1 queue with Weibull service time approximated by PH distributions (x-axis reports number of clients in the queue) with two utilization levels: 0.7 on the left and 0.9 on the right. (Since all approximations match the mean service time exactly, the relative error of the probability that the queue is empty is zero.)

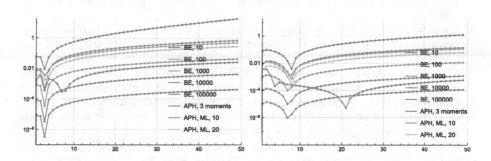

Fig. 6. Relative error of steady state queue length distribution of an M/G/1 queue with uniform service time approximated by PH distributions (x-axis reports number of clients in the queue) with two utilization levels: 0.7 on the left and 0.9 on the right.

6   Conclusions

We introduced Bernstein exponentials as a transformation of Bernstein polynomials that maintains the favorable properties of Bernstein polynomials and enables approximation over unbounded intervals. Bernstein exponentials turn out to be a subclass of acyclic phase type distributions, which can thus be effectively integrated into several solution methods for the approximate analysis of non-Markovian models. We showed that the queue length distribution of an M/PH/1 queue can be obtained with linear time complexity in the number of phases of the service time distribution if it is given by a Bernstein exponential.

We experimented Bernstein exponentials in distribution fitting and in the context of approximate analysis of M/G/1 queues. The obtained results highlight the issue of slow convergence which is inherited from Bernstein polynomials; however this is balanced by the simplicity of derivation of Bernstein exponentials which allows the construction of high-degree approximants with an extremely low computational cost.

References

1. Asmussen, S., Nerman, O.: Fitting phase-type distributions via the EM algorithm. In: Proceedings: "Symposium i Advent Statistik", Copenhagen, pp. 335–346 (1991)
2. Bobbio, A., Cumani, A.: ML estimation of the parameters of a PH distribution in triangular canonical form. In Balbo, G., Serazzi, G., (eds.) Computer Performance Evaluation, pp. 33–46. Elsevier Science Publishers (1992)
3. Bobbio, A., Horváth, A., Telek, M.: Matching three moments with minimal acyclic phase type distributions. Stoch. Model. **21**, 303–326 (2005)
4. Bobbio, A., Telek, M.: A benchmark for PH estimation algorithms: result for acyclic-PH. Stochastic Models **10**(661–677) (1994)
5. Carnevali, L., Grassi, L., Vicario, E.: State-density functions over dbm domains in the analysis of non-markovian models. IEEE Trans. on SW Eng. **35**(2), 178–194 (2009)
6. Carnevali, L., Sassoli, L., Vicario, E.: Using stochastic state classes in quantitative evaluation of dense-time reactive systems. IEEE Trans. on SW Eng. **35**(5), 703–719 (2009)
7. Cumani, A.: On the canonical representation of homogeneous Markov processes modelling failure-time distributions. Microelectron. Reliab. **22**, 583–602 (1982)
8. Cumani, A.: Esp - A package for the evaluation of stochastic petri nets with phase-type distributed transition times. In: Proceedings International Workshop Timed Petri Nets, Torino (Italy), pp. 144–151 (1985)
9. Feldman, A., Whitt, W.: Fitting mixtures of exponentials to long-tail distributions to analyze network performance models. Perform. Eval. **31**, 245–279 (1998)
10. Horváth, A., Telek, M.: Approximating heavy tailed behavior with Phase-type distributions. In: Proceedings of 3rd International Conference on Matrix-Analytic Methods in Stochastic models, Leuven, Belgium (June 2000)
11. Horváth, A., Telek, M.: PhFit: a general phase-type fitting tool. In: Field, T., Harrison, P.G., Bradley, J., Harder, U. (eds.) TOOLS 2002. LNCS, vol. 2324, pp. 82–91. Springer, Heidelberg (2002). https://doi.org/10.1007/3-540-46029-2_5
12. Horváth, A., Telek, M.: Matching more than three moments with acyclic phase type distributions. Stoch. Model. **23**(2), 167–194 (2007)
13. Johnson, M.A., Taaffe, M.R.: Matching moments to Phase distributions: nonlinear programming approaches. Stoch. Model. **6**, 259–281 (1990)

14. Latouche, G., Ramaswami, V.: Introduction to matrix analytic methods in stochastic modeling. SIAM (1999)
15. Lorentz, G.G.: Bernstein Polynomials. University of Toronto Press (1953)
16. Neuts, M.: Probability distributions of phase type. In: Florin, E.H. (ed.) Liber Amicorum Prof. pp. 173–206. University of Louvain (1975)
17. Neuts, M.F.: Matrix Geometric Solutions in Stochastic Models. Johns Hopkins University Press (1981)
18. Neuts, M.F.: Matrix Geometric Solutions in Stochastic Models. Johns Hopkins University Press, Baltimore (1981)
19. Scarpa, M., Bobbio, A.: Kronecker representation of stochastic Petri nets with discrete PH distributions. In: International Computer Performance and Dependability Symposium - IPDS 1998, pp. 52–61. IEEE CS Press (1998)
20. Telek, M., Horváth, G.: A minimal representation of Markov arrival processes and a moments matching method. Perform. Evaluat. 64(9–12), 1153–1168 (2007)

Bayesian Networks as Approximations of Biochemical Networks

Adrien Le Coënt[1]([✉]), Benoît Barbot[1], Nihal Pekergin[1], and Cüneyt Güzeliş[2]

[1] Université Paris Est Créteil, LACL, 94010 Creteil, France
[2] Department of Electrical-Electronics Engineering, Yaşar University, Izmir, Turkey
{adrien.le-coent,benoit.barbot}@u-pec.fr

Abstract. Biochemical networks are usually modeled by Ordinary Differential Equations (ODEs) that describe time evolution of the concentrations of the interacting (biochemical) species for specific initial concentrations and certain values of the interaction rates. The uncertainty in the measurements of the model parameters (*i.e.* interaction rates) and the concentrations (*i.e.* state variables) is not an uncommon occurrence due to biological variability and noise. So, there is a great need to predict the evolution of the species for some intervals or probability distributions instead of specific initial conditions and parameter values. To this end, one can employ either phase portrait method together with bifurcation analysis as a dynamical system approach, or Dynamical Bayesian Networks (DBNs) in a probabilistic domain. The first approach is restricted to the case of a few number of parameters, while DBNs have recently been used for large biochemical networks. In this paper, we show that time-homogeneous ODE parameters can be efficiently estimated with Bayesian Networks. The accuracy and computation time of our approach is compared to two-slice time-invariant DBNs that have already been used for this purpose. The efficiency of our approach is demonstrated on two toy examples and the EGF-NGF signaling pathway.

Keywords: Ordinary Differential Equations based models · Markov Chains · Bayesian Networks · Biochemical Networks · Time Homogeneous Systems

1 Introduction

In-silico analyses of biochemical networks based on their ODE and Chemical Master Equation (CME) models are valuable instruments for developing the methods for diagnosis, prognosis, drug discovery, therapeutic procedures in the clinical domain as well as for efficient and reliable experiment design. ODE models describe the deterministic future of the species' concentrations given the initial concentrations and the interaction rates. On the other hand, CMEs provide predictions for the stochastic future of the probabilities of the molecule numbers

This work was financed by the join ANR-JST project CyPhAI.

for species. Both have advantages on their sides; ODE models are appropriate when molecular species are available in large numbers whilst CMEs are more appropriate for low molecule numbers.

As a consequence of biological variability within a population of biochemical species of the same kind as well as measurement errors, environmental noise and estimation errors in determining values of unmeasurable parameters by means of directly measured ones, there is always uncertainty in the reaction rates [12].

Although both kinds of models for biochemical networks have been well-established, their analytical solutions are rarely available and their numerical analysis is computationally expensive and may become intractable due to the combinatorial explosion of the state space for CMEs. Stochastic Simulation Algorithms (SSAs) [9] for CMEs overcome this problem, however, simulations have to be repeated several times to provide accurate estimations making this approach excessively time consuming. In the literature, approximations and learning based methods have become a popular approach to accelerate the CME based models [1,19].

The ODE models of biochemical networks consist, in general, of many variables and parameters. Their analysis becomes computationally expensive since a large number of numerical simulations must be performed for numerous values of initial conditions and parameters. Approximations of ODEs are usually done with numerical solvers [5,6], but they usually provide approximations for a given initial condition and parameter combination value. The first probabilistic approximation of ODEs, aimed at capturing the behaviour of the evolution of the solution, and not just approximating a single simulation, has been proposed in [20] from researchers in the theoretical physics community. It has been complemented over the years [22], but their applications have been limited to very small dimensional systems, even though the dynamics can be very complex (nonlinear).

Probabilistic approximation methods have been proposed to tackle with the above-mentioned uncertainty in the ODE model parameters and also in the initial concentration levels of species. Although these approximations provide efficient solutions for low dimensional models, computing them becomes time consuming for large numbers of species and reaction rates. A recent study [14] proposes to use DBNs to efficiently encode the probabilistic approximations built, which at the same time allows to use standard Bayesian inference methods for computing marginal distributions of species, learning unobserved parameters, and performing sensitivity analysis.

In this paper, we focus on the parameter identification problem, and show that using BNs instead of DBNs allows better accuracy of the identification given the same computational power. The description of the construction of the approximations in [14] being very succinct, we first explain how to construct them with as much details as possible, so that interested readers can implement their own approximation methods easily in the future. We furthermore discuss how the structure of the approximations (*i.e.* the structure of the networks, the Markov chain time steps, and the size of the discrete states) play an important role in the accuracy of the identification problems. We finally exhibit that the BN approach is more efficient than the DBN one.

To show that the BN method provides an effective solution to the parameter identification problem also in high-dimensional networks, the EGF-NGF signaling pathway [3] is considered as a case study. The rationale behind this choice lies in the consideration of a case study on the same application area where the method desired to be surpassed in terms of performance was applied before. In this regard, the existing works [13, 14] that introduce DBN approximations of the ODE models, were initially aimed at analysing EGF-NGF signaling pathways.

The paper is organized as follows: Sect. 2 is devoted to the considered problem setting and to introduce briefly the applied models. In Sect. 3, we detail the proposed construction of the BN and emphasize that the structure of the network depends also the method used to simulate the ODEs. The numerical examples are given in Sect. 4, and finally we conclude and present our perspectives.

2   Problem Setting

As in [13, 14], we consider biochemical networks modeled by ODEs. However a large number of numerical simulations for all possible initial conditions and parameter values is necessary to perform parameter estimation or sensitivity analysis in such systems. The main idea in [13, 14] is to generate several trajectories of the ODE model by sampling different initial conditions and to construct an approximate discrete-time and discrete-valued stochastic model. Dynamic Bayesian inference methods are then applied to compute marginal distributions of some species, to provide parameter estimation, sensitivity analysis of large biochemical networks.

In this paper we propose a similar approach. However, we advocate that Bayesian networks would be sufficient if the underlying ODEs are autonomous (*i.e.* they do not explicitly depend on the time variable) which is hypothesized in [13, 14]. In the following subsections we briefly define the considered deterministic and stochastic models, and then explain the probabilistic approximation of ODE based models.

2.1   ODE Based Continuous-Valued Deterministic Models

The biochemical systems we consider are modeled by ODEs written under the following form:

$$\dot{x}_i(t) = f_i(x(t), k), \tag{1}$$

the variables $x(t) = x_1(t), x_2(t), ..., x_n(t)$ are real-valued concentrations of species at time t. The set $k_1, k_2, ..., k_m$ is the set of (positive) real-valued parameters, they are supposed to be constant over time, but unknown. We are interested in studying the system with various combinations of parameter values. The variables $x_i(t)$ for $i \in \{1, ..., n\}$ and the parameters k_j for $j \in \{1, ..., m\}$ are supposed to take values in products of intervals $\mathbf{X} = [x_1^{min}, x_1^{max}] \times \cdots \times [x_n^{min}, x_n^{max}]$ and $\mathbf{K} = [k_1^{min}, k_1^{max}] \times \cdots \times [k_m^{min}, k_m^{max}]$.

The functions f_i are issued from mass action kinetics, we assume them to be continuously differentiable. This ensures that the flows (vector fields) that are solutions of the ODE are measurable functions. We refer the reader to [7,10,14] for more details on the mathematical soundness ensured by these assumptions.

Our main motivating case study is the EGF-NGF network, described in [14], but our methods are illustrated on the following simpler examples.

Example 1. Our first example is a toy with two variables and one parameter:

$$\dot{x}_1 = kx_1 \qquad\qquad (2)$$
$$\dot{x}_2 = -0.9kx_1$$

Example 2. Our second example comes from a typical biochemical enzyme cat-alyzed reaction. Its continuous dynamics is given by

$$\dot{x}_1 = -k_1 x_1 x_2 + k_2 x_3 \qquad\qquad (3)$$
$$\dot{x}_2 = -k_1 x_1 x_2 + (k_2 + k_3)x_3$$
$$\dot{x}_3 = k_1 x_1 x_2 - (k_2 + k_3)x_3$$
$$\dot{x}_4 = k_3 x_3$$

In the following, we need to designate how one variable affects the others in the continuous dynamics. We thus define a function pa, which stands for parents, as follows.

Definition 1. *The function $pa_x(f_i)$ (resp. $pa_k(f_i)$) maps to the set of indices of variables (resp. parameters) on which variable i depends. Each variable depends at least on itself.*

For instance, for Example 2, $pa_x(f_1) = pa_x(f_2) = pa_x(f_3) = \{1, 2, 3\}$, $pa_x(f_4) = \{3, 4\}$ and $pa_k(f_1) = \{1, 2\}$, $pa_k(f_2) = pa_k(f_3) = \{1, 2, 3\}$ and $pa_k(f_4) = \{3\}$. We will denote $x_{|pa(j)}$ the projection of x where only the coordinates belonging to the parents of j remain, for example in the previous example $x_{|pa_x(f_4)} = (x_3, x_4)$.

2.2   Discrete-Valued Probabilistic Models

The first step of the approximation is the discretization of the value intervals (*i.e.* **X** and **K**) for each variable and parameter leading them to take values in a finite set of sub-intervals. Since each sub-interval is considered as a discrete state, the model is now discrete-valued.

Let us emphasize that it is not irrelevant to consider value intervals instead of precise values in biochemical systems, since model parameters are often subject to uncertainties. Furthermore it is assumed that the system is not observed continuously but at discrete time instants, the time is thus also discretized. The discrete time instants are an increasing sequence $(t_i)_{i=0}^{\tau}$ with $t_0 = 0$, $t_1 = \delta$, ..., and $t_\tau = T = \delta\tau$ with time step δ. The flow of the differential equation thus induces a discrete-time Markov chain (DTMC) by assuming a prior distribution of initial values [21].

Definition 2. *A time-homogeneous discrete-time Markov chain (DTMC) with finite or countable state space $\mathcal{H}$ is a sequence $X_0, X_1, \ldots$ of $\mathcal{H}$-valued random variables such that for all states $i, j, x_0, x_1, \ldots$ and all times $t = 0, 1, 2, \ldots$,*

$$Pr(X_{n+1} = j | X_n = i, X_{n-1} = x_{n-1}, \ldots) = Pr(X_{n+1} = j | X_n = i) = p(i,j)$$

where the transition probability $p(i,j)$ depends only on the states i, j, and not on the time nor the previous states $x_{n-1}, x_{n-2}, \ldots$.

The estimation of transition probabilities for the approximate DTMC by the statistical analysis of ODE trajectories will be described in the following sections. BN models will be used for the compact representation of the approximate DTMC of the underlying ODE model. We refer to [18] for further information.

Definition 3. *A Bayesian network (BN) is a finite acyclic directed graph $BN = (V, E)$. For each node $v \in V$, a finite-valued random variable X_v and a conditional probability table CPT_v are associated. The entries in CPT_v are of the form $Pr(X_v = x | X_{v_1} = x_1, X_{v_2} = x_2, ..., X_{v_j} = x_j)$ where $v_1, v_2, ..., v_j$ is the set of parents of v given by $pa(v) = \{u | (u,v) \in E\}$.*

BN represents the joint probability distribution over the random variables $\{X_v\}$, $1 \leq v \leq n$ given by: $Pr(\mathbf{X}) = \prod_{v=1}^{n} Pr(X_v | pa(X_v))$.

Dynamic Bayesian Networks (DBNs) are Bayesian Networks that allow us to model the temporal evolution of the system.

Definition 4. *A Dynamic Bayesian Network (DBN) is a tuple $(B_0, \{B_\rightarrow^j\}_{j=1}^T)$, where B_0 defines the initial probability distributions of the random variables, and $\{B_\rightarrow^j\}$ are two-slice temporal Bayesian networks for the transition from time step $j-1$ to j: $Pr(\mathbf{X}^j | \mathbf{X}^{j-1}) = \prod_{v=1}^{n} Pr(X_v^j | pa(X_v^j))$. In the case of time-invariant DBNs, there exists one two-slice BN, $B_\rightarrow$ whatever the time step j is.*

DBNs are useful for computing joint distributions of species after several time steps, or estimating initial distributions. In the case of parameter identification for time-homogeneous systems, the transition model (CPT) from time j to $j+1$ should be constant. In this paper, contrary to [14] where they use DBNs, we propose to use BNs for parameter inference.

2.3   Probabilistic Approximation of ODE Based Models

The main idea of the approximation is to estimate approximate transition probabilities from the numerical simulation of ODEs. The distribution of initial values is sampled several times and for each sampled initial value, the trajectory of the ODE is constructed. A sufficiently large set of such trajectories provides a good approximation of the dynamics of the ODE model. The transition probabilities of the underlying approximate Markov chain are estimated by analyzing the statistical properties of the ODE trajectories.

The probabilistic approximation of ODEs proposed in [14] consists of modeling the approximate DTMC by a two-slice time-variant DBN. Our observation is that the ODE considered in [14] is autonomous, and the parameters are considered constant over time. Transition probabilities should not vary with time. Contrary to [14], we propose to use Bayesian networks to store the transition probabilities, whatever the time step is.

As that will be explained in the next section, the underlying graph is constructed by exploiting the structure of the ODEs, as well as the continuous simulation method used to generate the sample set. The entries of the CPT are specified by the probability transition probabilities of the time-homogeneous approximate Markov chain. Using BN inference techniques for parameter marginal distribution computations leads to the following main computational improvements: the inference algorithms are less time consuming for BNs than DBNs; for large networks, one can use exact inference algorithms for BNs, while they become intractable for DBNs. The discrete probabilistic model still requires numerous simulations to compute the probability tables whatever is the kind of network. Let us emphasize that more simulations used in the BN or DBN construction lead to more accurate parameter estimation.

3   Construction of the Probabilistic Approximation

3.1   Computation of the BN Model

The BN model is built using two main steps which are the construction of both spatially and temporally discretized trajectories and the parameter estimation of the approximate Markov chain from these trajectories.

Construction of Discretized Trajectories. First step is the discretization of the continuous state spaces in which the state variable and parameter values lie: $\mathbf{X} = [x_1^{min}, x_1^{max}] \times \cdots \times [x_n^{min}, x_n^{max}]$ for variables and $\mathbf{K} = [k_1^{min}, k_1^{max}] \times \cdots \times [k_m^{min}, k_m^{max}]$ for parameters. The value interval for each variable i, $[x_i^{min}, x_i^{max}]$ is divided in L_i sub-intervals: $[x_i^{min}, x_i^1), [x_i^1, x_i^2), \ldots, [x_i^{L_i-1}, x_i^{max})$. Similarly in M_j sub-intervals for each parameter j: $[k_j^{min}, k_j^1), [k_j^1, k_j^2), \ldots, [k_j^{M_j-1}, k_j^{max})$.

- We define the discretization function d_{L_i} (resp. d_{M_i}) such that $d_{L_i}(x_i) = a$ iff $x_i \in [x_i^{a-1}, x_i^a)$ the function which maps the ith state variable (resp. parameter) to the index a of the corresponding sub-interval with $1 \leq a \leq L_i$ (resp. $1 \leq a \leq M_i$). Therefore for the vector of a continuous state $(\boldsymbol{x}, \boldsymbol{k})$ we define the vector of the corresponding discrete state $(\boldsymbol{d_L}(\boldsymbol{x}), \boldsymbol{d_M}(\boldsymbol{k}))$.
- A number of (continuous) simulations of the ODE (Eq. (1)) is then performed. Any simulation method yielding approximations of the continuous trajectories can be used, Runge-Kutta schemes are used here. These simulations are performed from random initial conditions in $\mathbf{X}$, and using random parameter values in $\mathbf{K}$. In order to get a probabilistic model that approximates the continuous equation as closely as possible for all the possible initial conditions,

the random initial conditions need to cover the initial sets as uniformly as possible. To this end, we make use of Halton sequences [8] to choose the initial conditions.
- The continuous simulations are converted spatially in discrete state simulations by applying d_L. More precisely, consider a given continuous trajectory (or an approximation of it) $x(t)$, $0 \le t \le T$. It is converted in a sequence of discrete states $d_L(x(t))$, $0 \le t \le T$.
- This sequence of discrete states is also converted temporally by specifying an increasing sequence of time $(t_j)_{j=0}^{\tau}$ with $t_0 = 0$ and $t_\tau = T$.
- The continuous parameters values are only converted spatially since parameters do not depend on time. By applying $d_M(k)$, we obtain the corresponding discrete parameter values.

Thus, for a trajectory (simulation) with the initial values $x(t_0)$, and parameter values (k), we produce the pair $\langle (d_L(x(t_j)))_{j=0}^{\tau}; d_M(k) \rangle$ containing the sequence of variables x at time t_j and parameter values k. In the following, we note this pair for a discretized (both spatially and temporarily) trajectory of Eq. (1) by $\langle (l^j)_{j=0}^{\tau}; m \rangle$.

Transition Probabilities Estimation for the Markov Chain. In the second step, we have a set S of discretized trajectories with elements of the form $\langle (l^j)_{j=0}^{\tau}; m \rangle_S$ of the Eq. (1). We estimate the transition probabilities of the approximate Markov chain (see Definition 2) by counting which transitions are taken, compared to the total possible transitions from the set S.

- Given S the set of discretized trajectories, $NT(S, l^{pred}, l^{next}, m, i)$ denotes the number of transitions taken during all of the trajectories in S which lead to a state where $l_i = l_i^{next}$ for a given variable index i from a state of l^{pred}. The transition may happen during time step j to time step $j+1$ whatever j is. However at time step j of the trajectory, the values for the parents of variable i denoted by i: $pa_x(f_i)$ (see Definition 1) must match those in l^{pred}. Similarly the parameters in the set $pa_k(f_i)$ must match with m.

$$NT(S, l^{pred}, l^{next}, m, i) = \sum_{\langle (l^j)_{j=0}^{\tau}; m' \rangle \in S}$$
$$|\{0 \le j < \tau \text{ s.t. } l^{pred}_{|pa_x(f_i)} = l^j_{|pa_x(f_i)} \wedge l_i^{next} = l_i^{j+1} \wedge m_{|pa_k(f_i)} = m'_{|pa_k(f_i)}\}| \tag{4}$$

The transition probabilities are estimated from the number of transitions for each variable as follows:

$$p(\langle l^{pred}; m \rangle, \langle l^{next}; m \rangle) = \prod_{i=1}^{n} \frac{NT(S, l^{pred}, l^{next}, m, i)}{\sum_{\langle l; m \rangle \in S} NT(S, l^{pred}, l, m, i)} \tag{5}$$

In Fig. 1, we illustrate a continuous simulation of the simple Example 1 and a realization of the approximate Markov chain.

– The approximate Markov chain built is finally efficiently stored as a BN. Its structure depends on the ODE and the numerical scheme used for Eq. (1). The structure of the BN to be used is discussed in the following section.

Inference. Once a BN is computed, various algorithms can be used for inference, parameter learning, structure learning, sensitivity analysis etc. Whether BNs or DBNs are considered, parameter inference is performed with some evidence that ideally consists of real life measurements. Here, we generate some continuous simulations with given parameter values and observe how accurately we recover these parameter values. For DBNs, the evidence takes the form of a sequence of variable values (species concentrations). For BNs, the same simulations are used, except that the sequence of variable values is simply decomposed in multiple couples (values at t and $t + \delta$). For BNs, we use a standard junction tree algorithm implemented in the BNT toolbox [16]. For DBNs, we use the DBN version of junction trees for exact inference, and the the Boyen and Koller fully factored (BKFF) algorithm [2] for approximate inference.

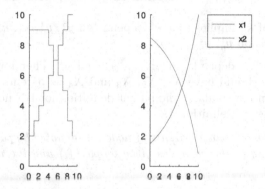

Fig. 1. Left: one realization of the approximate DTMC with 10 sub-intervals for x and 2 sub-intervals for k; Right: the continuous simulation of the ODE from the initial condition $(x_1^0, x_2^0) = (1.5, 8.5)$, using parameter $k = 0.19$. The initial sets of the continuous systems are $[0, 10] \times [0, 10]$ for xs and $[0, 0.2]$ for k.

3.2   Structure of the Bayesian Network

Let us now discuss the structure of the networks that should be used. If we consider ODE simulations performed with a simple Euler scheme, for which the time-step corresponds to Markov chain time step (*i.e.* with the abuse of notation $\delta = 1$, the solution of Eq. (1) is approximated by the series $y_{n+1} = y_n + f(y_n, p)$), then the structure proposed in [14] can be used. However, if a higher order scheme is used, then more edges should be added to the network.

Let us recall the principle of explicit S-stage Runge-Kutta (RK) schemes [4]:

$$y_{n+1} = y_n + h \sum_{i=1}^{S} b_i \kappa_i \tag{6}$$

where

$$\kappa_1 = f(t_n, y_n),$$
$$\kappa_2 = f(t_n + c_2 h, y_n + (a_{21}\kappa_1)h),$$
$$\kappa_3 = f(t_n + c_3 h, y_n + (a_{31}\kappa_1 + a_{32}\kappa_2)h),$$
$$\vdots$$
$$\kappa_S = f(t_n + c_S h, y_n + (a_{S1}\kappa_1 + a_{S2}\kappa_2 + \cdots + a_{S,S-1}\kappa_{S-1})h).$$

Let us now consider Example 2, and add an exponent $i \in \{1,2,3,4\}$ to the variables y_n and κ_j to denote its ith component (dimension). Application of an explicit 4-stage RK scheme to this example exhibits that:

- κ_1^3 depends on y_n^1, y_n^2, y_n^3; κ_1^4 depends on y_n^3, y_n^4 (parents in ODE as defined in Definition 1);
- the definition of κ_2^4 implies that κ_2^4 depends on $y_n^3, y_n^4, \kappa_1^3, \kappa_1^4$;
- then κ_2^4 depends on $y_n^1, y_n^2, y_n^3, y_n^4$.

And in the end, y_{n+1}^4 depends on y_n^1, y_n^2, y_n^3 and y_n^4. Therefore, as illustrated in Fig. 2, node X_3' should have X_1, X_2, X_3 and X_4 as parents in the BN. The following definition reformulates the parent definition for BN nodes, it was used earlier for variables in Definition 1.

Definition 5. *The one-stage parents of node X_i, denoted by $pa_{BN}(X_i)$, are the nodes corresponding to the variables given by $pa_x(f_i)$ and the parameters given by $pa_k(f_i)$.*

Definition 6. *The s-stage parents of node X_i, denoted by $pa_{BN}^s(X_i)$ for any $s \geq 1$, are defined recursively as follows :*

$$pa_{BN}^1(X_i) = pa_{BN}(X_i),$$
$$pa_{BN}^{s+1}(X_i) = \bigcup_{X \in pa_{BN}^s(X_i)} pa_{BN}(X).$$

In the above definitions, if $pa_x(f_i) = \{1,2\}$ and $pa_k(f_i) = \{1\}$, then $pa_{BN}(X_i) = \{X_1, X_2, K_1\}$. The 2-stage parents $pa_{BN}^2(X_i)$ are the one stage parents of the one stage parents of X_i. Such node connections are then used to build the BN structure depending on the numerical scheme used. If an explicit Euler scheme with time step δ was used as the continuous simulation scheme, then the structure of the BN would be given by $pa_{BN}(X_i)$ for all i (this is the BN structure suggested in [14], it is represented in Fig. 2(a) for Example 2). The following theorem formally states the graph structure yielding the highest accuracy depending on the numerical scheme used to generate the continuous trajectories.

Theorem 1. *Consider the system of Eq. (1) simulated with time step δ using an S-stage RK scheme. The most accurate BN approximation is obtained with the following graph structure. Node X_i has parents:*

$$pa_{BN_S}(X_i) = \bigcup_{s=1}^{S} pa_{BN}^s(X_i)$$

where $pa_{BN}^s(X_i)$ are the s-stage parents of node X_i as defined in Definition 6.

The above theorem formalizes that computing an S-stage RK step requires at least S repeated applications of function f of Eq. (1), which the BN node connections should reflect. Application of this theorem to Example 2 yields the structure represented in Fig. 2(b). One can observe that, contrary to Fig. 2(a), the nodes X'_1, X'_2, X'_3, X'_4 all depend on all the parameter nodes K_1, K_2, K_3. Note however that the graph is not fully connected, repeated applications of f do not connect X_4 to the other nodes. Once a graph structure is chosen, computation of the transition probabilities should be performed with modified functions pa_x and pa_k that take into account node dependencies and not continuous variable dependencies of Definition 1.

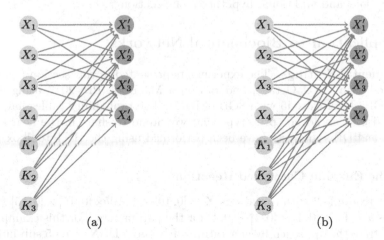

(a)   (b)

Fig. 2. The BN structure used in [14] (left) and the BN structure we suggest for an RK4 scheme (right).

3.3   Time and Space Discretization Discussion

The time and space discretization parameters used in the present method play an important role in the accuracy of the parameter learning and the computation times, and their choice heavily depends on each other. Let us first observe that over a short discrete time step δ, very small changes in the continuous state are made. If these changes are small enough, the corresponding discrete state does not change. On the other hand, if the space discretization is fine enough, discrete state changes will be observed in the approximating Markov chain.

As a general guideline, we propose to choose the discrete time step, δ and the number of sub-intervals for variable i, L_i such that there exists at least one $i \in \{1, \ldots, n\}$ such that:

$$\frac{\int_{y_0 \in \mathbf{X}} \int_{k \in \mathbf{K}} \|y_\delta^i(y_0, k) - y_0^i\|}{|\mathbf{X}| \times |\mathbf{K}|} \geq \frac{x_i^{max} - x_i^{min}}{L_i}$$

where $|\mathbf{X}| = (x_1^{max} - x_1^{min}) \times \cdots \times (x_n^{max} - x_n^{min})$, $|\mathbf{K}| = (k_1^{max} - k_1^{min}) \times \cdots \times (k_m^{max} - k_m^{min})$, and $y_\delta(y_0, k)$ is (the Runge-Kutta approximation of) the solution at time δ of Eq. (1) from the initial condition y_0 at $t = 0$ and using parameter k. This formula ensures that, on average, the continuous state variation over a time step δ is greater than the size of at least one sub-interval considered for the Markov chain, thus inducing at least one discrete state change over time step δ. This formula cannot be computed exactly, but a few Monte-Carlo simulations should yield a satisfying first estimate of the time and space discretization parameters to use. Note that we do not have a proof that such a parameter combination exists, but we hope to have one in a future work using uniform discretizations and additional hypotheses on function f.

4   Application to Biochemical Networks

Experimental Settings. The experiments presented in this section have been performed with GNU Octave, and ran on a MacBook Pro 2017 using a dual core 2.3GHz Intel Core i5 with 8GB of RAM. The probability tables have been generated with an ad hoc prototype that we intend to share in the near future. The BN and DBN analyses have been performed using the BNT toolbox [16].

4.1   The Enzyme Catalyzed Reaction

Consider Example 2 with initial sets $\mathbf{X} = [0, 10] \times [0, 10] \times [0, 15] \times [0, 15]$ for the variables and $\mathbf{K} = [0, 1] \times [0, 1] \times [0, 1]$ for the parameters. For this example, we first illustrate the approach by generating a BN and a DBN using 5 sub-intervals per variable and parameter. *I.e.*, the interval $[0, 10]$ for variable x_1 is divided in $[0, 2), [2, 4), [4, 6), [6, 8), [8, 10]$.

To test the accuracy of the BN and DBN learning methods, we generate a sample set, *i.e.* some continuous trajectories for a given parameter combination and some random initial conditions. They are taken as evidence to learn some parameter probability distributions. We generate 10 continuous samples using $(k_1, k_2, k_3) = (0.9, 0.5, 0.1)$, and random initial conditions in $\mathbf{X}$. We suppose that parameter k_2 is observed (taken as evidence), the distributions for k_1 and k_3 have to be estimated.

An example of parameter combination learned is given in Table 1. For samples generated using $k_1 = 0.9$ and $k_3 = 0.1$, k_2 being known, the highest probability for parameter k_1 estimation is 0.45 (given in bold). Similarly k_3 is estimated to be in $[0.0, 0.2)$ with a 67% probability in the BN approach. With the DBN approach, the highest probability for k_1 estimation is 0.58 for interval $[0.6, 0.8)$ while the estimation to be in $[0.8, 1.0]$ is with a 31% probability. And k_3 is estimated to be in $[0.0, 0.2)$ with a 59% probability. We have tested the approach for different sets of samples and parameter combinations. We obtain similar results using 7 sub-intervals per parameter and variable. Note that the results obtained highly depend on the sample set. On average, the BN approach yields right parameter intervals more often than the DBN approach. Interestingly, the DBN approach gives results with higher confidence (probability), even though these results might be wrong. Depending on the sample set, the BN learning time is between 25% and 95% faster than the DBN learning time.

Table 1. An example of parameter marginal distribution learned with the BN and DBN approaches, using 5 sub-intervals per parameter. The highest probability is in bold for each method.

original parameters	sub-interval	BN marginal distribution	DBN marginal distribution
	$[0, 0.2)$	0.0106	0.0000
	$[0.2, 0.4)$	0.1034	0.0039
$k_1 = 0.9$	$[0.4, 0.6)$	0.1480	0.0993
	$[0.6, 0.8)$	0.2889	**0.5855**
	$[0.8, 1.0]$	**0.4501**	0.3114
	$[0, 0.2)$	**0.6770**	**0.5940**
	$[0.2, 0.4)$	0.2582	0.2845
$k_3 = 0.1$	$[0.4, 0.6)$	0.0456	0.1215
	$[0.6, 0.8)$	0.0072	0.0000
	$[0.8, 1.0]$	0.0120	0.0000

We also compared the accuracy of the parameter identification using the BN structures suggested in [14] and our proposition (Theorem 1). We built two BNs with the structures given in Fig. 2(a) and Fig. 2(b), with 5 sub-intervals per variable and parameter. The sample set was generated using $k_1 = 0.9$ and $k_3 = 0.1$, k_2 being known. As expected, the graph that is more connected yields more accurate results, as seen in Table 2. The computation time was slightly higher with the more connected graph of Fig. 2(a) with $3.86s$, compared to $3.32s$ with the graph of Fig. 2(b), which is a 16% increase in computation time. This is also expected since the junction tree algorithm is polynomial in the number of cliques [11], which is increased by the number of edges in the graph.

Finally, the parameter learning accuracy has been tested with different state and time discretizations. Very fine time discretization ($\delta < 0.1$) led to very random results since the BN barely captures any dynamical behaviour, this is due to the number discrete state transitions being very small compared to the number of trajectories used to generate the BNs, and confirms our suggestions

Table 2. An example of parameter marginal distribution learned with the BN with our suggested graph connection and the graph connection suggested in [14], computed with 5 sub-intervals. The highest probability is in bold for each method.

original parameters	sub-interval	BN marginal distribution with structure of Theorem 1	BN marginal distribution with structure of [14]
$k_1 = 0.9$	[0, 0.2)	0.0000	0.0000
	[0.2, 0.4)	0.0806	0.0947
	[0.4, 0.6)	0.1875	0.2075
	[0.6, 0.8)	0.3159	**0.5249**
	[0.8, 1.0]	**0.4160**	0.1730
$k_3 = 0.1$	[0, 0.2)	**0.9001**	**0.8462**
	[0.2, 0.4)	0.0362	0.1502
	[0.4, 0.6)	0.0638	0.0036
	[0.6, 0.8)	0.0000	0.0000
	[0.8, 1.0]	0.0000	0.0000

in Sect. 3.3. On Example 2, we observe that state discretization has a noticeable influence on computation time, as shown in Table 3.

Table 3. Average parameter marginal distribution learning times (in seconds) for the BN and DBN approaches, using different numbers of sub-intervals per parameter and variable.

number of sub-intervals	BN	DBN
3	28 s	35 s
5	37 s	45 s
7	83 s	108 s

4.2 The EGF-NGF Model

The EGF-NGF signaling pathway is a large biochemical network that has been extensively studied in [3,14]. The model under consideration, which is described by the ODEs in Appendix in (7), is depicted as a hybrid functional Petri net (HFPN) [15] in Fig. 3. The original model has 32 variables and 48 parameters, 28 of which are known, the remaining ones needing to be learned. A first formal analysis of the network allowed us to identify that only 14 of the 20 unknown parameters are essential to fully grasp the dynamics of the system.

We have successfully computed a BN and a DBN approximation of the model in 53 hours with a very straightforward implementation. They have been constructed with 10^6 trajectories, using 5 sub-intervals per variables and unknown parameter. The number of trajectories is very small compared to what is used in [14]. To test the learning accuracy, we generated 5 trajectories of length 10 taken as evidence, for given randomly selected parameter values, and observed how often the learned parameters were in the right sub-interval. Note that exact inference methods quickly become intractable for large DBNs, and approximate inference methods were required to test the DBN learning approach. We used here the Boyen and Koller fully factored (BKFF) algorithm [2] which is already implemented in the BNT toolbox. More time-efficient algorithms such as the factored frontier algorithm could be considered [17], but the accuracy of the BN approach would still be higher since we can use an exact inference algorithm (the junction-tree algorithm). Our approach, despite the small number of trajectories used for generating the BN, recovers the right parameter sub-intervals for 7 of the 14 parameters on average, the others being close, compared to 5 for the DBN. Using exact inference, the learning time is around 15 seconds for the BN, and 3 hours for the DBN. Using the BKFF algorithm, the computation time is lowered to 3 minutes for the DBN, with a minor accuracy decrease.

5   Conclusions and Future Work

Throughout this paper, we have shown that using BNs as probabilistic approximations for biochemical networks yields better accuracy and computation times than using DBNs for parameter learning, provided that the ODEs are autonomous. We provided a detailed explanation for building the BN approximation. We compared our approach to the DBN approximations suggested in [14]. We furthermore proposed a BN structure that, depending on the numerical schemes used, provides the highest possible accuracy. We compared the accuracy of the parameter identification with BN and DBN approximations on a simple biochemical system, and computed parameters on the real life sized case study of the EGF-NGF signaling pathway model.

Our future work will be devoted to using model reduction techniques as a preliminary step before building the BN approximations, in order to use more computational power on the learning part and less on the BN construction part. Experimental measurements are potentially performed at times that are not in a perfect time sequence (e.g. some species concentrations are obtained at times 0, 5, 20, 50, ...). To use such experimental measurements, our method should be extended to BNs that allow learning parameters with samples that have missing time instants.

Appendix

$$\dot{x}_1 = -k_1 * x_1 * x_3 + k_2 * x_4$$

$$\dot{x}_2 = -k_3 * x_2 * x_5 + k_4 * x_6$$

$$\dot{x}_3 = -k_1 * x_1 * x_3 + k_2 * x_4$$

$$\dot{x}_4 = k_1 * x_1 * x_3 - k_2 * x_4$$

$$\dot{x}_5 = -k_3 * x_2 * x_5 + k_4 * x_6$$

$$\dot{x}_6 = k_3 * x_2 * x_5 - k_4 * x_6$$

$$\dot{x}_7 = k_9 * x_{10} * x_8/(x_8 + k_{10}) - k_5 * x_4 * x_7/(x_7 + k_6) - k_7 * x_6 * x_7/(x_7 + k_8)$$

$$\dot{x}_8 = -k_9 * x_{10} * x_8/(x_8 + k_{10}) + k_5 * x_4 * x_7/(x_7 + k_6) + k_7 * x_6 * x_7/(x_7 + k_8)$$

$$\dot{x}_9 = -k_{27} * x_{21} * x_9/(x_9 + k_{28})$$

$$\dot{x}_{10} = k_{27} * x_{21} * x_9/(x_9 + k_{28})$$

$$\dot{x}_{11} = -k_{11} * x_8 * x_{11}/(x_{11} + k_{12}) + k_{13} * x_{13} * x_{12}/(x_{12} + k_{14})$$

$$\dot{x}_{12} = k_{11} * x_8 * x_{11}/(x_{11} + k_{12}) - k_{13} * x_{13} * x_{12}/(x_{12} + k_{14})$$

$$\dot{x}_{13} = 0$$

$$\dot{x}_{14} = -k_{15} * x_{12} * x_{14}/(x_{14} + k_{16}) + k_{45} * x_{32} * x_{15}/(x_{15} + k_{46})$$
$$+ k_{35} * x_{25} * x_{15}/(x_{15} + k_{36})$$

$$\dot{x}_{15} = k_{15} * x_{12} * x_{14}/(x_{14} + k_{16}) - k_{45} * x_{32} * x_{15}/(x_{15} + k_{46})$$
$$- k_{35} * x_{25} * x_{15}/(x_{15} + k_{36})$$

$$\dot{x}_{16} = -k_{43} * x_{29} * x_{16}/(x_{16} + k_{44}) + k_{47} * x_{32} * x_{17}/(x_{17} + k_{20})$$

$$\dot{x}_{17} = k_{43} * x_{29} * x_{16}/(x_{16} + k_{44}) - k_{47} * x_{32} * x_{17}/(x_{17} + k_{20})$$

$$\dot{x}_{18} = -k_{17} * x_{15} * x_{18}/(x_{18} + k_{18}) - k_{19} * x_{17} * x_{18}/(x_{18} + k_{48})$$
$$+ k_{21} * x_{31} * x_{19}/(x_{19} + k_{22})$$

$$\dot{x}_{19} = k_{17} * x_{15} * x_{18}/(x_{18} + k_{18}) + k_{19} * x_{17} * x_{18}/(x_{18} + k_{48})$$
$$- k_{21} * x_{31} * x_{19}/(x_{19} + k_{22})$$

$$\dot{x}_{20} = -k_{23} * x_{19} * x_{20}/(x_{20} + k_{24}) + k_{25} * x_{31} * x_{21}/(x_{21} + k_{26})$$

$$\dot{x}_{21} = k_{23} * x_{19} * x_{20}/(x_{20} + k_{24}) - k_{25} * x_{31} * x_{21}/(x_{21} + k_{26})$$

$$\dot{x}_{22} = -k_{29} * x_4 * x_{22}/(x_{22} + k_{30}) - k_{31} * x_{12} * x_{22}/(x_{22} + k_{32})$$

$$\dot{x}_{23} = k_{29} * x_4 * x_{22}/(x_{22} + k_{30}) + k_{31} * x_{12} * x_{22}/(x_{22} + k_{32})$$

$$\dot{x}_{24} = -k_{33} * x_{23} * x_{24}/(x_{24} + k_{34})$$

$$\dot{x}_{25} = k_{33} * x_{23} * x_{24}/(x_{24} + k_{34})$$

$$\dot{x}_{26} = -k_{37} * x_6 * x_{26}/(x_{26} + k_{38})$$

$$\dot{x}_{27} = k_{37} * x_6 * x_{26}/(x_{26} + k_{38})$$

$$\dot{x}_{28} = -k_{39} * x_{27} * x_{28}/(x_{28} + k_{40}) + k_{41} * x_{30} * x_{29}/(x_{29} + k_{42})$$

$$\dot{x}_{29} = k_{39} * x_{27} * x_{28}/(x_{28} + k_{40}) - k_{41} * x_{30} * x_{29}/(x_{29} + k_{42})$$

$$\dot{x}_{30} = 0$$

$$\dot{x}_{31} = 0$$

$$\dot{x}_{32} = 0$$

(7)

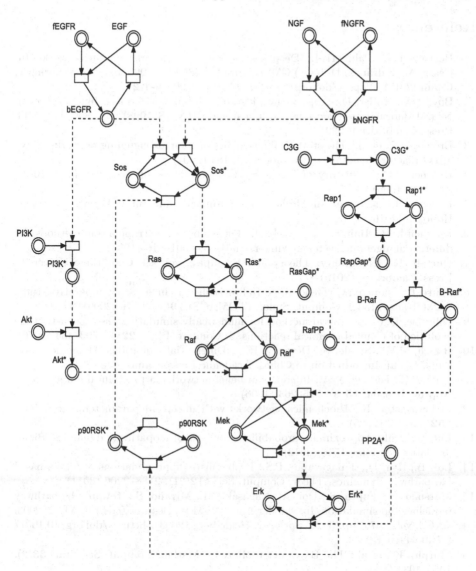

Fig. 3. The HFPN model of the EGF-NGF signaling pathway [14]. The HFPN representation of chemical networks is detailed in [15]. In a few words, circles (places) represent species, squares (transitions) represent reactions, and arrows represent how the species are involved in the reactions, dashed arrows mean that the species are not consumed (like in enzyme reactions).

References

1. Bortolussi, L., Palmieri, L.: Deep abstractions of chemical reaction networks. In: Češka, M., Šafránek, D. (eds.) CMSB 2018. LNCS, vol. 11095, pp. 21–38. Springer, Cham (2018). https://doi.org/10.1007/978-3-319-99429-1_2
2. Boyen, X., Koller, D.: Approximate learning of dynamic models. In: Advances in Neural Information Processing Systems, vol. 11(NIPS 1998), pp. 396–402. MIT Press, Cambridge (1999)
3. Brown, K.S., et al.: The statistical mechanics of complex signaling networks: nerve growth factor signaling. Phys. Biol. 1(3), 184 (2004)
4. Butcher, J.C.: A history of Runge-Kutta methods. Appl. Numer. Math. 20(3), 247–260 (1996)
5. Butcher, J.C.: Numerical Methods for Ordinary Differential Equations. Wiley, Hoboken (2016)
6. Deuflhard, P., Hairer, E., Zugck, J.: One-step and extrapolation methods for differential-algebraic systems. Numer. Math. 51, 501–516 (1987)
7. Durrett, R.: Probability: Theory and Examples, vol. 49. Cambridge University Press, Cambridge (2019)
8. Faure, H., Lemieux, C.: Generalized Halton sequences in 2008: a comparative study. ACM Trans. Model. Comput. Simul. (TOMACS) 19(4), 1–31 (2009)
9. Gillespie, D.T.: A general method for numerically simulating the stochastic time evolution of coupled chemical reactions. J. Comput. Phys. 22(4), 403–434 (1976)
10. Hirsch, M.W., Smale, S., Devaney, R.L.: Differential Equations, Dynamical Systems, and an Introduction to Chaos. Academic Press, Cambridge (2012)
11. Huang, C., Darwiche, A.: Inference in belief networks: a procedural guide. Int. J. Approx. Reason. 15(3), 225–263 (1996)
12. Liebermeister, K.: Biochemical networks with uncertain parameters. Syst. Biol. 152, 97–107 (2005)
13. Liu, B., et al.: Approximate probabilistic analysis of biopathway dynamics. Bioinformatics 28, 1508–1516 (2012)
14. Liu, B., Hsu, D., Thiagarajan, P.S.: Probabilistic approximations of ODEs based bio-pathway dynamics. Theor. Comput. Sci. 412(21), 2188–2206 (2011)
15. Matsuno, H., Fujita, S., Doi, A., Nagasaki, M., Miyano, S.: Towards biopathway modeling and simulation. In: van der Aalst, W.M.P., Best, E. (eds.) ICATPN 2003. LNCS, vol. 2679, pp. 3–22. Springer, Heidelberg (2003). https://doi.org/10.1007/3-540-44919-1_2
16. Murphy, K., et al.: The Bayes net toolbox for Matlab. Comput. Sci. Stat. 33(2), 1024–1034 (2001)
17. Murphy, K., Weiss, Y.: The factored frontier algorithm for approximate inference in DBNs. In Proceedings of the 17th Conference on Uncertainty in Artificial Intelligence, pp. 378–385 (2001)
18. Murphy, K.P.: Probabilistic Machine Learning: An Introduction. MIT Press, Cambridge (2022)
19. Schnoerr, D., Sanguinetti, G., Grima, R.: Approximation and inference methods for stochastic biochemical kinetics-a tutorial review. J. Phys. A: Math. Theor. 50(9), 093001 (2017)

20. Skilling, J.: Bayesian solution of ordinary differential equations. Maximum Entropy Bayesian Meth. Seattle **1991**, 23–37 (1992)
21. Trivedi, K.: Probability and Statistics with Reliability, Queuing and Computer Science Applications. Wiley, Hoboken (2001)
22. Tronarp, F., Kersting, H., Särkkä, S., Hennig, P.: Probabilistic solutions to ordinary differential equations as nonlinear Bayesian filtering: a new perspective. Stat. Comput. **29**, 1297–1315 (2019)

Performance of Genetic Algorithms in the Context of Software Model Refactoring

Vittorio Cortellessa(ID), Daniele Di Pompeo(✉)(ID), and Michele Tucci(ID)

University of L'Aquila, L'Aquila, Italy
{vittorio.cortellessa,daniele.dipompeo,michele.tucci}@univaq.it

Abstract. Software systems continuously evolve due to new functionalities, requirements, or maintenance activities. In the context of software evolution, software refactoring has gained a strategic relevance. The space of possible software refactoring is usually very large, as it is given by the combinations of different refactoring actions that can produce software system alternatives. Multi-objective algorithms have shown the ability to discover alternatives by pursuing different objectives simultaneously. Performance of such algorithms in the context of software model refactoring is of paramount importance. Therefore, in this paper, we conduct a performance analysis of three genetic algorithms to compare them in terms of performance and quality of solutions. Our results show that there are significant differences in performance among the algorithms (*e.g.,* `PESA2` seems to be the fastest one, while `NSGAII` shows the least memory usage).

Keywords: Performance · Multi-Objective · Refactoring · Search-Based Software Engineering

1 Introduction

Multi-objective optimization techniques proved to be effective in tackling many model-driven software development problems [21,25,29,31]. Such problems usually involve a number of quantifiable metrics that can be used as objectives to drive the optimization. Problems related to non-functional aspects undoubtedly fit into this category, as confirmed by the vast literature in this domain [1,2,23]. Most approaches are based on evolutionary algorithms [6], which allow exploring the solution space by combining solutions.

The improvement of software models quality through refactoring is a kind of task that can be carried out by multi-objective optimization. However, multi-objective algorithms demand a lot of hardware resources (*e.g.,* time, and memory allocation) to search the solution space and generate a (near-)optimal Pareto frontier. Therefore, the actual performance of multi-objective algorithms in software model refactoring is of paramount importance, especially if the goal is to integrate them into the design and evolution phases of software development.

M. Iacono et al. (Eds.): EPEW/ASMTA 2023, LNCS 14231, pp. 234–248, 2023.
https://doi.org/10.1007/978-3-031-43185-2_16

For this reason, in this paper, we compare the performance in terms of execution time, memory allocation, and quality of Pareto frontiers of the NSGAII, SPEA2, and PESA2 multi-objective algorithms within the context of software model refactoring. We have selected NSGAII due to its extensive use in the context of software refactoring, SPEA2 because it has already been compared with NSGAII in other domains [8,18,22], and PESA2 because it uses a different technique (*i.e.*, hyper-grid crowding degree operator) to search the solution space.

We have evaluated the performance of each algorithm by using a reference case study presented in [15]. To achieve this, we have executed *30* independent runs, as suggested in [3], for each algorithm by varying the number of iterations for each run, and we have collected execution time and memory usage. We provide a replication package of the experimentation presented in this study.[1]

We aim at answering the following research questions:

- RQ_1: *How do NSGA-II, SPEA2, and PESA2 compare in terms of execution time?*
- RQ_2: *How do NSGA-II, SPEA2, and PESA2 compare in terms of memory usage?*
- RQ_3: *How do NSGA-II, SPEA2, and PESA2 compare in terms of multi-objectindicatorsive optimization indicators?*

Our experimentation showed that PESA2 is the algorithm whose executions last quite less than the NSGAII and SPEA2 ones. Furthermore, PESA2 generates Pareto frontiers that showed better solutions in terms of reliability and performance. NSGAII, instead, consumed less memory than SPEA2, and PESA2. However, it generated less densely populated Pareto frontiers. Finally, SPEA2 showed worse performance and Pareto frontiers than PESA2, and NSGAII.

The remaining of the paper is structured as follows: Section 2 reports related work; Section 3 introduces the algorithms subject of the study; Section 4 briefly introduces the case studies; Section 5 discusses results and findings. Section 6 describes takeaways from the study; Section 7 discussed threats to validity. Section 8 concludes the paper.

2   Related Work

Genetic algorithms are exploited in different domains to identify alternatives of the initial problem that show at least one better attribute (*i.e.*, at least on objective). In particular, studies have analyzed the performance in building Pareto frontiers in heterogeneous domains, which span from automotive problems to economic ones [11,20,22,32]. In this paper, instead, we analysed performance in terms of hardware consumption needed to search the solution space for software model refactoring. In the context of software architecture, studies have investigated how multi-objective optimization can improve quality of software architectures.

---

[1] Replication   package: https://github.com/danieledipompeo/replication-package_Perf-Comp-GA-4-Multi-Obj-SW-Model-Ref.

For example, Cortellessa and Di Pompeo [8] studied the sensitivity of multi-objective software architecture refactoring to configuration characteristics. They compared two genetic algorithms in terms of Pareto frontiers quality dealing with architectures defined in Æmilia, which is a performance-oriented Architecture Descrption Language (ADL). In this paper, we propose a performance comparison between NSGAII, SPEA2, and PESA2 to identify which algorithm needs less resources to search the solution space.

Aleti et al. [1] have presented an approach for modeling and analyzing Architecture Analysis and Design Language (AADL) architectures [17]. They have also introduced a tool aimed at optimizing different quality attributes while varying the architecture deployment and the component redundancy. Instead, our work relies on UML models and offers more complex refactoring actions as well as different target attributes for the fitness function. Besides, we investigate the role of performance antipatterns in the context of multi-objective software architecture refactoring optimization.

Menascé et al. [27] have presented a framework for architectural design and quality optimization, where architectural patterns are used to support the searching process (e.g., load balancing, fault tolerance). Two limitations affects the approach: the architecture has to be designed in a tool-related notation and not in a standard modelling language (as we do in this paper), and it uses equation-based analytical models for performance indices that might be too simple to capture architectural details and resource contention. We overcome the possible the Menascé et al. limitation by employing Layred Queueing Network (LQN) models to estimate performance indices.

Martens et al. [26] have presented PerOpteryx, a performance-oriented multi-objective optimization problem. In PerOpteryx the optimization process is guided by tactics referring to component reallocation, faster hardware, and more hardware, which do not represent structured refactoring actions, as we employ in our refactoring engine. Moreover, PerOpteryx supports architectures specified in Palladio Component Model (PCM) [5] and produces, through model transformation, a LQN for of performance analysis.

Rago et al. have presented SQuAT [30], an extensible platform aimed at including flexibility in the definition of an architecture optimization problem. SQuAT supports models conforming to PCM language, exploits LQN for performance evaluation, and PerOpteryx tactics for architecture.

A recent work compares the ability of two different multi-objective optimization approaches to improve non-functional attributes [28], where randomized search rules have been applied to improve the software model. The study of Ni et al. [28] is based on a specific modelling notation (i.e., PCM) and it has implicitly shown that the multi-objective optimization problem at model level is still an open challenge. They applied architectural tactics, which in general do not represent structured refactoring actions, to find optimal solutions. Conversely, we applied refactoring actions that change the structure of the initial model by preserving the original behavior. Another difference is the modelling notation, as we use UML with the goal of experimenting on a standard notation instead of a custom DSL.

3   Algorithms

NSGA-II The Non-dominated Sorting Algorithm II (NSGAII), introduced by Deb et al. [13], is widely used in the software engineering community due to its good performance in generating Pareto frontiers. The algorithm, randomly generates the initial population P_0, shuffles it and applies the *Crossover* operator with probability $P_{crossover}$, and the *Mutation* operator with probability $P_{Mutation}$ to generate the Q_t offspring. Thus, the obtained $R_t = P_t + Q_t$ mating pool is sorted by the *Non-dominated sorting* operator, which lists Pareto frontiers with respect to considered objectives. Finally, a *Crowding distance* is computed and a new family (*i.e.*, P_{t+1}) is provided to the next step by cutting the worse half off.

SPEA2. Strength Pareto Evolutionary Algorithm 2 (SPEA2) has been introduced by Zitzler et al. [34]. Differently from NSGAII, SPEA2 does not employ a non-dominated sorting process to generate Pareto frontiers.

SPEA2 randomly generates an initial population P_0 and an empty archive $\bar{P}_0$ in which non-dominated individuals are copied at each iteration. For each iteration $t = 0, 1, \ldots, T$, the fitness function values of individuals in P_t and $\bar{P}_t$ are calculated. Then non-dominated individuals of P_t and $\bar{P}_t$ are copied to $\bar{P}_{t+1}$ by discarding dominated individuals or duplicates (with respect to the objective values). In case size of $\bar{P}_{t+1}$ exceeds $\bar{N}$, *i.e.*, the size of the initial population, the *Truncation* operator drops exceeded individuals by preserving the characteristics of the frontier, using the *k-th nearest neighbor* knowledge. In case size of $\bar{P}_{t+1}$ is less than $\bar{N}$, dominated individuals from P_t and $\bar{P}_t$ are used to fill $\bar{P}_{t+1}$. The algorithm ends when a stopping criterion is met, *e.g.*, the iteration t exceeds the maximum number of iterations T, and it generates the non-dominated set A in output.

PESA2. The Pareto Envelope-based Selection Algorithm 2 (PESA2) is a multi-objective algorithm, introduced by Corne et al. [7] that uses two sets of population, called internal (*IP*) and external (*EP*). The internal population is often smaller than the external one and it contains solution candidates to be included in the external population. Furthermore, the external population is generally called *archive*. The selection process is driven by a hyper-grid crowding distance degree. The current set of *IP* are incorporated into the *EP* one by one if it is non-dominated within *IP*, and if is not dominated by any current member of the *EP*. Once a candidate has entered the *EP*, members of the *EP* which it dominated (if any) will be removed. If the addition of a candidate renders the *EP* over-full, then an arbitrary chromosome which has the maximal squeeze factor in the population of *EP* is removed. Also, the squeeze factor describes the total number of other chromosomes in the archive which inhabit the same box. The PESA2 crowding strategy works by forming an implicit hyper-grid which divides the solution space into hyper-boxes. Furthermore, each chromosome in the *EP* is associated with a particular hyper-box in solution space. Then, the squeeze factor is assigned to each hyper-box, and it is used during the searching phase.

4   Case Study

In this section, we apply our approach to the Train Ticket Booking Service (TTBS) case study [15,33], and to the well-established model case study CoCOME, whose UML model has been derived by the specification in [19].

Train Ticket Booking Service. Train Ticket Booking Service (TTBS) is a web-based booking application, whose architecture is based on the microservice paradigm. The system is made up of 40 microservices, and it provides different scenarios through users that can perform realistic operations, *e.g.,* book a ticket or watch trip information like intermediate stops. Our UML model of TTBS is available online.[2] The static view is made of **11** UML Components, where each component represents a microservice. In the deployment view, we consider **11** UML Nodes, each one representing a docker container. We selected these three scenarios because they commonly represent performance-critical ones in a ticketing booking service.

CoCOME. CoCOME describes a Trading System containing several stores. A store might have one or more cash desks for processing goodies. A cash desk is equipped with all the tools needed to serve a customer (e.g., a Cash Box, Printer, Bar Code Scanner). CoCOME describes 8 scenarios involving more than 20 components. From the CoCOME original specification, we analyzed different operational profiles, *i.e.,* scenarios triggered by different actors (such as Customer, Cashier, StoreManager, StockManager), and we excluded those related to marginal parts of the system, such as scenarios of the *EnterpriseManager* actor. Thus, we selected **3** UML Use Cases, **13** UML Components, and **8** UML Nodes from the CoCOME specification.

5   Results

In this section, we compare execution times, memory consumption, and quality of Pareto frontiers across the considered algorithms and case studies.

---

[2] https://github.com/SEALABQualityGroup/2022-ist-replication-package/tree/main/case-studies/train-ticket.

5.1 RQ_1: How Do NSGAII, SPEA2, and PESA2 Compare in Terms of Execution Time?

In order to answer to the RQ_1 we collected execution time of each algorithm 30 times. Based on the results of our experimentation, we can state that the PESA2 algorithm showed the best execution time with respect to NSGAII and SPEA2 in both case studies. Also, it appears as complexity and size of the case study plays an important role in determining execution time and its variability across iterations.

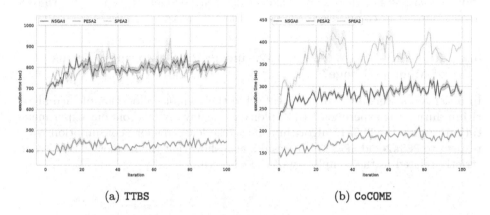

(a) TTBS   (b) CoCOME

Fig. 1. Comparison of algorithms execution time.

Figure 1 compares NSGAII, SPEA2, and PESA2 in terms of their execution times for TTBS and CoCOME, respectively. Darker lines report the mean over 30 runs for each iteration, while the bands represent 95% confidence intervals for the mean, and are computed for the same runs. Our results show substantial differences in the execution times of the algorithms, both on the same case study, and across them.

It is easy to notice that, regardless of the algorithm, the search is twice as fast in CoCOME that it is in TTBS, as it is obvious when observing the scale on the y-axis. PESA2 is clearly the fastest algorithm in both cases (around 400 sec in TTBS, and 180 sec in CoCOME). However, when it comes to comparing NSGAII and SPEA2, their execution time, while consistently larger than PESA2, is almost on par in TTBS, and noticeably apart in CoCOME. This suggests that the execution time might very well be dependent on the complexity and size of the specific case study. For instance, it looks like the more complex the case study is, the slower SPEA2 is. Therefore, it appears evident that the search policy used by SPEA2, *i.e.,* the dominance operator, is slower than the crowing distance used by NSGAII. Moreover, the search policy employed by PESA2, *i.e.,* the hyper-grid crowding distance, seems to be faster than the ones used by NSGAII and SPEA2, as it lasts half the time of the other two techniques.

Another interesting point, could be the stability of execution times, as it appears that the three algorithms exhibit different variability. For instance, PESA2 and NSGAII showed a more stable execution time in both the case studies, while SPEA2 showed a quite stable execution time with TTBS, and a considerably larger variability with CoCOME, with some abrupt changes. This might be due to the usage of the archive for storing generated solutions. When the case study is more complex, as it is the case for TTBS, the usage of the archive seems to help find a Pareto frontier, while the usage of two archives with a less complex case study results in prolonged executions. In fact, when a higher number of different solutions are found, these slower executions may be caused by the fact that a higher number of comparisons are needed to fill the two archives.

5.2 RQ_2: How Do NSGAII, SPEA2, and PESA2 Compare in Terms of Memory Usage?

In order to answer to the RQ_2 we collected the memory allocation of each algorithm during the experiments by exploiting the Java API. From our experimentation results, the NSGAII algorithm shows the least memory consumption with respect to PESA2 and SPEA2. Our results also show that the memory usage is not strictly related to the complexity of the case study.

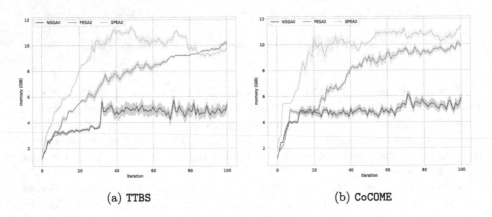

(a) TTBS   (b) CoCOME

Fig. 2. Comparison of algorithms memory usage.

Figure 2 shows the memory allocation of the three algorithms. NSGAII, SPEA2, and PESA2 occupy the same quantity of memory showing an increase trend of the memory usage around the 20 iterations, then NSGAII becomes almost flat. Moreover, SPEA2 shows a steep memory usage, and it occupies all the available memory after 40 iterations, while PESA2 showed a smooth but linear increase of the memory, and it filled the available memory after 80 iteration.

Undoubtedly, the NSGAII search policy is the least memory demanding among the three analyzed in our study, and it requires around 5 GiB when it stabilizes.

SPEA2 and PESA2, on the other hand, occupy almost all the available memory (*i.e.*, 12 GiB).

SPEA2 shows a different behavior in the two case studies, in our results. In the case of CoCOME, we can see an almost flat memory consumption around the 12 Gib after 20 iteration, while in TTBS we can observe a reduction of the memory allocation after 80 iterations. Combining the latter with the overall quality of the generated Pareto fronts (see Sect. 5.3), we can assume that SPEA2 cannot find better solutions after 80 iterations, thus any new solution was probably already stored in the two archives.

Finally, PESA2 showed an interesting trend, as it allocated more memory almost linearly. This might be due to the search policy of splitting the solution space in hyper-grids that will require to store new solutions when other locations of the solution space will be investigated by longer iterations. Therefore, we can expect that PESA2 will likely exceed the 12 GiB with longer iterations.

5.3 RQ_3: How Do NSGAII, SPEA2, and PESA2 Compare in Terms of Multi-objective Optimization Indicators?

In order to answer to the RQ_3 we graphically compare properties of the Pareto frontiers computed by each algorithm, and we use well-known indicators to estimate the quality of Pareto frontiers. From our results, the PESA2 algorithm is the best to search the solution space, with solutions closest to the reference Pareto in TTBS and CoCOME. Also, NSGAII generates solutions with the highest variability in both case studies. Finally, SPEA2 did not show any quality indicators with higher quality.

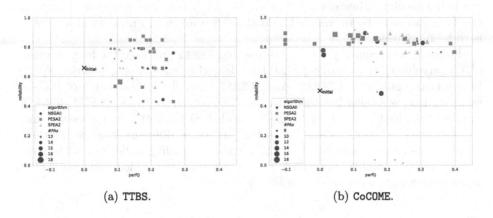

(a) TTBS.   (b) CoCOME.

Fig. 3. Comparison of reference Paretos.

The overall quality of computed Pareto frontiers (PF^c) is one of the most critical parameters to consider when comparing genetic algorithms. Fig. 3 depicts the PF^c generated by the three genetic algorithms for TTBS and CoCOME, where, in each plot, the top right quadrant is the optimal location for the optimization. Furthermore, we measure the quality of PF^c through the quality indicators listed in Table 1.

From Figs. 3a and 3b, we can clearly deduce that none of the subject algorithms shows the ability of finding solutions towards the top right quadrant in both the case studies. In fact, we can see that the solutions are organized in a vertical cluster in Fig. 3a, and in a horizontal one in Fig. 3b. Also, it appears that the optimization process selects similar refactoring actions, therefore generating almost identical solutions within the frontiers.

Furthermore, we can observe a different behavior of each algorithm in TTBS, and CoCOME. For example, PESA2 found the best solutions for CoCOME, in terms of *reliability* and *perfQ* (*e.g.*, see the rightmost squares in Fig. 3b), while this is not the case for TTBS, where, instead, PESA2 found the best solution in terms of *perfQ*, with worse *reliability* than the initial solution (see the square near the point (0.3, 0.4) in Fig. 3a).

Besides the graphical analysis, we performed a study of the quality of PF^c in both case studies, by exploiting established quality indicators for multi-objective optimization. It is important to recall that an indicator estimates the quality of a specific property of PF^c with respect to the reference Pareto frontier (PF^{ref}). Since PF^{ref} has not yet been defined for our case studies, we estimated the PF^{ref} as the set of non-dominated solutions produced by any algorithm. In particular, we computed the *Hypervolume (HV)* [35], *Inverted Generational Distance + (IGD+)* [24], *GSPREAD* [16], and *Epsilon (EP)* [16] quality indicators. We listed quality indicators *(Q Ind)* in Table 1, where the up arrow (↑) means the indicator is to be maximized, and the down arrow (↓) means the indicator is to be minimized.

From our experimental results, we see that PESA2 produced the highest value of Hypervolume, thus proving that the algorithm covered the solution space better than NSGAII and SPEA2. Also, PESA2 showed the best value of IGD+, meaning that solutions belonging to the PF^c are closer to the PF^{ref}. NSGAII produced the best value of generalized spread (GSPREAD), thus indicating that the solutions in NSGAII Pareto frontiers are more different from each other. Finally, our results prove that SPEA2 computes quality indicators with good quality only for CoCOME. Therefore, it seems that SPEA2 is able to find good PF^c when case studies with lower complexity.

Table 1. Quality indicators to establish the overall quality of Pareto frontiers.

Q Ind	# iter	NSGAII		PESA2		SPEA2	
		TTBS	CoCOME	TTBS	CoCOME	TTBS	CoCOME
HV ($\uparrow$)	102	0.22433	0.07022	0.50909	0.44431	0.14467	0.36521
IGD+ ($\downarrow$)	102	0.11221	0.06005	0.04683	0.04046	0.10270	0.06620
GSPREAD ($\downarrow$)	102	0.16013	0.12675	0.38391	0.52451	0.39153	0.33592
EP ($\downarrow$)	102	0.33333	0.20339	0.20000	0.10000	0.50000	0.36191

6   Lesson Learned

Genetic algorithms have proved to help optimize quantifiable metrics, such as performance and reliability. Their ability to search for optimal solutions is influenced by several configuration parameters. In our experience, we noticed that each configuration parameter has a different impact on the overall performance, *e.g.*, the population size impacts the execution time and the memory usage. A wider initial population size requires longer execution times to generate individuals, and it might produce stagnation during the optimization [4] that, in turn, might hamper the quality of Pareto frontiers. Furthermore, in model-based software refactoring, a wider initial population also implicates a higher memory consumption, because entire models need to be loaded in memory for the refactoring to be performed. Hence, it is crucial to find the optimal trade-off between the configuration parameters and the quality of the Pareto frontiers.

Besides the initial population, crossover and mutation operators might impact the execution time. For instance, a higher mutation probability will obviously produce more frequent mutations within the population. The more mutations are produced, the higher the probability of having an invalid individual, thus requiring additional time to check for feasibility, repair or even change the individual entirely. The crossover probability, instead, impacts the execution time since combinations of individuals are more frequent. Furthermore, the crossover operator requires time to perform the combination and might also generate invalid individuals. Therefore, using the right crossover and mutation probabilities is crucial for the time and quality of subsequent populations. This is clearly an opportunity for further research on heuristic to estimate some configuration values, since it would be impractical to evaluate every parameter combination. In future work, we plan to examine how different configurations affect the resulting quality and performance of different genetic algorithms.

We cannot guarantee that our analysis can be generalized in other domains or with other modeling notations. However, in the context in which we performed our study, there is not an in-depth analysis of performance traits of genetic algorithms. We believe this study might open a research direction on improving genetic algorithm performance for model-based refactoring optimization. For example, in a recent work, Di Pompeo and Tucci [14] studied the possibility of reducing the search time by limiting it with a budget. Also, knowing how algorithms compare in terms of performance might open to even using them in an interactive optimization process, where the designer could be involved in a step-by-step decision process aided by the automation provided by the algorithms, but bounded in time. In such a scenario, the designer could be at the core of the process, potentially making optimization trade-offs at each step.

7   Threats to Validity

Our results may be affected by threats related to the specific versions of Java and the JMetal library. We mitigated these threats by using the same configuration for the Java Virtual Machine and the same JMetal version in each run. In particular, we used the OpenJDK 11 with default configuration, and we built the implementation on JMetal v5.10.

Multiple factors may influence the performance measurements gathered during the experiments and, therefore, could undermine our conclusions. However, we mitigated external influences by disabling any other software running on the same server, and we repeated each experiment 30 times as suggested in [3].

Also, the study we conducted in this paper may be affected, as for any performance analysis experiment, by the input data. Although we use two case studies, our deductions may change if other case studies, *i.e.*, different software models, are employed as inputs to the optimization problem. However, we considered two models presented in [15,19] that has been already used in other performance analysis problems [9,10,14]. To the best of our knowledge, there are no previous studies that analyzed and compared performance of multi-objective algorithms in the context of software model refactoring, as we did in this study. Therefore, although the paper may suffer from this threat to conclusion validity, it represents a first investigation in this direction.

The overall quality of Pareto frontiers generated by each algorithm has been estimated through well-known quality indicators. These indicators leverage the estimation by comparing a Pareto frontier to problem-specific reference points. Since in our experimentation these reference points are not yet available, we computed them as the non-dominated points within every Pareto frontier of each run of each algorithm. Therefore, the reference points might affect the overall quality computation, and we further investigate the usage of more appropriated reference points.

Finally, our results may be affected by threats related to the configurations of the genetic algorithms. For example, the number of iterations can influence performance results. We cannot be sure to have effectively mitigated these threats because of the long execution time required to run each configuration. Such long execution times make trying many alternative configurations unfeasible. For this reason, we used a configuration in an attempt to detect performance flaws that may only manifest during longer executions.

8   Conclusion

This study presented a performance comparison of three genetic algorithms, *i.e.*, NSGAII, SPEA2, and PESA2. We selected those algorithms due to their wide usage in the software refactoring context and their search algorithm characteristics.

We compared the execution time, the memory allocation, and the quality of the produced Pareto frontiers. We collected performance metrics by using two case studies presented in [15, 19] by executing 30 different runs, and we compared the overall quality of Pareto frontiers through specific quality indicators, such as Hypervolume and IGD+. Our analysis can summarize that PESA2 is the fastest algorithm, and NSGAII is the least memory-demanding algorithm. Finally, SPEA2 has shown the worst memory usage as well as the worst execution time. We will further investigate memory consumption by employing a more sophisticated memory profiling, which might introduce an overhead within the measurements.

Concerning the overall quality of the produced Pareto frontiers, we found that PESA2 produced the most densely populated Pareto frontiers, while NSGAII generated the least densely populated frontiers. PESA2 has also shown a linear memory consumption, thus we intend to further analyze the trend by exploring longer execution in terms of the number of iterations.

Furthermore, we intend to investigate if our findings can be generalized to other case studies, different algorithms, and different kinds of refactoring actions, as those aimed at other non-functional properties, such as availability [12].

Acknowledgments. Daniele Di Pompeo and Michele Tucci are supported by European Union - NextGenerationEU - National Recovery and Resilience Plan (Piano Nazionale di Ripresa e Resilienza, PNRR) - Project: "SoBigData.it - Strengthening the Italian RI for Social Mining and Big Data Analytics" - Prot. IR0000013 - Avviso n. 3264 del 28/12/2021.

References

1. Aleti, A., Björnander, S., Grunske, L., Meedeniya, I.: ArcheOpterix: an extendable tool for architecture optimization of AADL models. In: ICSE 2009 Workshop on Model-Based Methodologies for Pervasive and Embedded Software. MOMPES 2009, May 16, 2009, Vancouver, Canada, pp. 61–71. IEEE Computer Society, Washington, DC, USA (2009)
2. Aleti, A., Buhnova, B., Grunske, L., Meedeniya, I.: Software architecture optimization methods: a systematic literature review. IEEE Trans. Software Eng. **39**(5), 658–683 (2013)
3. Arcuri, A., Briand, L.C.: A practical guide for using statistical tests to assess randomized algorithms in software engineering. In: Taylor, R.N., Gall, H.C., Medvidovic, N. (eds.) Proceedings of the 33rd International Conference on Software Engineering, ICSE 2011, Waikiki, Honolulu, HI, USA, May 21–28, 2011, pp. 1–10, ACM (2011). https://doi.org/10.1145/1985793.1985795
4. Arcuri, A., Fraser, G.: Parameter tuning or default values? an empirical investigation in search-based software engineering. Empir. Softw. Eng. **18**(3), 594–623 (2013). https://doi.org/10.1007/s10664-013-9249-9

5. Becker, S., Koziolek, H., Reussner, R.H.: The Palladio component model for model-driven performance prediction. J. Syst. Softw.**82**(1), 3–22 (2009)
6. Blum, C., Roli, A.: Metaheuristics in combinatorial optimization: overview and conceptual comparison. ACM Comput. Surv. **35**(3), 268–308 (2003). https://doi.org/10.1145/937503.937505
7. Corne, D.W., Jerram, N.R., Knowles, J.D., Oates, M.J.: Pesa-ii: Region-based selection in evolutionary multiobjective optimization. In: GECCO, pp. 283–290 (2001), ISBN 1558607749, https://doi.org/10.5555/2955239.2955289
8. Cortellessa, V., Di Pompeo, D.: Analyzing the sensitivity of multi-objective software architecture refactoring to configuration characteristics. Inf. Softw. Technol. **135**, 106568 (2021). https://doi.org/10.1016/j.infsof.2021.106568
9. Cortellessa, V., Di Pompeo, D., Eramo, R., Tucci, M.: A model-driven approach for continuous performance engineering in microservice-based systems. J. Syst. Softw. **183**, 111084 (2022). https://doi.org/10.1016/j.jss.2021.111084
10. Cortellessa, V., Di Pompeo, D., Stoico, V., Tucci, M.: On the impact of performance antipatterns in multi-objective software model refactoring optimization. In: 47th SEAA 2021, Palermo, Italy, September 1–3, 2021, pp. 224–233, IEEE (2021). https://doi.org/10.1109/SEAA53835.2021.00036
11. Cortellessa, V., Di Pompeo, D., Stoico, V., Tucci, M.: Many-objective optimization of non-functional attributes based on refactoring of software models. Inf. Softw. Technol. **157**, 107159 (2023). https://doi.org/10.1016/j.infsof.2023.107159
12. Cortellessa, V., Eramo, R., Tucci, M.: Availability-driven architectural change propagation through bidirectional model transformations between UML and petri net models. In: IEEE International Conference on Software Architecture, ICSA 2018, Seattle, WA, USA, April 30 - May 4, 2018, pp. 125–134, IEEE Computer Society (2018). https://doi.org/10.1109/ICSA.2018.00022
13. Deb, K., Pratap, A., Agarwal, S., Meyarivan, T.: A fast and elitist multiobjective genetic algorithm: NSGA-II. IEEE Trans. Evol. Comput. **6**(2), 182–197 (2002)
14. Di Pompeo, D., Tucci, M.: Search budget in multi-objective refactoring optimization: a model-based empirical study. In: 48th Euromicro Conference on Software Engineering and Advanced Applications, SEAA 2022, pp. 406–413, IEEE (2022). https://doi.org/10.1109/SEAA56994.2022.00070,to appear
15. Di Pompeo, D., Tucci, M., Celi, A., Eramo, R.: A microservice reference case study for design-runtime interaction in MDE. In: STAF 2019 Co-Located Events Joint Proceedings: 1st Junior Researcher Community Event, 2nd International Workshop on Model-Driven Engineering for Design-Runtime Interaction in Complex Systems, and 1st Research Project Showcase Workshop co-located with Software Technologies: Applications and Foundations (STAF 2019), Eindhoven, The Netherlands, July 15–19, 2019, CEUR Workshop Proceedings, vol. 2405, pp. 23–32, CEUR-WS.org (2019)
16. Durillo, J.J., Nebro, A.J.: jmetal: A Java framework for multi-objective optimization. Adv. Eng. Softw. **42**(10), 760–771 (2011)
17. Feiler, P.H., Gluch, D.P.: Model-Based Engineering with AADL - An Introduction to the SAE Architecture Analysis and Design Language. SEI series in software engineering, Addison-Wesley (2012)
18. Gadhvi, B., Savsani, V., Patel, V.: Multi-objective optimization of vehicle passive suspension system using NSGA-II, SPEA2 and PESA-II. Procedia Technol. **23**, 361–368 (2016)

19. Herold, S., et al.: Cocome - the common component modeling example. In: The Common Component Modeling Example: Comparing Software Component Models, LNCS, vol. 5153, pp. 16–53 (2008) https://doi.org/10.1007/978-3-540-85289-6_3

20. Hiroyasu, T., Nakayama, S., Miki, M.: Comparison study of spea2+, spea2, and NSGA-II in diesel engine emissions and fuel economy problem. In: CEC, pp. 236–242 (2005). https://doi.org/10.1109/CEC.2005.1554690

21. Kessentini, M., Sahraoui, H.A., Boukadoum, M., Benomar, O.: Search-based model transformation by example. J. Softw. Syst. Model. **11**(2), 209–226 (2012). https://doi.org/10.1007/s10270-010-0175-7

22. King, R.A., Deb, K., Rughooputh, H.: Comparison of NSGA-II and SPEA2 on the multiobjective environmental/economic dispatch problem. Univ. Mauritius Res. J. **16**(1), 485–511 (2010)

23. Koziolek, A., Koziolek, H., Reussner, R.H.: PerOpteryx: automated application of tactics in multi-objective software architecture optimization. In: 7th International Conference on the Quality of Software Architectures, pp. 33–42, ACM, New York, New York, USA (2011)

24. López, E.M., Coello, C.A.C.: An improved version of a reference-based multi-objective evolutionary algorithm based on igd$^+$. In: Aguirre, H.E., Takadama, K. (eds.) GECCO, pp. 713–720 (2018). https://doi.org/10.1145/3205455.3205530

25. Mariani, T., Vergilio, S.R.: A systematic review on search-based refactoring. J. Inform. Softw. Technol. **83**, 14–34 (2017)

26. Martens, A., Koziolek, H., Becker, S., Reussner, R.H.: Automatically improve software architecture models for performance, reliability, and cost using evolutionary algorithms. In: ICPE 2010 - Proceedings of the 1st ACM/SPEC International Conference on Performance Engineering, pp. 105–116 (2010). https://doi.org/10.1145/1712605.1712624

27. Menascé, D.A., Ewing, J.M., Gomaa, H., Malek, S., Sousa, J.P.: A framework for utility-based service oriented design in SASSY. In: Adamson, A., Bondi, A.B., Juiz, C., Squillante, M.S. (eds.) Proceedings of the First Joint WOSP/SIPEW International Conference on Performance Engineering, pp. 27–36 (2010). https://doi.org/10.1145/1712605.1712612

28. Ni, Y., et al.: Multi-objective software performance optimisation at the architecture level using randomised search rules. Inf. Softw. Technol. **135**, 106565 (2021). https://doi.org/10.1016/j.infsof.2021.106565

29. Ouni, A., Kessentini, M., Inoue, K., Cinnéide, M.Ó.: Search-based web service antipatterns detection. IEEE Trans. Serv. Comput. **10**(4), 603–617 (2017). https://doi.org/10.1109/TSC.2015.2502595

30. Rago, A., Vidal, S.A., Diaz-Pace, J.A., Frank, S., van Hoorn, A.: Distributed quality-attribute optimization of software architectures. In: Proceedings of the 11th Brazilian Symposium on Software Components, Architectures and Reuse, SBCARS, pp. 7:1–7:10 (2017). https://doi.org/10.1145/3132498.3132509

31. Ramírez, A., Romero, J.R., Ventura, S.: A survey of many-objective optimisation in search-based software engineering. J. Syst. Softw. **149**, 382–395 (2019)

32. Zhao, F., Lei, W., Ma, W., Liu, Y., Zhang, C.: An improved spea2 algorithm with adaptive selection of evolutionary operators scheme for multiobjective optimization problems. Math. Probl. Eng. **2016**, 1–20 (2016)

33. Zhou, X., et al.: Fault analysis and debugging of microservice systems: industrial survey, benchmark system, and empirical study. TSE **47**(2), 243–260 (2021). https://doi.org/10.1109/TSE.2018.2887384

34. Zitzler, E., Laumanns, M., Thiele, L.: SPEA 2: Improving the strength pareto evolutionary algorithm. TIK-report 103, Swiss Federal Institute of Technology (ETH) Zurich (2001)
35. Zitzler, E., Thiele, L.: Multiobjective optimization using evolutionary algorithms - A comparative case study. In: Eiben, A.E., Bäck, T., Schoenauer, M., Schwefel, H. (eds.) Parallel Problem Solving from Nature, LNCS, vol. 1498, pp. 292–304 (1998). https://doi.org/10.1007/BFb0056872

An Approach Using Performance Models for Supporting Energy Analysis of Software Systems

Vincenzo Stoico[1]([✉])[iD], Vittorio Cortellessa[1][iD], Ivano Malavolta[2][iD],
Daniele Di Pompeo[1][iD], Luigi Pomante[1][iD], and Patricia Lago[2][iD]

[1] Department of Information Engineering, Computer Science and Mathematics,
University of L'Aquila, L'Aquila, Italy
vincenzo.stoico@graduate.univaq.it
{vittorio.cortellessa,daniele.dipompeo,luigi.pomante}@univaq.it
[2] Department of Computer Science, Vrije Universiteit Amsterdam, Amsterdam,
The Netherlands
{i.malavolta,p.lago}@vu.nl

Abstract. Measurement-based experiments are a common solution for assessing the energy consumption of complex software systems. Since energy consumption is a metric that is sensitive to several factors, data collection must be repeated to reduce variability. Moreover, additional rounds of measurements are required to evaluate the energy consumption of the system under different experimental conditions. Hence, accurate measurements are often unaffordable because they are time-consuming. In this study, we propose a model-based approach to simplify the energy profiling process and reduce the time spent performing it. The approach uses Layered Queuing Networks (LQN) to model the scenario under test and examine the system behavior when subject to different workloads. The model produces performance estimates that are used to derive energy consumption values in other scenarios. We have considered two systems while serving workloads of different sizes. We provided 2K, 4K, and 8K images to a Digital Camera system, and we supplied bursts of 75 to 500 customers for a Train Ticket Booking System. We parameterized the LQN with the data obtained from short experiment and estimated the performance and energy in the cases of heavier workloads. Thereafter, we compared the estimates with the measured data. We achieved, in both cases, good accuracy and saved measurement time. In case of the Train Ticket Booking System, we reduced measurement time from 5 h to 35 min by exploiting our model, this reflected in a Mean Absolute Percentage Error of 9.24% in the estimates of CPU utilization and 8.72% in energy consumption predictions.

Keywords: Layered Queuing Networks · Performance Analysis · Energy Consumption · Software

© The Author(s), under exclusive license to Springer Nature Switzerland AG 2023
M. Iacono et al. (Eds.): EPEW/ASMTA 2023, LNCS 14231, pp. 249–263, 2023.
https://doi.org/10.1007/978-3-031-43185-2_17

1   Introduction

The ubiquity of ICT devices triggered a continuous digitalization of information, thus facilitating access, storage, and manipulation of data. ICT brought several benefits to society, such as monitoring the health conditions of people in real-time or having almost universal access to educational content. However, continuous digitization has also downsides. A considerable amount of information demands expensive resources for processing and storage, with a consequent rising need for energy to build and power ICT devices. As energy demand increases, the impact of ICT in terms of carbon dioxide (CO_2) emissions becomes significant [21]. Belkhir et al. estimate that ICT devices will produce 14% of global CO_2 emissions by 2040 [5].

As already discussed in 2018 by Georgiou et al. [15] in the context of IoT systems, technology has made considerable advancements for increasing hardware power consumption savings, which however can be undermined by poor design decisions at the software level. Software energy optimization is a hard endeavor, where multiple (and frequently conflicting) design and implementation decisions can influence the energy footprint of the software [21]. Making developers aware of the impact of their decisions on the energy consumption of their software is fundamental to cutting it down [10]. However, energy optimization cannot be pursued in isolation, because it may negatively impact on other non-functional attributes of software and, in particular, on performance. Hence, software design decisions have to induce acceptable tradeoffs between the satisfaction of performance requirements and power consumption savings [12]. Energy/performance tradeoffs can be analyzed by measurement-based experiments, although they can be very time-consuming and they need contextual conditions to be taken under control (e.g., temperature of devices) for achieving reliable results. Modeling is often a valuable alternative to measurements, especially in cases where enough information about the software system and its context is known. Obviously, energy/performance models have to hold an appropriate level of abstraction that allows practitioners not to miss relevant aspects of the system and, at the same time, to keep an acceptable complexity of the model evaluation process [6].

This study explores the combination of measurement-based experiments and modeling in the context of energy/performance analysis of software systems, in order to benefit from the advantages of both of them. In particular, we investigate how to exploit performance models (specifically, Layered Queuing Networks – LQNs) to reduce the experimentation time while keeping a high accuracy in the energy consumption estimate of a software system. Although LQNs are a well-known modeling notation for software performance analysis [13,20], their adoption for energy consumption analysis has yet to be developed. LQNs fit our purposes as they represent system resources as time-consuming entities. As energy consumption is a time-based metric, it is possible to define a relationship between performance and energy consumption according to resource utilization (namely, the amount of time a resource is busy). We reduce the reality gap between the LQN model and the system under analysis by systematically refining the LQN model using data obtained from a small-scale measurement-based experiment. After achieving satisfactory accuracy, the LQN can be used to study the system under different workloads and get corresponding resource utilization estimates. These estimates can be multiplied by the energy consumed per second by each resource to obtain the total energy consumed while the resources are busy. We tested our approach on two different systems: Digital Camera, which we employ as a running example, and Train Ticket Booking System. The former is an image processing application that we deployed on an embedded platform, while the latter is a container-based web application for managing train bookings. For the Digital Camera, the supplied workloads correspond to batches of images of different resolutions, namely 2K, 4K, and 8K. Instead, for Train Ticket Booking System the workloads consists of bursts of 75, 150, 225, 300, 375, 450, and 500 customers. We parametrized the LQN with data measured for the batch of 2K images and the 75-customer burst, respectively, then we estimated resource utilization and energy consumption for different scenarios. Promising results emerged by comparing the measured data with the estimates. The Mean Absolute Percentage Error (MAPE) obtained for Train Ticket Booking System equals 9.24% for CPU Utilization and 8.47% for energy consumption. At the same time, we reduced experimentation time from 5 h to 35 min.

Hence, the main contributions of this study are: (i) an approach for using LQNs to make accurate energy estimations of a software system, (ii) a preliminary empirical evaluation of the proposed approach on two different software systems across different domains, (iii) a replication package for the independent verification and replication of the performed evaluation[1]. The paper is structured as follows. Section 2 presents energy consumption basics and describes the Layered Queuing Networks. Section 3 delves into the approach and shows it through a running example: the Digital Camera case study. Section 4 outlines the experiments and the results achieved on the Train Ticket Booking System case study. Threats to validity and related work are discussed, respectively, in Sect. 5 and Sect. 6. The paper ends with conclusions in Sect. 7.

2   Background

2.1   Software Energy Measurement

The physical quantities used for expressing the energy consumed by software executions are electrical energy and electrical power. Electrical energy quantifies the amount of work needed to drive current through a circuit, while electrical power refers to the rate the energy is consumed by the circuit at any instant. Energy is commonly measured in joule (J), which is defined in the International System of Units (SI). Power, unlike energy, expresses a rate. Indeed, power is defined as the total energy consumed over time measured in joule per second ($\frac{J}{s}$) or, following the standard SI, in watt (W).

Literature includes a plethora of tools for measuring software energy consumption [9]. We distinguish between energy profilers and power monitors.

---

[1] https://doi.org/10.5281/zenodo.7877782.

Energy profilers are software tools providing an estimation of the energy consumed by a running application. Compared with power monitors, energy profilers are easy to set up but are less accurate as they provide estimates. Among the most popular ones we have: `perf` and `powerstat`. Power Monitors are hardware devices wired directly to the system to profile, e.g., to the battery of the system. Therefore, they are more accurate but also more complex to set up. In this work, we exploit two power monitors for our experiments: the Monsoon [19] and the Watts up Pro? [22]. Instead of reporting total energy usage in joules, several power monitors report the power consumption in watts. They read, at each instant t, the current intensity (I) and voltage (V) and calculate the power as $P = I \times V$. The total energy consumption (E) can be derived from the power consumption. When the power consumption is constant, the total energy spent is proportional to the observation interval Δt, that is $E = P \times \Delta t$. As previously mentioned, some power monitors calculate power values querying the system every instant t over an observation period Δt. This process results in a dataset, where each row is formed by the power value in watts and the timestamp of the reading. This dataset describes the distribution of power values consumed during Δt. Since power corresponds to the rate at which energy is consumed over time, it is possible to retrieve the total amount of energy spent between two instants t_0 and t_n, by calculating the area bounded by these two instants underneath the distribution. Formally, this area can be calculated by integrating the power consumption values over t_0 and t_n.

2.2   Layered Queuing Network

Layered Queuing Networks (LQNs) are used to describe and analyze the performance of a system [24]. An LQN captures the behavior of a system as a set of interacting entities sending and servicing requests. Incoming requests generate the workload that is handled by system resources such as CPU or Disk. If a request comes to a resource that is already busy, the request is queued. Such a model of computation is peculiar to ordinary queuing models. LQNs extend ordinary queuing models by implementing simultaneous resource possession. Simultaneous resource possession occurs when a resource is blocked waiting for another to finish serving a request. Figure 3 shows the LQN used for analyzing a Train Ticket Booking System. The root task may be used to specify the workload which the system will undergo or receive requests from the environment. In the former case, the task is named reference task and represents the number of users in the system, while in the second case, requests arrive following a rate specified with the λ parameter. System entities are shown as parallelograms and are called tasks. A task provides service through one or multiple entries and has a single host processor. Processors are represented with circles connected to a task and embody system resources. Thus, processors handle the workload generated by the entries. Entries are represented as rectangles within tasks and specify a demand corresponding to the mean time the processor is busy serving the entry. Communication among tasks is described by arrows connecting the entries. These arrows are labeled with the mean number of requests the client task sends to the server task. Carleton University provides a suite of tools including modeling languages and an analytic solver to create and retrieve performance metrics [7].

3   Our Approach

As mentioned in the introduction, we exploit performance models to assist designers in making energy-related decisions. For this purpose, we have conceived a modeling process to reach a good trade-off between abstraction and accuracy of estimates. Developing models of existing systems may help to understand them better, which includes identifying flaws and finding opportunities for improvement. At design time, models are used to describe design choices or to verify conformance to the requirements. Considering that implementation details highly influence the energy consumption of the software, we chose to proceed bottom-up and exploit models for studying existing software systems. This implies the use of models that can be simulated or solved analytically to study "what if" cases and gain fast insights into the system under study. In addition, we see modeling as an opportunity to reduce the complexity and time of measurements. Our approach combines both strengths of measurements and abstraction.

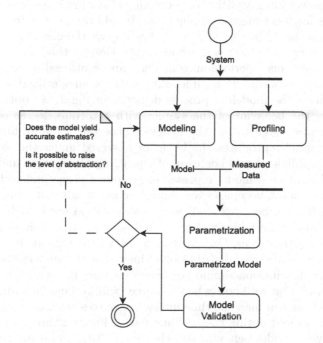

Fig. 1. The process underlying our approach.

Our modeling approach is schematically represented in Fig. 1. Each rounded box represents an activity, while labels on the edges embody the exchanged artifacts. Initially, an existing system is modeled and profiled. Profiling means measuring the characteristics of the system that we planned to evaluate using the models, in this case performance and energy consumption. Modeling and profiling activities can be conducted in parallel. After the end of these two activities, the measured data are used to parametrize the model. Taking performance as an example, we can look at the rate at which requests arrive at the system, or how long it takes software components to handle a request. Parameters such as arrival rate, or service time are the input parameters of the model. Once the model is parametrized, it is validated and possibly refined. In the validation step, the correctness of the model and the accuracy of the estimations are considered. Moreover, during this phase, designers refine the model removing unnecessary details and simplifying it. The process stops when consensus about model accuracy and abstraction level is reached. We envisage that, under a set of assumptions, it is possible to exploit the flexibility of performance models for estimating energy consumption. Briefly speaking, since energy consumption is a metric based on time, we can relate energy consumption and resource utilization over a fixed observation time. Our approach, at the moment, considers only the cases in which *energy consumption grows linearly with execution time*. However, this is not always true. As reported by Cruz et al. [9], some mobile architectures have fast but power-hungry CPUs for processing heavy tasks and slower but more efficient CPUs for less time-consuming tasks. In addition, CPUs frequency scaling mechanisms should be considered, as they impact the non-linearity between workload intensity and power consumption [18]. Despite this, the approach has significant benefits since performance models can be utilized to scale workload and retrieve energy consumption values along with resource utilization estimates.

As a result of the modeling process depicted in Fig. 1, we obtain a model approximating the behavior of the system with a certain degree of accuracy. This approximation stems from the behavior observed while profiling the system. So, the model reproduces the behavior observed under the experimental setup of the profiling phase. For example, profiling the system under a given workload will produce a model representing resource usage under that specific workload. This aspect becomes even more important when it comes to power consumption. Indeed, behavior and power consumption are heavily connected. Different experimental settings will result in different system behaviors and thus different power distributions. On the other hand, we can exploit the relationship between behavior and power distribution to infer that the modeled behavior will induce a power distribution similar to the system one. Indeed, from the performance model solution we know when resources will be busy handling requests. For this reason, we can map the time interval resources are busy onto the power distribution measured during the profiling phase. Figure 2 shows an example of mapping between model behavior and the power distribution measured on the system. In this example, the power distribution is measured while running several times a software on a server. The blue cross on the x-axis represents the end of a single execution. Requests arrive periodically and require CPU and disk at the same time. The CPU, depicted by the blue area, is occupied for a longer interval than the disk, which is depicted in orange. White areas represent the time interval when resources are idle. Each colored segment underneath the power distribution represents the energy consumed by a resource during that time interval (see Sect. 2.1). Therefore, it is possible to obtain the energy consumption of a resource by integrating the areas where the resource was busy (1). Thus, given a resource, the sum of all the intervals in which a resource was busy (i.e., same color intervals in Fig. 2) represents the total energy demand (i.e., $ED(res)$) during the observation time (Eq. 2). Accordingly, we can derive the average consumption per visit of each resource, namely $E(res)$ (Eq. 3). This value is tied to a visit, so it reflects the energy spent serving a specific software task. For example, if the CPU spent 2 s serving a visit, the energy spent per visit refers to the consumption during two seconds. So, the energy consumed per visit is measured in $\frac{Joule}{Visit}$. The Joule consumed per second (i.e., $e(res)$) can be obtained from $E(res)$, which is unbound to the size of a software task. We calculated $e(res)$ by dividing $E(res)$ by the average time the resource is busy performing a software task (Eq. 4). Since $e(res)$ is decoupled from the size of the task visiting a resource, we can use it to estimate the energy consumed by a resource busy serving a task with whatever size. In other words, since $e(res)$ is measured in $\frac{Joule}{s}$, it is untied from the workload and becomes a property of the resource. Thus, we assume that it does not sensibly vary by scaling the workload. In this way, we can use $e(res)$ as a multiplier of the busy time of a resource and estimate the energy consumption in case of larger workload sizes. This method allows designers to scale up the estimations retrieved during low-effort experiments to predict the energy and performance of more complex scenarios. In this way, designers can very quickly obtain estimates for complex cases and thus avoid laborious and complex experiments.

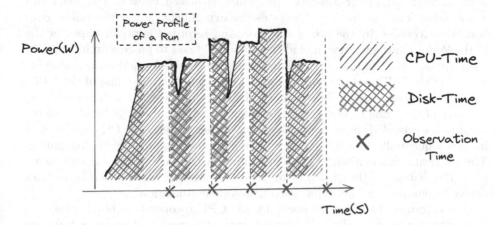

Fig. 2. Sample Power Profile highlighting CPU and Disk busy time, respectively, in blue and orange. (Color figure online)

$$E(res, i) = \int_{t0,i}^{S_{res}} P(t)\, dt [\frac{Joule}{Visit}] \tag{1}$$

$$ED(res) = \sum_{i=1}^{\#Visit} \int_{t0,i}^{S_{res,i}} P(t)\, dt [Joule] \tag{2}$$

$$E(res) = \frac{ED(res)}{\#Visit} [\frac{Joule}{Visit}] \tag{3}$$

$$e(res) = \frac{E(res)}{S(res)} [\frac{Joule}{s}] \tag{4}$$

where:

res = a resource

$t0$ = the instant a resource starts to be busy

i = ith visit to a resource

$\#Visit$ = total number of visits to a resource

S_{res} = the average time the resource spends serving a software task

Running Example: The Digital Camera

We deployed the application of a Digital Camera (DC) on a BeagleBone Black (BBB) [4] development platform. The BBB is a ARM-based single core platform equipped with Linux Debian, which, in our setting, executes only the services of the operating system and the DC. As our approach envisions (see Sect. 3), we set an experiment in which the DC is subject to a synthetic workload. While the DC processes the workload, we measure the power required by the BBB using a Monsoon Power Monitor [19], which is placed between the BBB and a notebook. The notebook orchestrates the experiment, stores the power consumption recorded by the power monitor, and records the performance of the BBB. Performance measures include the time DC takes to process an image (i.e., the response time) and resources utilization. The notebook queries the operating system of the BBB and retrieves, for each execution, the busy time of the CPU.

The workload consists of a stream of image batches. A batch contains 30 pictures of the same format chosen between 2K, 4K, and 8K. A total of thirty batches are provided to the application, i.e., 10 per format. The DC processes all images sequentially in a batch, then pauses for two minutes before continuing. Their arrival is randomized to avoid any influence of image size on measurements [23]. The batch size is set to 30 to achieve statistical significance of metrics derived from observed behavior, such as resource utilization.

We determined the energy spent by the CPU to operate a batch of a particular format based on Eq. 3. Therefore, using the measured response time, we calculated the e multiplier using Eq. 4, which represents the average power in $\frac{Joule}{s}$ spent by the CPU. As Sect. 3 remarks, e is a property of the resource, which is detached from the characteristics of the visiting task and therefore has the same value across scaled workloads. If we know or estimate the $S(res)$ variable of Eq. 4, we can use the e profiled for the batches of 2K images to discover the average energy spent for batches of format 4K and 8K. Consequently, we obtain estimates of the energy consumed for batches of 4K and 8K images without taking any measurements. We used an LQN to simulate the arrival of 30 images of 4K and 8K and estimate the response time for the corresponding batch. The LQN has a single task, representing the DC, connected to a single processor, which embodies the CPU. Incoming images arrive following a rate (i.e., λ). By varying the λ parameter, we replicated sequential arrival of 4K and 8K images, while changing the service time of the DC task, we set the average CPU processing time per image. In fact, the service time of the CPU differs based on image size. The model yields estimations concerning both the time it takes the CPU to handle a batch of 30 images and CPU usage.

Table 1 provides the results according to the image format of a batch. The table includes the arrival rate λ, the time spent handling a batch, i.e. the response time, CPU utilization, and energy consumed per batch. Columns containing two values show the measured value on the left and the corresponding estimates on the right. The estimates for the energy consumed per batch are calculated multiplying $1.57\frac{J}{s}$, which corresponds to the e calculated for batches of 2K images, by $240.30s$ and $960.60s$, which is the estimated response time for batches of 4K and 8K images. By comparing estimates to the measurements, it can be concluded that the results are promising. Finally, the BBB data sheet reports a range of $1.04\frac{J}{s}$ to $2.3\frac{J}{s}$ consumed by the platform when subject to various load [4]. The $e(CPU)$ multiplier of the DC falls within this range. This observation confirms the reliability of the $e(CPU)$ used for the analysis and the value of the approach for evaluating particular combinations of hardware and software. The small complexity of this case study and the availability of the DC source code, have simplified the mapping process between power distribution and the busy time of the CPU. Indeed, the DC executes only few functions in sequence, and from the source code we could see when the CPU operations were performed. In addition to the plethora of analyses that can be done by varying LQN parameters, we also gain benefits in terms of time. In fact, we obtained energy estimations without performing experiments in the 4K and 8K cases.

4 Evaluation

In light of the promising results obtained from the experiments with the Digital Camera, we decided to validate the method using a more widely used case study: Train Ticket Booking System (TTBS) [14], which is an application comprised of 68 Docker containers that manages bookings of a railway system.[2]

Consistently with the method described in Sect. 3, we set up a testbed for profiling TTBS and supplied a synthetic workload to the application. Therefore,

---

[2] For our measurements, we used release 0.0.4: https://github.com/FudanSELab/train-ticket/tree/release-0.0.4.

Table 1. Results for the Digital Camera and Train Ticket Booking System according to the size of the input. Legend: IS: Input Size; λ: Arrival Rate; R: Response Time; U: CPU Utilization; e: Average Power Consumption; EC: Average Energy Consumed. Columns with double values indicate measured and estimated values (on the right).

IS	λ ($\frac{images}{s}$)	R (s)	U (%)	e ($\frac{J}{s}$)	EC (J)
Digital Camera					
2K	0.48	60.30 - 60.30	96.30 - 96.48	1.57	95.27 -95.16
4K	0.12	240.36 - 240.30	96.76 - 96.12	1.59	382.46 - 379.24
8K	0.03	960.73 - 960.60	97.39 - 96.06	1.59	1537.96 - 1516.04
Train Ticket Booking System					
75	6.45	4.09 - 4.63	35.96 - 39.86	78.56	321.99 - 364.17
150	9.17	8.89 - 9.27	50.72 - 56.73	80.89	719.82 - 728.34
225	10.32	13.86 - 13.90	57.88 - 63.77	82.45	1143.19 - 1092.51
300	11.02	18.96 - 18.54	61.70 - 68.10	82.28	1560.54 - 1456.68
375	11.34	24.95 - 23.17	64.91 - 70.08	82.02	2047.20 - 1820.85
450	11.51	29.89 - 27.81	66.32 - 71.13	82.77	2474.40 - 2185.03
500	11.64	33.73 - 30.90	67.67 - 71.94	82.67	2788.76 - 2427.81

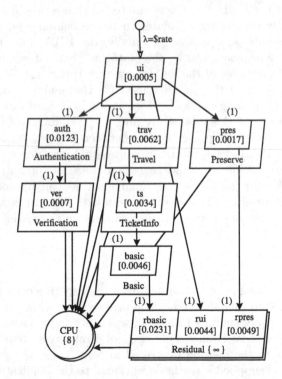

Fig. 3. Layered Queuing Network of Train Ticket Booking System. The variable $rate varies based on the workload to provide to the model.

we modeled TTBS through an LQN (see Fig. 3) and parametrized the model with the measurements retrieved during the fastest experiment. For the sake of space, we do not show the iterations, described in Sect. 3, performed to obtain the LQN. Each task in the model embodies a Docker container, which has its service time indicated in square brackets. We remark that this value stands for the average time a container kept the CPU busy. We included a task called Residual that models the time spent on the CPU by unrepresented containers. Thus, this task can be seen as a delay that is triggered as requests arrive.

The testbed consists of two machines, M1 and M2, used, respectively, to record the measurements and run TTBS. Both machines run Ubuntu Linux. M2 has 32 GB of RAM and 8 CPUs. It is worth to remark that using two different machines we reduce the perturbation on the machine where the application is running. Hence, we collected results as cleaner as possible. We provide TTBS with a variable-sized burst of customers and measure the performance and the energy expense of M2 for each burst. The bursts can have size equal to 75 (i.e., the one we used to parameterize the LQN model), 150, 225, 300, 375, 450, 500 and were randomly supplied to TTBS for 30 times. This means for each size we collected 30 readings for a total of 210 executions. Randomization is necessary to remove the burst size from the factors that might influence the readings [23]. Further, we inserted a one-minute pause between executions to allow M2 to cool and thus prevent subsequent executions from affecting the profiled data. We empirically validated the pause we need to have a fresh machine. Measurements are coordinated through a bash script running on M1. The script runs JMeter [2], which generates a burst of customers operating on TTBS. At the same time, it records application and M2 performance from the operating system of M2, while the power consumption of TTBS is recorded from a Wattsup Pro power monitor [22] connected to M2. Additionally, we measured disk utilization, but excluded it from the analysis because it was too low to be meaningful.

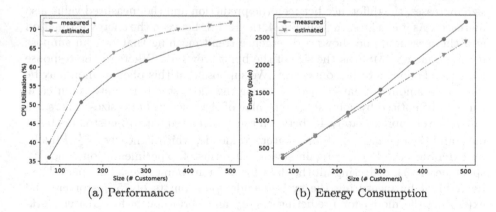

(a) Performance   (b) Energy Consumption

Fig. 4. Comparison between measured and estimated results.

We set the arrival rate, i.e., the λ parameter of the LQN in Fig. 3, and the service time of each task according to the data collected while supplying 75-customer bursts. Therefore, we incremented the λ parameter to replicate the arrival of 150, 225, 300, 375, 450, and 500 customers. The LQN returns CPU utilization predictions that we compared against the measurements. Table 1 reports performance and energy metrics for each batch size. It provides the arrival rate, i.e., the one we also supplied to the LQN, the response time, CPU utilization, the average power consumed by the CPU, i.e., the e multiplier, and the average energy consumed per batch. The columns containing two values show the measured value on the left and the corresponding predicted value on the right.

As expected, the CPU utilization rises according to the size of the burst. In our experiment, CPU utilization ranges from 35.96% when supplying a burst of 75 customers to 67.67% with a burst size of 500 customers. Figure 4a shows the distance between CPU utilization estimates and measurements varying burst size. The predictions slightly overestimate the measurements. This overestimation is quantified by the Root Mean Squared Error (RMSE) which equals 5.27%. Moreover, we obtained a Mean Absolute Percentage Error (MAPE) of 9.24%, which confirms the accuracy of the CPU utilization predictions. Besides utilization, the LQN returns response time estimates for each burst size. Therefore, it is possible to estimate the average energy consumed for a given burst by combining the corresponding response time prediction with the e multiplier calculated for the 75-customer burst, i.e., $\frac{E(CPU)}{S(CPU)} = \frac{321.99J}{4.098s} = 78.56\frac{J}{s}$. For example, given the response time estimation for a 500-customer burst, i.e., 30.90 s, we calculated the corresponding energy consumption by multiplying it by e using Eq. 4. We obtained an estimation of 2427.81 J, which is lower than the measured energy consumption, i.e., 2788.76 J. Figure 4b summarizes energy consumption predictions for each tested burst size. The energy consumption estimates are quite accurate as also evidenced by the RMSE and the MAPE which are equal to 200.16 J and 8.72%, respectively. However, we can observe that as the burst size increases, the difference between the prediction and the measured value also increases. As it can be seen from Fig. 4, the RMSE between the energy estimation and the measured one shows a divergence trend meaning that we can suppose having a larger RMSE as the size of the burst grows up. Whereas, the response time trend appears to be convergent. We suppose that this phenomenon may be due to the amount of data collected during the shortest experiment. In our case, this can be noticed by comparing the value of the e across burst sizes. There is a difference of approximately $4\frac{J}{s}$ between the e calculated with 75-customers burst data and the one measured for greater workloads, which is nearly $82\frac{J}{s}$. Finally, by exploiting the LQN, we gain savings in terms of experimentation time. We spent nearly 5 h collecting all the data for different burst sizes. This period can be reduced by measuring the system undergoing bursts of 75 customers and exploiting the model for predicting energy and performance for greater workloads. By doing so, we would have spent only 35 min for experimentation versus 5 h for measuring the 7 cases.

5   Threats to Validity

This discussion of threats to validity follows the classification made by Wohlin et el. [23]. The results of the study might be affected by *Conclusion Validity* threats due to the low significance of the sample collected during the lower-effort experiment. In fact, due to the short duration of this experiment, the sample collected may not be enough to accurately characterize either the LQN parameters (e.g., service time of containers) or the energy data (e.g., value of $e\frac{J}{s}$). This inaccuracy affects the estimates since we use this dataset to parameterize the LQN and derive energy consumption values for heavier workloads. Moreover, we considered a small set of data points, i.e., 3 image formats for the Digital Camera and 7 different burst sizes for Train Ticket Booking System. This limitation hampers generality and could influence the results, as the linearity between energy consumption and burst size for Train Ticket Booking System. In both case studies, we consider only the load handled by the CPU and scaled workloads. The findings might not be confirmed in situations involving more resources and different types of workload. Therefore, they might not be generalizable and the work might be affected by *Construct Validity* threats. Finally, we do not consider a broad sample of hardware/software systems. For example, we do not examine battery-powered systems, which may have power-saving modes. These characteristics might impact the measurements of energy consumption and performance. As a result, the study might be affected by *External Validity* threats, making it difficult generalize the findings to all types of systems.

6   Related Work

To the best of our knowledge, this is the first study investigating and quantitatively evaluating how performance models (specifically, LQNs) can be exploited to make accurate energy estimations of software systems. Moreover, in our case, the models are used to support measurement-based experiments and reduce experimentation time, thus assessing situations that designers aimed to measure. Several papers in the literature use queuing models to define energy-aware behaviors. These works come from different domains, such as robotics [11], wireless sensor networks (WSNs) [16,17,25], or cloud computing [3]. Cerotti et al. [8] use a queuing model to improve the utilization of the servers in a data center. Indeed, depending on the workload, some servers may be subject to long periods of low utilization, which still generate significant energy consumption. The queuing model incorporates a controller which manages the incoming workload so that servers maximize their throughput and resource utilization. So, each request will require less energy to serve, reducing the total energy consumption. Marsan and Meo [1] apply queuing models to optimize the energy footprint of a university WLAN. The authors consider the areas covered by multiple access points (APs). Co-located APs can be turned on/off, depending on the capacity of the group of APs to provide service and the number of active users accessing the WLAN. This situation is modeled with a queuing system which outputs the number of APs of a group that should be active to handle a given workload. In some of the situations, the authors manage to save even more than half of the energy usually expended to power the WLAN.

7 Conclusions

In this paper we have introduced a model-based approach for simplifying the energy consumption estimation of software systems. We have exploited the linear dependency of energy consumption and performance to extrapolate estimations of the former one in scenarios that would require high measurement times in practice. We tested the approach using a running example: Digital Camera, then validated the results on a more complex application, Train Ticket Booking System. The experimental results are quite promising, thus we plan to apply our approach to larger-size energy-critical software systems. Besides, we intend to examine the performance and energy consumption of resources other than the CPU, such as the disk and network. Although the approach has been implemented on top of LQN model, the whole process is independent of the modeling notation adopted for sake of performance analysis. Therefore, as further future work, we plan to consider different modeling notations that could be more suitable in specific application domains.

References

1. Ajmone Marsan, M., Meo, M.: Queueing systems to study the energy consumption of a campus WLAN. Comput. Netw. **66**, 82–93 (2014). https://doi.org/10.1016/j.comnet.2014.03.012
2. Apache Software Foundation: Apache JMeter. https://jmeter.apache.org, Accessed 02 Apr 2023
3. Balde, F., Elbiaze, H., Gueye, B.: GreenPOD: leveraging queuing networks for reducing energy consumption in data centers. In: 2018 21st Conference on Innovation in Clouds, Internet and Networks and Workshops (ICIN). pp. 1–8 (2018). https://doi.org/10.1109/ICIN.2018.8401602
4. BeagleBoard.org Foundation: The BeagleBone Black Development Platform. https://beagleboard.org/black, Accessed: 11 Nov 2022
5. Belkhir, L., Elmeligi, A.: Assessing ICT global emissions footprint: trends to 2040 & recommendations. J. Cleaner Prod. **177**, 448–463 (2018)
6. Brambilla, M., Cabot, J., Wimmer, M.: Model-Driven Software Engineering in Practice, 2nd edn. Springer International Publishing, Synthesis Lectures on Software Engineering (2017)
7. Carleton University Software Performance Research Group: layered queuing network solver. https://github.com/layeredqueuing, Accessed 23 Mar 2023
8. Cerotti, D., Gribaudo, M., Piazzolla, P., Pinciroli, R., Serazzi, G.: Multi-class queuing networks models for energy optimization. In: Proceedings of the 8th International Conference on Performance Evaluation Methodologies and Tools. p. 98–105. VALUETOOLS '14, ICST (Institute for Computer Sciences, Social-Informatics and Telecommunications Engineering), Brussels, BEL (2014). https://doi.org/10.4108/icst.Valuetools.2014.258214

9. Cruz, L., Abreu, R.: Performance-based guidelines for energy efficient mobile applications. In: 2017 IEEE/ACM 4th International Conference on Mobile Software Engineering and Systems (MOBILESoft). pp. 46–57 (2017). https://doi.org/10.1109/MOBILESoft.2017.19

10. Eder, K., et al.: ENTRA: whole-systems energy transparency. Microprocessors Microsyst. **47**, 278–286 (Nov2016)

11. Ekren, B.Y., Akpunar, A.: An open queuing network-based tool for performance estimations in a shuttle-based storage and retrieval system. Appl. Math. Model. **89**, 1678–1695 (2021). https://doi.org/10.1016/j.apm.2020.07.055

12. Esmaeilzadeh, H., Cao, T., Yang, X., Blackburn, S., McKinley, K.: What is happening to power, performance, and software? IEEE Micro **32**(3), 110–121 (2012). https://doi.org/10.1109/MM.2012.20

13. Franks, G., Al-Omari, T., Woodside, M., Das, O., Derisavi, S.: Enhanced modeling and solution of layered queueing networks. IEEE Trans. Softw. Eng. **35**(2), 148–161 (2009). https://doi.org/10.1109/TSE.2008.74

14. Fudan Software Engineering Laboratory: Train Ticket Booking System. https://github.com/FudanSELab/train-ticket, Accessed 12 Apr 2023

15. Georgiou, K., Xavier-de Souza, S., Eder, K.: The IoT energy challenge: a software perspective. IEEE Embed. Syst. Lett. **10**(3), 53–56 (2018)

16. Ghosh, S., Unnikrishnan, S.: Reduced power consumption in wireless sensor networks using queue based approach. In: 2017 International Conference on Advances in Computing, Communication and Control (ICAC3). pp. 1–5 (2017). https://doi.org/10.1109/ICAC3.2017.8318794

17. Jiang, F.C., Huang, D.C., Wang, K.H.: Design approaches for optimizing power consumption of sensor node with n-policy m/g/1 queuing model. In: Proceedings of the 4th International Conference on Queueing Theory and Network Applications. QTNA '09, Association for Computing Machinery, New York, NY, USA (2009). https://doi.org/10.1145/1626553.1626556

18. Marinescu, D.C.: Cloud computing: theory and practice. Morgan Kaufmann (2022)

19. Monsoon Solutions: monsoon power monitor. https://www.msoon.com/, Accessed 26 Sep 2021

20. Tribastone, M., Mayer, P., Wirsing, M.: Performance prediction of service-oriented systems with layered queueing networks. In: Margaria, T., Steffen, B. (eds.) Leveraging Applications of Formal Methods, Verification, and Validation, pp. 51–65. Springer, Berlin Heidelberg, Berlin, Heidelberg (2010)

21. Verdecchia, R., Lago, P., Ebert, C., De Vries, C.: Green it and green software. IEEE Software **38**(6), 7–15 (2021)

22. WattsUp: Watts up? pro power monitor. https://github.com/isaaclino/wattsup, Accessed 05 Apr 2023

23. Wohlin, C., Runeson, P., Höst, M., Ohlsson, M.C., Regnell, B., Wesslén, A.: Experimentation in Software Engineering. Springer, Berlin, Heidelberg (2012). https://doi.org/10.1007/978-3-642-29044-2

24. Woodside, M., Franks, G.: Tutorial introduction to layered modeling of software performance (2002)

25. Zhang, Y., Li, W.: Modeling and energy consumption evaluation of a stochastic wireless sensor network. EURASIP J. Wireless Commun. Netw. **2012**(1), 282 (2012). https://doi.org/10.1186/1687-1499-2012-282

An Implementation Study of the Impact of Batching on Replication Protocols Performance

Christopher Johnson(✉) and Paul Ezhilchelvan

Newcastle University, Newcastle upon Tyne NE4 5TG, UK
{c.johnson14,paul.ezhilchelvan}@newcastle.ac.uk

Abstract. Raft is a well-known, distributed ordering protocol used for replicating services on multiple servers for crash-tolerance. Chain Paxos is a recent development that is shown to offer a much higher throughput than many known ordering protocols. This superior performance is attributed to different dissemination structures employed between the lead server and the rest, while the comparative evaluation experiments do not employ batching of service requests for ordering. We observe here, through a range of experiments, that the throughput differences indeed vary from being significant to negligible when batching is allowed and batch size varied. This behaviour is explained by modelling the ordering process at the lead server and deriving stability conditions at various stages.

Keywords: replication · total ordering · batching · throughput · queueing theory · implementation

1 Introduction

Uniform Total Order (UTO) protocols are at the heart of building replicated state machines (RSMs). They ensure that updates are applied in the same order at each machine to maintain the abstraction of a single, reliable server. This abstraction simplifies application logic since clients can now interact with a distributed replicated system as if it were a single system.

UTO protocols, such as Paxos [5], are generally complex, involving potentially multiple rounds of communication among servers; the leader based ones are relatively simple to understand and implement, and have only a 2 hop latency. Among them, Raft [6] is commonly used in practice as its implementation is well structured, making it easier to operate on.

However, the Raft leader is a performance bottleneck, as it does most of the heavy-lifting: receives operations from clients, sequences and unicasts the operations to all other servers, called *followers*, waits to receive acknowledgements (*acks* for short) back from followers, applies the acked operations in sequence to local state machine and finally sends responses back to clients.

M. Iacono et al. (Eds.): EPEW/ASMTA 2023, LNCS 14231, pp. 264–278, 2023.
https://doi.org/10.1007/978-3-031-43185-2_18

Many protocol variations have been proposed to avoid this leader bottleneck. Notably, some propose a unidirectional ring or chain dissemination pattern: starting with the leader, the sequenced operation is transmitted only to one immediate neighbour (if any); acks are also passed on in a hop-by-hop manner. This has the following benefits of (i) *CPU*: The leader composes one sequenced message and processes one ack; (ii) *network*: The leader only transmits one message and receives one ack. This dissemination pattern however results in potentially higher latency as messages must travel the length of the chain. Additionally, the chain configuration can not mask slow or faulty servers. A server crashing temporarily halts the flow of messages until a reconfiguration can occur, and performance is also limited in gray failures where a server suffers a degradation in performance but does not completely fail.

Assessing the performance impact of various dissemination patterns employed is thus interesting and necessary for system developers. However, many such studies ignore the possibility of batching client operations, especially at large request arrival rates. Such set-ups constitute an unrealistic context for evaluating relative protocol performances.

This paper seeks to address these drawbacks by running experiments using a realistic cluster-based implementation. Clients are thus real computer nodes interacting with remote servers that execute UTO protocols and batch the requests as any real world application would seek to do. Recent years have seen rapid improvements in networking capability, even the most basic general purpose VMs in public clouds are allocated 10Gbps of network bandwidth. As a result in a wide range of scenarios CPU, rather than network bandwidth, becomes the limiting factor for performance. This aspect is also factored into the way our experiments are run.

We compare Raft and Chain Paxos [4] in a genuine real world setting on a cloud cluster. In Raft, leader directly interacts with each follower and Chain Paxos is chosen as a representative of protocols that employ alternative forms of interaction between leader and followers. This choice is made for two reasons. First, Chain Paxos [4] is the latest work on leader-based ordering that is demonstrated to offer higher throughput and scalability compared to many other leader-based counterparts; these benefits are attributed to the dissemination structure employed together with many optimisation features. Secondly, the experiments of [4] do not consider batching.

The wide range of experiments we have carried out show that when throughput is the performance metric of interest, Raft can remain competitive with Chain Paxos even at extreme loads through careful tuning of batching. This means that the choice between Raft and Chain Paxos is not so obvious when the network bandwidth is not a limiting factor.

Our second main contribution is in explaining these relative performance behaviours of these two protocols. We model the leader activities as a series of *logical* servers and, using queueing theory principles, establish stability condition for each. We observe that the Raft leader load due to unicasting sequenced batches to several followers is reduced when the batch size is sufficiently large and hence the leader can remain stable even when arrival rates are very high.

2   System Description

We first describe the execution of Raft in its basic configuration, and then describe the differences made in Chain Paxos.

Raft: The system is comprised of $N = 2f + 1$ servers where f is the maximum number of servers that can crash.

Clients connect to the leader and send an operation to be executed on the replicated state machine (RSM). However, to fulfill the requirements of 1-server abstraction, the operation cannot be immediately applied and must first be sequenced so all servers apply it in an identical order.

Operations from clients may first be batched. The leader assigns a sequence number to each batch of operations and sends a message containing one batch and the sequence number assigned to it, to each follower. Figure 1 depicts Raft dissemination structure with server 0 acting as the leader.

Followers append the received batch to their log and send an ack message back to the leader to confirm they have received the batch. The leader processes the acks and *commits* the corresponding batch once it can confirm that at least $f + 1$ servers, i.e., at least f followers, have received and logged that batch.

Committed batches are passed to the *apply* service in their sequenced order. The apply service takes up each batch in their order and serially applies operations within each batch in their batching order. Once applied, the outcome of that apply operation is sent to the client that sent that operation.

The leader also piggybacks the sequence number of the most recently committed batch with all messages to followers. This allows followers to also apply operations to their state machine.

If the leader becomes unresponsive the followers consider it failed and *elect* a new leader. Clients connect to the new leader and continue. Details of leader election are not relevant here as we will compare only crash-free performances.

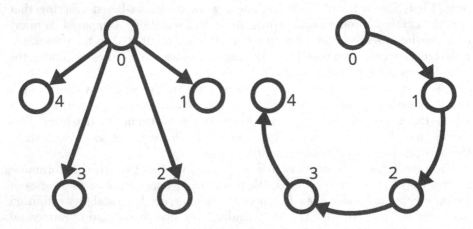

Fig. 1. Message dissemination pattern of Raft (left) and Chain Paxos (right).

Chain Paxos: Chain Paxos has several execution differences compared to Raft. The leader organises all N servers as a chain as shown in Fig. 1 where server 0 is the leader; server $i, i > 0$, receives sequenced batches from $i - 1$ and forwards them to $i+1, i+1 < N$. As the batch is forwarded along the change, an attached ack counter is incremented. Once a batch has been forwarded f times, subsequent servers, i.e., from server f to server $N - 1$, can commit the batch immediately upon receiving it. The final server in the chain, server $N - 1$, sends an ack back to the leader so the leader can also commit the batch. Note that the Chain Paxos leader is not the first server in the system to commit a batch that it had earlier sequenced.

Followers in the first half of the chain, i.e., servers up to server $f - 1$, can commit batches when they receive a commit piggybacked to subsequent batches forwarded by the leader. By default, clients connect to any server and each server creates operation batches that are then sent directly to the leader.

For comparison, we also allow Raft followers to receive client requests, batch them, forward the batches to the leader, and respond to clients when a directly received operation is applied locally. Similarly, for Chain Paxos, we consider also the case of clients sending their requests only to the leader. Thus, we compare performances when clients connect (i) only to the leader and (ii) to all servers. To assess the protocols at different scales we compare performance on clusters of servers of size N=3, 5, 7 for both configurations.

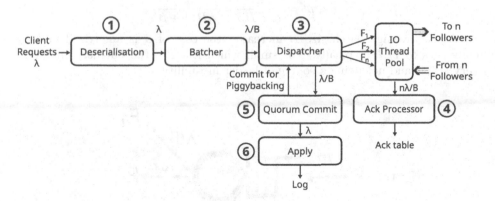

Fig. 2. Six main service stages of a client request operation at the leader.

3   Clients Connecting only to Leader

Suppose that all clients connect only to the leader. The logical arrangement of services that make up the processing pipeline within the leader is shown in Fig. 2. We will first derive the stability conditions for each of the six main services. All services need to be stable for the leader server as a whole to be stable.

3.1   Stability Conditions

Deserialisation. Client requests arrive at the leader at the rate λ and are deserialised by a pool of γ IO threads and added to the batcher queue. Deserialisation process is modelled as a shared queue with γ threads, each able to deserialise requests at a rate of μ_1^t. The stability condition for deserialisation is

$$\lambda < \mu_1 = \gamma \mu_1^t \tag{1}$$

This service can be scaled by increasing the number of deserialisation threads γ which is ultimately limited by the amount of CPU resources available.

Batcher. Deserialised client requests are converted into batches by the *batcher* service. The batcher creates batches with an average size of B requests, controlled by the run time configuration parameters of the maximum batch size B_m and batch timeout Δ. We assume that the service time for constructing one batch is comprised of a fixed cost α time units for creating a batch and adding it to the dispatcher queue, and a variable cost of β time units for adding a client request into a batch. Thus, the average batching rate μ_2 is B requests in $(\alpha + B\beta)$ time units.

This gives the stability condition of the batcher as

$$\lambda < \mu_2 = \frac{B}{\alpha + B\beta} = \frac{1}{\alpha B^{-1} + \beta} \tag{2}$$

As the average batch size B increases, the cost of α is amortized and μ_2 converges to $1/\beta$. Typically, β is small which leads to the batcher remaining very stable and not being a source of leader instability.

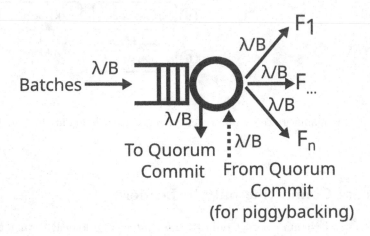

Fig. 3. Dispatcher. Batches are sequenced and sent to followers for replication. Batches passed to quorum commit service to await ack quorum.

Dispatcher. Batches of client requests are processed by the dispatcher to assign sequence numbers and send to followers for replication. The number of followers the leader directly sends batches to is denoted by n. In Raft $n = N - 1$, as the leader is responsible for sending the batch to each follower. In Chain Paxos $n = 1$, as the leader only sends the batch to one follower and relies on the followers to forward the batch along the chain. The work done by the dispatcher is the same regardless of the size of the batch, as it acts on the batch as a whole rather than the individual client requests. The stability condition for the dispatcher is

$$\lambda < \frac{B\mu_3}{n} \tag{3}$$

where μ_3 is the service rate for preparing a batch to send to a follower, and n in the number of followers the batch is sent to.

Forwarded batches also contain the highest committed sequence number so followers can apply committed batches. Once a batch is forwarded to followers, it is added to the commit queue where batches are held until a quorum of acks is received.

Ack Processor. Acks are received from followers and stored in the ack table to confirm that batches have been received by followers. When acks have been received from a majority of servers the batch is marked as stable and can be released by the quorum commit service. In Raft, individual acks are received from each server resulting in n acks per batch. In Chain Paxos, all acks are accumulated as the batch is forwarded along the chain, so only one ack is received. The stability condition is therefore

$$\frac{n\lambda}{B} < \mu_4$$
$$\lambda < \frac{B\mu_4}{n} \tag{4}$$

Quorum Commit. Batches are held in the quorum commit queue until each batch is considered ready for commit i.e., once a quorum of servers have acked that batch. The service monitors the ack table, and releases a batch once a quorum of acks for the batch are added to the ack table. Both protocols ensure acks from a follower arrive in the same sequence order of the batches, so batches are not blocked by other batches in the queue. We assume the work done by the queue is negligible as the queue simply releases the batch to the apply queue once a quorum of acks have been recorded. Additionally, the arrival rate to quorum commit is limited by the rate the dispatcher can process new batches.

The time a batch remains in the quorum commit queue is the time difference between the batch being sequenced by the dispatcher, and the batch being committed after a quorum of acks are processed by the ack processor. We can model this as a queue with a sojourn time of

$$W_q = \phi d + W_a \tag{5}$$

$$\phi = \begin{cases} 2 & \text{for Raft,} \\ N & \text{for Chain Paxos.} \end{cases} \qquad (6)$$

where W_q is the average total time spent in the quorum commit queue , d is the transmission delay between any two servers, and W_a is the sojourn time for an ack to be processed by the ack processor.

If the network transmission is stable, variance in transmission delays is small with a constant mean. If the network is bursty, this assumption no longer holds and successive transmissions will have a longer delay. However, under partial synchrony [2] assumptions, which are often used for these protocols, we assume that network delay has an unknown finite upper bound. Therefore we assume bursty behaviour is transient and stabilises at a maximum delay of d_{max}, so d is at most d_{max} and therefore bounded.

If the ack processor is stable, quorum commit will also be stable with the average number of batches in the queue L_q being

$$L_q \leq (\phi d_{max} + W_a)\lambda/B \qquad (7)$$

Apply. The apply service is responsible for applying batches of operations to the RSM in the order determined by the dispatcher. Operations within each batch are applied to the RSM in order, resulting in each server applying the same operations in the same order. Applying operations to the RSM is inherently serial and application dependent, so this can become a bottleneck to the system. After each operation is applied a response is generated for the client that sent the original request. The stability condition for the apply service is

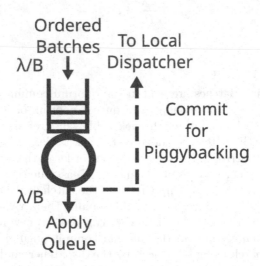

Fig. 4. Quorum commit. A batch is released after a quorum of acks have been received.

$$\lambda < B\mu_6 = \frac{B}{B(S_o + S_r)} = \frac{1}{S_o + S_r} \tag{8}$$

where S_o is the average service time for applying an operation to the RSM, and S_r is the average service time for creating a response to the client.

3.2 Summary

The overall stability of the system can be summarised as stability at services that scale with batching and those that don't. Without batching the dispatcher or ack processor is likely to reach saturation first, as for each client operation it must do work proportional to n. When batching is employed these services scale with B/n, so this bottleneck can be relieved provided sufficiently large batches can be created in a reasonable time. The ultimate saturation point of the system is determined by the service rates of the deserialisation and apply services, $\min\{\mu_1, \mu_6\}$, as they are not affected by batching.

3.3 Experimental Results

The evaluation was carried out on Microsoft Azure with D16ds_v5 instances (16 vCPU, 64GiB RAM, 12.5Gbps network bandwidth). This was chosen as the closest available configuration to the test environment in the original Chain Paxos paper (18 CPU, 64GB RAM, 25Gbps network bandwidth).

For the comparison we used the Chain Paxos implementation from the original paper [4], and implemented a Raft equivalent in the same framework [3] to ensure language or architecture differences would not influence results. For Raft writing to disk was disabled as Chain Paxos does not write to disk and it would become the bottleneck over CPU. This was done in the interest of running a fair comparison that compares the CPU bottleneck given the different dissemination patterns.

Experiments were carried out on N=3, 5, 7 servers each on separate VMs colocated in the same region and availability zone, with load being generated by 5 separate VMs running the YCSB [1] benchmark across a varying number of threads in a closed loop, also in the same availability zone. Latency was measured at the clients, recording the total time from a request being sent to a response being received. We repeat experiments while varying the number of client threads, and maximum number of operations per batch B_m. The number of client threads ranged from 5 to 15000 sending operations of 100 bytes, and B_m was set to 1, 10, 50, 100, 200 and 500. The batch timeout Δ was set to 1ms. An initial warming up period was used before running the workloads to ensure results were not affected by JIT optimisation between runs.

The results for $N = 5$ for up to 5000 clients are shown in detail in Fig. 5 and Fig. 6, showing the plot of throughput and latency against number of client threads respectively. The results of other experiments are shown as a plot of latency against throughput in Fig. 7 for the sake of space. The results of the

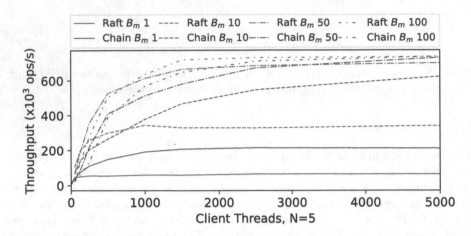

Fig. 5. Comparison of number of concurrent clients with throughput with 5 servers and varying batch sizes.

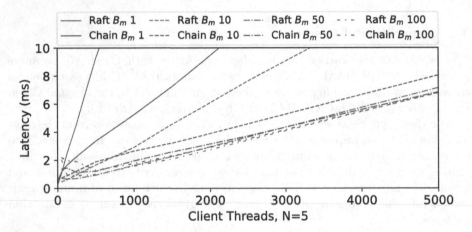

Fig. 6. Comparison of number of concurrent clients with latency with 5 servers and varying batch size.

experiment can be seen in Fig. 9, with Raft in red and Chain Paxos in blue. Note that the results for runs with some values of B_m were omitted for clarity when they did not change the overall results of the experiment. In Fig. 7 each line is the connected points from successive runs with increasing number of concurrent clients to increase load on the system. As the number of concurrent clients increases so does load, resulting in the system reaching saturation at one of the logical services described in 3.1. This results in a increase in latency without a notable increase in throughput from further increases in the number of concurrent clients. In some cases increasing the number of clients causes a decrease in throughput while increasing latency. This is due to the extra load negatively affecting performance while not increasing throughput due to saturation.

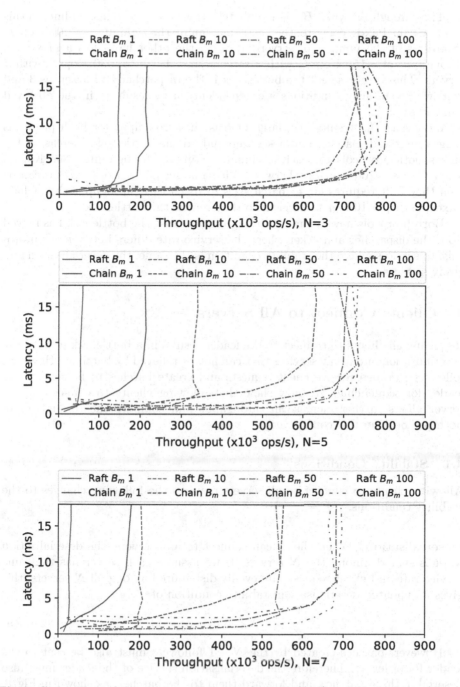

Fig. 7. Comparison of throughput and latency for Raft (Red) and Chain Paxos (Blue) for varying max batch sizes when clients connect to leader. (Color figure online)

The throughput when B_m is 1 (unbatched) is shown with a solid line. In this setup Chain Paxos consistently has a maximum throughput of ∼200 Kops/s regardless of the number of servers. Raft on the other hand has a maximum throughput of ∼150 Kops/s with 3 servers, and drops to ∼30 Kops/s with 7 servers. This suggests that the bottleneck is the dispatcher and acker, as 3 and 4 show the stability conditions are dependent on n, resulting in the observed behaviour.

Increasing B_m results in a large increase in throughput for both protocols suggesting the dispatcher and acker were indeed the bottleneck. Increasing B_m allows both protocols to reach a similar maximum throughput; ∼800 Kops/s when $N = 3$, ∼700 Kops/s otherwise. With a larger number of servers it can be seen that Raft requires larger batch sizes to reach this throughput than Chain Paxos, as $B_m = 10$ is not sufficient to fully relieve the bottleneck.

Both protocols have a similar throughput because the bottleneck has moved from the dispatcher and acker where the service rate differs between the protocols, to services where the stability condition is the same, the deserialisation and apply services.

4   Clients Connect to All Servers

Requiring all clients to connect to the leader results in a bottleneck in both the deserialisation and apply services that can not be relieved by batching. However, followers can also receive client requests and create batches to be sent to the leader for sequencing. We now show that allowing clients to connect to any server allows further increase in throughput and compare the throughput of both protocols in this configuration.

4.1 Stability Conditions

Allowing clients to connect to all servers requires the following changes to the stability conditions.

Deserialisation. When clients can connect to any servers, the deserialisation work is shared among the N servers. If we assume each server has the same service rate and client requests are evenly distributed among all N servers, this gives an updated deserialisation stability condition of

$$\lambda < N\mu_1 \tag{9}$$

However, batches created by the $N - 1$ followers must also be sent to the leader for ordering. Therefore the deserialisation service of the leader must also deserialise these batches, and forward them to the batcher, as shown in Fig. 8. Deserialisation of a batch of B messages can be faster than deserialising B individual requests due to the overhead cost per message. Let k be the ratio of the time taken to deserialise a batch of B requests over the time taken to deserialise

B individual requests, $0 < k \leq 1$. The stability condition for the deserialisation service at the leader then becomes

$$\lambda < \mu_1[\frac{1}{N} + k\frac{N-1}{N}]^{-1} \tag{10}$$

with the service rate tending to μ_1/k as N increases.

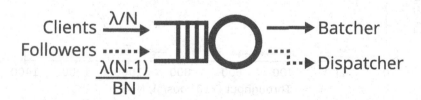

Fig. 8. Deserialisation service when client requests can be received by all servers.

Batcher. The work of batching is shared among the N servers. Assuming all servers have the same arrival and service rate, the stability condition becomes

$$\lambda < N\mu_2 = \frac{NB}{\alpha + B\beta} = \frac{N}{\alpha B^{-1} + \beta} \tag{11}$$

Apply. Each server must still apply every operation to the RSM regardless of which process the client connects to in order to maintain consistent state. However a server must only create a response for operations originally submitted to that server. Therefore the stability condition becomes

$$\frac{\lambda}{B} < \mu_6 = \frac{1}{B(S_o + \frac{S_r}{N})} \tag{12}$$

4.2   Experimental Results

For this configuration, client threads connected to servers in a round robin fashion to evenly distribute the load among all servers. The number of concurrent clients again ranged from 5 to 15000.

The results of the experiment can be seen in Fig. 9, with Raft in red and Chain Paxos in blue.

The unbatched throughput remains similar to that of when all clients connect to the leader. This is because without batching the bottleneck is still the dispatcher and acker of the leader, which is experiencing the same load.

When batching is considered, the maximum throughput of both protocols improves to $\sim$1200 Kops/s for $N = 3, 5$ compared to $\sim$700–800 Kops/s when

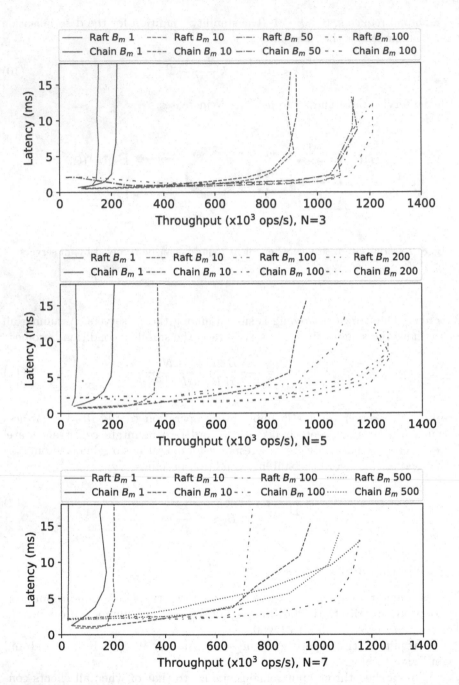

Fig. 9. Comparison of throughput and latency for Raft (Red) and Chain Paxos (Blue) for varying max batch sizes over when clients can connect to any server. (Color figure online)

all clients connected to the leader. This is because deserialising incoming client operations and creating batches is divided among all servers, and likewise sending responses to clients is also shared among all servers.

When $N = 7$, Raft's throughput is slightly behind that of Chain Paxos, with an observed throughput of 1075 Kops/s compared to 1158 Kops/s with 15000 clients, even with a maximum batch size of 500. Increasing the number of clients to 25000 resulted in a throughput of 1132 Kops/s for Raft and 1287 Kops/s for Chain Paxos.

5 Discussion

The results show that batching is an effective way to control the bottleneck of sequencer based UTO systems. Although this is intuitive, it is surprising that it can have a large effect on the relative performance of different protocols. When compared in a scenario where operations aren't batched, Raft's maximum throughput is only 16% of Chain Paxos's throughput with 7 replicas, but with batching it is 88%. With lower number of replicas, which are more common, Raft was found to have similar throughput. Both protocols require batching to reach maximum throughput.

It is worth noting that a relatively high level of batching is required for Raft to reach comparable throughput to Chain Paxos - especially with larger cluster sizes. This becomes impractical as the size of client operations increases, as there is a limit to the packet size.

Additionally, when client connections are distributed between all servers the B/n scaling factor required to keep the dispatcher and ack processor of the leader stable becomes hard to maintain for large cluster sizes. Not only does n increase with N for Raft, but also B will decrease due to each server receiving client operations at a rate λ/N, restricting the size of batches that can be created in a timely manner.

It is also clear that tuning of batching parameters is important in achieving good performance with these systems. If batch size is large in times of low load unnecessary time is spent waiting at the batcher. If batch size is set too small throughput is limited. A model for tuning these parameters was proposed in [7] but this assumes sequential execution on a single core machine which is not representative of modern multi-core machines. Further work in this area could offer practical performance increases in real systems.

6 Conclusion

We have shown that batching is an important factor when considering the relative performance of consensus protocols. Batching amortizes the cost of achieving consensus, alleviating the most immediate bottleneck of the system. Batching can further improve performance by utilising followers as proxy batchers, reducing the load generated by client requests at the leader. However, this can be sensitive to the workload and batcher parameters, so careful tuning is required for optimal performance.

References

1. Cooper, B.F., Silberstein, A., Tam, E., Ramakrishnan, R., Sears, R.: Benchmarking cloud serving systems with YCSB. In: Proceedings of the 1st ACM symposium on Cloud Computing, pp. 143–154 (2010)
2. Dwork, C., Lynch, N., Stockmeyer, L.: Consensus in the presence of partial synchrony. J. ACM (JACM) 35(2), 288–323 (1988)
3. Fouto, P., Costa, P.Á., Preguiça, N., Leitão, J.: Babel: a framework for developing performant and dependable distributed protocols. In: 2022 41st International Symposium on Reliable Distributed Systems (SRDS), pp. 146–155. IEEE (2022)
4. Fouto, P., Preguiça, N., Leitão, J.: High throughput replication with integrated membership management. In: 2022 USENIX Annual Technical Conference (USENIX ATC 22), pp. 575–592. USENIX Association, Carlsbad, CA (2022), https://www.usenix.org/conference/atc22/presentation/fouto
5. Lamport, L.: Paxos made simple. ACM SIGACT News (Distributed Computing Column) 32, 4 (Whole Number 121, December 2001) 51–58 (2001)
6. Ongaro, D., Ousterhout, J.: In search of an understandable consensus algorithm. In: 2014 USENIX Annual Technical Conference (USENIX ATC 14), pp. 305–319 (2014)
7. Santos, N., Schiper, A.: Tuning paxos for high-throughput with batching and pipelining. In: Bononi, L., Datta, A.K., Devismes, S., Misra, A. (eds.) ICDCN 2012. LNCS, vol. 7129, pp. 153–167. Springer, Heidelberg (2012). https://doi.org/10.1007/978-3-642-25959-3_11

Workflow Characterization of a Big Data System Model for Healthcare Through Multiformalism

Tancredi Covioli[1], Tommaso Dolci[1] iD, Fabio Azzalini[1] iD, Davide Piantella[1] iD, Enrico Barbierato[2(✉)] iD, and Marco Gribaudo[1] iD

[1] Politecnico di Milano, Piazza Leonardo da Vinci, 32, 20133 Milano, Italy
tancredi.covioli@mail.polimi.it, {tommaso.dolci,fabio.azzalini,
davide.piantella,marco.gribaudo}@polimi.it
[2] Università Cattolica del Sacro Cuore, Via della Garzetta 48, 25133 Brescia, Italy
enrico.barbierato@unicatt.it

Abstract. The development of technologies such as cloud computing, IoT, and social networks caused the amount of data generated daily to grow at an incredible rate, giving birth to the trend of Big Data. Big data has emerged in the healthcare field, thanks to the introduction of new tools producing massive amounts of structured and unstructured data. For this reason, medical institutions are moving towards a data-based healthcare, with the goal of leveraging this data to support clinical decision-making through suitable information systems. This comes with the need to evaluate their performance. One of the techniques commonly used is modeling, which consists in performing an evaluation of a model of the system under analysis, without actually implementing it. However, to make an adequate performance assessment of Big Data systems, we need a diversity of volumes and speeds that, due to the sensitivity of data concerning healthcare, is not available. While in other fields this problem is usually solved through the use of synthetic data generators, in healthcare these are few and not specialized in performance evaluation. Therefore, this work focuses on the creation of a synthetic data generator for evaluating the performance of a Big Data system model for healthcare. The dataset used as a reference for creating the generator is MIMIC-III, which contains the digital health records of thousands of patients collected over a time span of multiple years. First, we perform an analysis of the dataset, adopting multiple distribution fitting techniques (e.g., phase-type fitting) to model the temporal distribution of the data. Then, we develop a generator structured as a multi-module library to allow the customization of each component, specifically we propose a multiformalism model to reproduce the patient behavior inside the hospital. Finally, we test the generator by evaluating the performance in different scenarios. Through these experiments, we show the granular control that the generator offers over the synthetic data produced, and the simplicity with which it can be adapted to different uses.

Keywords: performance evaluation · synthetic data generation · Big Data · healthcare data

M. Iacono et al. (Eds.): EPEW/ASMTA 2023, LNCS 14231, pp. 279–293, 2023.
https://doi.org/10.1007/978-3-031-43185-2_19

1   Introduction

Big Data is flooding the healthcare field thanks to the introduction of new tools for continuous patient monitoring, producing massive amounts of structured and unstructured data every day. For this reason, medical facilities are moving towards data-driven healthcare, with the goal of leveraging this incredible source of information to support clinical decision-making and public health management. This naturally comes with the request to elaborate a multitude of heterogeneous data, resulting in the emergence of new system architectures and new methods to evaluate their performance. One of the evaluation techniques frequently used is modeling, which consists in creating a model of the system under analysis, and performing the evaluation on it instead of implementing and testing the actual system. In order to make an adequate performance assessment of Big Data–centered systems, we need a diversity of volumes and workloads that, due to the sensitivity of data related to healthcare, it is usually not available. In other fields, this problem is solved through the use of synthetic data generators, but in the field of healthcare there are just few available and not specialized for performance evaluation [6].

In this paper we create a synthetic data generator for the evaluation of performances of a Big Data system model focusing on healthcare data. Reference data for the generator is gathered from two sources: *a)* MIMIC-III dataset [14], a large freely-available database comprising de-identified health-related observations (called "events") associated with over forty thousand patients who stayed in intensive care units (ICU) at the Beth Israel Deaconess Medical Center between 2001 and 2012; *b)* MIMIC-III Waveform Matched Dataset [17], a dataset associated to the original MIMIC-III containing the digitized vital signals ("waveforms") recorded for all patients that stayed in ICU.

The rest of the paper is organized as follows: Sect. 2 describes the state of the art, Sect. 3 illustrates MIMIC-III and the analyses performed on it, Sect. 4 introduces our synthetic data generator, Sect. 5 shows the experimental evaluation of the generator, Sect. 6 sums up and concludes the paper.

2   Related Work

As previously mentioned, the advent of Big Data in healthcare led to the need of new systems for data processing and performance evaluate. The first challenge is finding datasets that contain both large volumes of structured and unstructured data, and with a fine enough temporal granularity to identify the peak usage in order to make a proper and complete performance assessment. The scarcity of publicly available datasets that meet these characteristics causes the need to consider generating synthetic data. The features sought for these datasets are:

- Data heterogeneity, namely the need to contain structured, partially structured and unstructured data.
- A high level of temporal granularity, namely the need to cover events that occur even in fairly short time intervals.
- A volume of data large enough to make the simulator realistic.

In literature, publicly available datasets with the characteristics sought are very few. One of them is the New Zealand National Minimal dataset, a collection of public and private hospital discharge information, including clinical information [13]. Although it meets the volume requirement, covering the discharges of all patients in New Zealand public hospitals since 1993, it is not considered granular enough for the purposes presented (it does not contain records collected about patients during their stay but only information recorded during their discharge) and not heterogeneous enough (containing only tabular data).

Another dataset that we considered is the ChestX-ray8 dataset, a collection of X-ray images of over 30000 patients [24]. Unfortunately, the radiological reports associated with each image are not included in the dataset, severely limiting its heterogeneity and, therefore, its usefulness. The only dataset deemed large, heterogeneous and granular enough to be used as a reference for the creation of the generator is MIMIC-III [14].

2.1 Synthetic Generators

A commonly used methodology related to machine learning for generating synthetic data in the healthcare field are Generative Adversarial Networks (commonly referred to as GANs). A GAN is a kind of artificial intelligence algorithm based on an adversarial training system, where two competing models (a generator model and a discriminator model) are trained against each other [10]. GANs been used for the development of medGAN [6], a generator that focuses on privacy preserving synthetic health data which, however, is only able to generate binary or at most integer values, severely limiting its usefulness. An evolution of medGAN is provided by healthGAN [25], which is capable of the simulation of real-distributed values, while remaining limited in the heterogeneity of the simulated data. In [18] is proposed SmoothGAN, a new approach for the generation of high quality synthetic data that maintains important relations and factors of the original data; it is not focused on following the time distribution of the original data though and, moreover, it does not seem to have a functioning implementation. GANs have also been used previously in [5] to model events contained in the MIMIC-III dataset, but their focus is again not on the timing of the data, which are apparently omitted from the output of the generator.

The most notable synthetic data generator that we found not based on GANs is provided by Synthea, a tool suite focusing on the generation of synthetic health records that cover the entire lifetime of the patients [23], but that lacks the granularity to be used for performance evaluation purposes.

2.2 Workflow Characterization by Multiformalism

The multiformalism modeling approach aims to facilitate the coordination and integration of models that focus on different aspects or components of a system by employing heterogeneous formalisms. This approach also enables the creation of macro-models by combining specific submodels that use the most appropriate formalism for the problem at hand, while ensuring unity and coherence in the overall model. Different approaches exist regarding the implementation of multiformalism modeling concepts. For instance, some approaches combine multiformalism with multisolution, while others allow the choice of solver to be decoupled from the formalism and bound to general characteristics of the model. Additionally, different multiformalism modeling frameworks vary in terms of the number and types of formalisms that are allowed. Some frameworks employ a fixed set of known formalisms, which simplifies the implementation of proper single or multiple solvers and enables significant optimizations. Other frameworks allow an unlimited number of formalisms and require means to incorporate new formalisms, solution processes, and solvers.

Scientific literature offers a vast family of multiformalism tools, such as SHARPE [22], SMART [7] and DEDS [2] where the type of formalisms is limited. In other tools, the limitation is overcome, as the range of formalisms can be augmented in different ways: see, for example, AToM^3 [15] and Möbius [8], OsMoSys [9], and finally SIMTHESys [1].

3   MIMIC-III

The analysis performed on MIMIC datasets are centered around gathering an understanding of the process behind the generation of its events, focusing on the interaction process between patients and hospital and its duration. MIMIC datasets contain data associated with thousands of patients collected over 10 years, including, in addition to a large amount of structured tabular data collected and placed temporally with minute accuracy, a collection of medical notes written in plain English and issued by the caregivers of the patients. Moreover, it is associated with a database containing a multitude of physiological signals (waveforms) recorded during the ICU stays of the patient, which further increases the heterogeneity of the dataset.

From now on, when talking about MIMIC, unless clearly specified, we will mean to consider the combination of both the original MIMIC-III dataset and the MIMIC-III Waveform Matched Dataset. The objective of this analysis is to find a set of distributions to be used to model the time of acquisition of the events described in MIMIC. To do so, we focus on the process of interaction between the patient and the system, specifically looking at the different events that are registered at each stage of such interaction process. Figure 1 shows the evolution of the number of patients inside the considered hospital per year and during a day. It is interesting that it does not show the classical self-similarity present in classical sources, like internet traffic, but it has a lot of variability during a year. Once these stages have been identified, we are able to look at the distribution of their duration and model it with some fitting techniques and tools. Finally, the same fitting procedures will be applied to model the inter-time between the registrations of the events contained in MIMIC.

Once the distributions of the duration of the interaction stages and of the inter-arrival time of the events for each stage are determined, we use them as the foundation for the synthetic data generator.

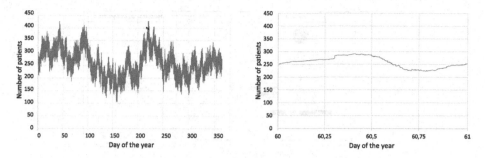

Fig. 1. Evolution of the number of patients during a year and during a day.

3.1 Design Decisions

Due to the sensitive nature of the data that comprises MIMIC, the authors needed to perform a de-identification procedure before the dataset was made available to the public. In particular, for each individual patient, the dates have been shifted into the future by a random offset. Nonetheless, all the time intervals between two entries relative to the same patient are kept intact; therefore, only a handful of time information associated to each timestamp are still valid after the random shift, i.e., the day of the week, the seasonality, and the time of the day. This procedure deeply influenced our methodology: the de-identification procedure made an analysis of the exchange of data between the patients as a whole and the hospital system impossible and, for this reason, we decided to focus on analyzing the duration of the interactions of the patients singularly.

3.2 Stages of Interaction with the Hospital

Figure 2 shows the possible temporal evolution of patients inside the hospital, focusing on their permanence in the Intensive Care Units. After multiple refinements, the identified stages of the interaction process, represented as time intervals to be modeled, are the following:

1. time spent in an ICU;
2. time interval after the hospital admission and before the start of the first ICU stay;
3. time spent in the hospital between two consecutive ICU stays;
4. time between the end of the last ICU stay and the end of the entire hospital stay;
5. total time in the hospital, which shall be considered as the sum of the times listed above (unless the patient is not admitted in an ICU, in which case the hospital time shall be computed separately);
6. time between the end of a hospital stay and the beginning of the next one.

MIMIC considers two main categories of events: those associated with the specific ICU stay of the patient (e.g., automatic measurements of their blood

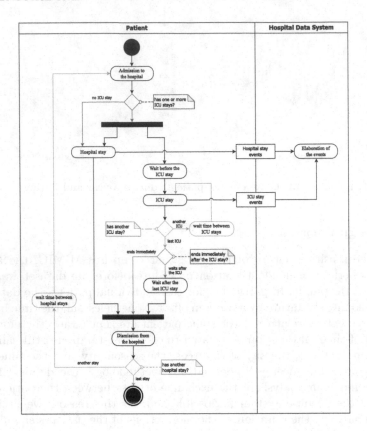

Fig. 2. Activity diagram representing the stages of the interaction process between the patient and the hospital system.

pressure made during their ICU permanence) and those associated with the general hospital stay of the patient (e.g., laboratory results collected from their cultures). The former are generated only during the first stage of the ones listed above, i.e., the time spent in ICU, while the latter are generated through all the other intervals except the last one, during which no event shall be generated, since the patient is not in the premise of the hospital. Each of these categories is then split into multiple kinds of event, like laboratory events or note events, each stored in its own table in the dataset.

3.3   Classification

Before focusing on the distributions that can be fit to model the interaction stages identified in the previous section, we decide to split both the events stored in MIMIC and the interactions into *classes*. This decision is necessary to avoid considering a single distribution to model all the events produced by all the patients during their stays, which would inevitably worsen the fitting results.

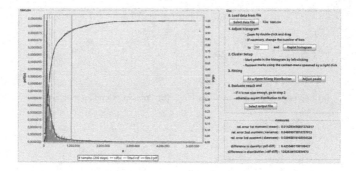

Fig. 3. Interface of HyperStar for distribution fitting.

The chosen classes are based on easily observable features of the data in order to maintain the grouping process simple. Such features, though, shall be relevant to the medical field and distinctive enough to split the dataset in comparable portions. For instance, the link between gender and health conditions is well known [16], and the link between ethnicity and health conditions, although not as common, has also been documented and studied for a long time [21]. Finally, the features chosen for the classification are the following:

- the gender of the patient (either Male or Female);
- the age of the patient (a categorical attribute with values 0–45, 45–65, 65–75, 75–100 and over 100 years old);
- the day of the week when the hospital stay begins.

The combination of these features leads to a total of 70 classes in which the patients, the admissions and the associated events can be divided.

3.4 Distribution Fitting

After the classification step, we are ready to fit some distributions to the interaction stages previously defined and to the events registered in MIMIC. Using HyperStar [19], a simple tool for distribution fitting, we fit the duration of each interaction stage to Phase-Type distributions [12], as shown in Fig. 3. They are commonly employed for performance evaluation thank to their adaptability to any kind of empirical distribution [4]. In particular, the time intervals are fitted to Hyper-Erlang distributions.

Since HyperStar requires the intervention of the user during the fitting process and we have 70 distributions to consider (one for each class) for each of the interactions identified, the time complexity would grow excessively. For this reason, we opt for bringing together those classes that have a similar empirical distribution. To find similar classes, we leverage the Kolmogorov-Smirnov test, commonly used to assess whether two samples come from the same distribution [3]. For each of the interaction stages identified, we perform the test on all possible couples of classes, and cluster together those that exhibit the same

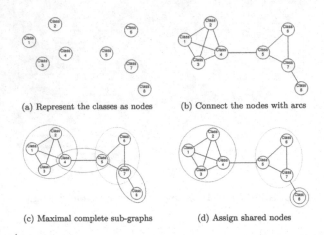

(a) Represent the classes as nodes   (b) Connect the nodes with arcs

(c) Maximal complete sub-graphs   (d) Assign shared nodes

Fig. 4. Visualization of the steps performed to make the groups.

distribution. In Fig. 4 is shown a sample of the result, where the nodes represent the different classes, the arcs connect the classes with the same distribution according to the Kolmogorov-Smirnov test, and the circles represent the groups formed. The events, on the other hand, are fitted using exponential distributions, one for each class and for each kind of event registered in MIMIC. The procedure chosen to fit an exponential to the inter-time between the recording of the events of each kind is based on the method of moments, which focuses on achieving the same average of the considered sample. Since most of the kinds of events registered in MIMIC are associated with one or more attributes that provide a clear indication of the time of their creation, this procedure is used for them. Events that require a specific procedure to be fit, due to the way they are structured, are considered separately and specific procedures are employed for each of them. Waveforms are also fitted using exponential distributions.

4   Development of the Generator

The objective features which we aim to obtain during the development of the generator are fine-tunability and adaptability, in line with the requirements of data and workload diversity for the performance evaluation of Big Data systems [11]. **Fine-tunability** is intended as as the possibility of changing the parameters of the identified distributions from those obtained from the analysis performed and the ability to control the synthetic data generated as output. **Adaptability** is intended as the ability to adapt the generator to different analysis procedures and, potentially, different datasets, without requiring excessive and complex modifications. The generator is divided into the following three modules:

– **Configuration module**, which contains the necessary components for reading and managing the outputs of the analysis.

– **Classification module**, containing the components intended to model the classification made during the analysis phase.
– **Generation module**, containing the components needed to generate the synthetic events.

4.1 Configuration Module

In order to allow for an easier customization of the parameters of the generator, the management of the information obtained from the analysis is centralized within a single module. The configuration module consists in two main components: *a)* the **Manager** class, responsible for providing the other components of the generator with the information obtained from the analysis; **b)** the *Configuration dictionary*, which contains the default file paths where to find the outputs of the analysis. To avoid the use of hard-coded strings in the other components of the generator, some enumerations are also introduced to enclose and group the keys of the previously mentioned dictionary. As we will see later on, they are also used to model the interaction stages identified during the analysis and the kinds of events associated with each of them.

4.2 Classification Module

The classification module collects the enumerations intended to model the groups obtained by the classification procedures previously described. To do so, it uses two enumerations, namely `PatientClass` (used to model the classifications based on the gender and the age of the patients) and `AdmissionClass` (which models the classifications based on the weekday at which the hospital stay started).

4.3 Generation Module

The generation module is the main module of the generator, in charge of the generation of the synthetic events from the distributions fitted during the analysis. Its components follows a layered structure. Each layer is comprised of a single class that implements the `EventsGenerator` interface, which defines the `get_events` and `get_waveforms` methods, used to retrieve the events and the waveforms that result from the generation. Each component, when asked to generate the events, creates an instance of the component of the following layer and routes the request to them, up until the `Interaction` layer is reached, which ultimately generates the events.

This structure is chosen to allow the user to finely control the events to be generated by choosing which layer to request the generation of the events to. For example, if the user is interested in the generation of events of multiple patients, regardless of their classification group, s/he can simply use the Hospital class, which is able to generate the required events, once provided with the number of patients to consider and the time distance between their first admission to the hospital. If instead the user wants to consider the events of the patients of a specific age group, the Patient class may be used. The classification group to consider is decided by the Hospital layer, that chooses the age and the gender of the patient, and by the Patient layer, that chooses the day of the week on which the admission begins. The chosen classification group is passed on to the following layers by the corresponding element of the enumeration discussed in Sect. 4.2.

The distributions to be used for the generation of the events and the other outputs of the analysis (such as the probability of the patient having a certain age or a certain gender) are requested to the Manager class by the components of the generation module. The Manager class is meant to be instantiated by the user and provided with the configuration dictionary to retrieve the outputs of the analysis; a default configuration dictionary, previously introduced in the configuration module, may also be used. The created instance can then be provided to the layer intended to use. During the creation of the following layers, the instance of the Manager is passed on to allow the generation of the events.

As previously stated, the only layer that effectively performs the generation of the events is the **Interaction** layer. This layer models the stages of the Interaction process identified previously and, differently to the other layers, contains more than one class. In particular, the Interaction layer contains two classes, both extending the abstract class *Interaction*:

- the StayInteraction class, which models all the interaction stages that generate the events associated with the entire hospital stay.
- the ICUInteraction class, which models all the interaction stages that generate the events associated with only the ICU stay.

To generate the synthetic events, the two classes request to the instance of the Manager class (either passed on by the previous layers or provided by the user) the parameters of the exponential distributions fitted during the analysis. The duration of the interaction is provided by the Admission layer, which requests the necessary parameters to the Manager to generate it according to the phase-type distributions fitted during the analysis; during this exchange, information about the specific interaction stage considered is communicated through the elements of the enumerations introduced in the configuration module. The events are represented in the generator by the Event class, regardless of their category or type. The waveforms are represented by an extension of such class.

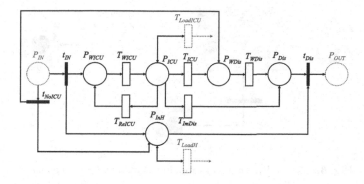

Fig. 5. The GSPN model for the activity diagram shown in Fig. 2.

5   Experimental Evaluation

To show the potentiality of the approach, we propose a multiformalism model which uses the GSPN shown in Fig. 5 to reproduce the patient behavior described in Fig. 2. In particular, new patients enters the hospital in place P_{IN}. Immediate transitions t_{IN} and t_{NoICU} determines whether the patient will need at least a visit to the ICU or if she will only require regular hospitalization. Waiting time to enter or re-enter the ICU is modeled by place P_{WICU} and timed transition T_{WICU}. Permanence in ICU is modeled by place P_{ICU}. In this case three possible events can occur: a patient can conclude her permanence in the ICU, and be immediately dismissed (transition T_{ImDis}), she can be hospitalized for a subsequent period (T_{ICU}), or re-enter the ICU after another period (T_{ReICU}). Normal hospitalization is modeled by place P_{WDis} and timed transition T_{WDis}, while the dismiss is modeled by place P_{Dis} and immediate transition t_{Dis}. Finally places P_{InH} and P_{OUT} account respectively for the total number of patients in the hospital (both ICU and non-ICU), and the patients that are leaving the hospital. Requests produced by patient in the ICU and in the hospital, are modeled by infinite server timed transitions $T_{LoadICU}$ and T_{LoadH}. The firing rate of both transitions is controlled by the marking of the place corresponding to the number of patients either in ICU (P_{ICU}), or inside the hospital (P_{InH}). The type of request is modeled by the color of the token being produced during firing of the corresponding transition: either ($H-jobs$) or ($ICU-Jobs$).

The information system of the hospital is modeled with the Queing Network model shown in Fig. 6, where the GSPN component shown if Fig. 5 is represented with a rounded box. There are two classes of jobs, representing respectively the requests incoming from normal patients ($H-jobs$) and from the ones in ICU ($ICU-Jobs$). Patients arrives to the hospital according to source λ_{Pat}, produces jobs with the GSPN submodel, and leaves the system through immediate transition T_{END} and sink σ_{Pat}. In this simplified scenario, the system is modeled by a C-server queue, with finite capacity K, and drop policy. In this context, all distributions are considered exponential, and queues works in FCFS order.

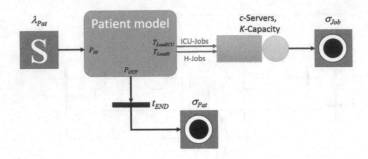

Fig. 6. The multiformalism model of the considered scenario.

The considered parameters are reported in Table 1. To avoid privacy issues, data used are not the exact ones computed by the fitting procedure and the generator previously described, but sufficiently similar to make results meaningful.

Table 1. Model parameters.

Parameter	Value	Parameter	Value	Parameter	Value
λ	5...10 p/h	C	40	K	100
S_{ICU}	20 s	S_H	30 s	t_{IN}	9
$\lambda_{LoadICU}$	30 r/h	λ_{LoadH}	6 r/h	t_{NoICU}	1
T_{WICU}	4 h	T_{ReICU}	120 h	T_{ImDis}	72 h
T_{ICU}	24 h	T_{WDis}	48 h		

The performance analysis and the design of the model is performed using *JMT*, a suite of tools for the performance evaluation of computer systems [20]. With the generated events, we are able to evaluate the system's performances, showing through an analysis of various performance indices (such as throughput, response time, and utilization), considering an increase in the arrival rate. Figure 7a shows the evolution of the arrival rate of requests: the flex at around $\lambda = 7$ patients per hour shows the possibility of having reached the maximum capacity of the system. This is confirmed by Fig. 7b, where throughput and drop rates are shown. As expected, with more than $\lambda > 7$ patients per hour, requests start being dropped, showing that the considered number of servers is not sufficient to handle this peak of requests. It is interesting to see that although the system is experiencing drops, it still does not reach 100% utilization, as shown in Fig. 7c: this is a consequence of the fact that the losses are due to the high variability of the input traffic: most of them occur during bursts, making then the system work at a very low utilization during normal operation times.

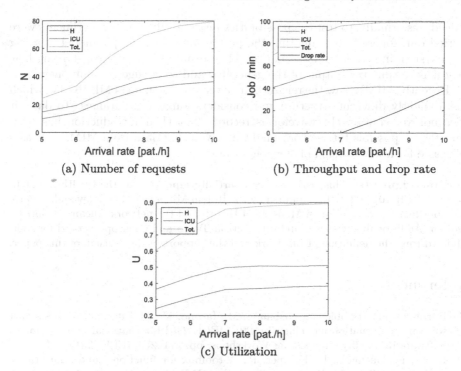

(a) Number of requests

(b) Throughput and drop rate

(c) Utilization

Fig. 7. The evolution of number of requests, throughput and drop rate, and utilization of the servers as function of the arrival rate λ.

6   Conclusions

In this work, we have developed a synthetic data generator to be used for the evaluation through modeling techniques of a Big Data System. As a reference for the creation of the generator, we relied on MIMIC, a publicly available dataset containing the measurements and observations made on patients of a hospital during a multiple years period and recorded by the data system of the hospital in question. With the aim of obtaining additional information on the temporal distribution of such events, we first analyzed the interaction process between the patients and the hospital system, focusing on the time periods during which the events described in MIMIC were recorded in the system. These time periods were then modeled through appropriate distributions, taking advantage of distribution fitting techniques and tools; these same techniques were then used to model the inter-times of each kind of event. The results obtained, such as the parameters of the fitted distributions and the structure identified for the interaction process between the patients and the hospital system, were used as the foundation for the generator's architectural choices. To allow a granular control over the output of the generator, a layered architecture was adopted during its development. Moreover, throughout development, the ability to customize the generator was foregrounded, to make it easy to adapt it to different analysis procedures. Finally, to test the possibilities just described, two experiments were carried out, focused on evaluating the performance of a simple model of a Big Data system in two different scenarios. They resulted in a success, showing how easy it is to tune the output of the generator and to change its components.

Recently, MIMIC saw the release of a newer version, named MIMIC-IV, which has a slightly different structure and considers longer time frames. To date, it does not yet contain the waveforms records, but their introduction has been initiated. A possible future work might involve using data from MIMIC-IV as a reference for a new version of the generator.

Acknowledgments. This work has been partially supported by the Health Big Data Project (CCR-2018-23669122), funded by the Italian Ministry of Economy and Finance and coordinated by the Italian Ministry of Health and the network Alleanza Contro il Cancro. Additionally, we are grateful to Letizia Tanca and Giuseppe Serazzi for their advice during the definition of this work and the support in the revision of this paper.

References

1. Barbierato, E., Bobbio, A., Gribaudo, M., Iacono, M.: Multiformalism to support software rejuvenation modeling. In: 2012 IEEE 23rd International Symposium on Software Reliability Engineering Workshops, pp. 271–276. IEEE (2012)
2. Bause, F., Buchholz, P., Kemper, P.: A toolbox for functional and quantitative analysis of DEDS. In: Puigjaner, R., Savino, N.N., Serra, B. (eds.) TOOLS 1998. LNCS, vol. 1469, pp. 356–359. Springer, Heidelberg (1998). https://doi.org/10.1007/3-540-68061-6_32
3. Berger, V.W., Zhou, Y.: Kolmogorov-Smirnov Tests. In: Encyclopedia of Statistics in Behavioral Science. John Wiley & Sons, Ltd. (2005)
4. Bladt, M.: A review on phase-type distributions and their use in risk theory. ASTIN Bull.: J. IAA **35**(1), 145–161 (2005)
5. Chin-Cheong, K., Sutter, T., Vogt, J.E.: Generation of heterogeneous synthetic electronic health records using GANs. In: Workshop on Machine Learning for Health at the 33rd Conference on Neural Information Processing Systems (2019)
6. Choi, E., Biswal, S., Malin, B., Duke, J., Stewart, W.F., Sun, J.: Generating multi-label discrete patient records using generative adversarial networks. In: Machine Learning for Healthcare Conference, pp. 286–305. PMLR (2017)
7. Ciardo, G., Jones, R.L., III., Miner, A.S., Siminiceanu, R.I.: Logic and stochastic modeling with SMART. Perform. Eval. **63**, 578–608 (2006)
8. Clark, G., et al.: The mobius modeling tool. In: Proceedings 9th International Workshop on Petri Nets and Performance Models, pp. 241–250 (2001)
9. Franceschinis, F., Gribaudo, M., Iacono, M., Mazzocca, N., Vittorini, V.: Towards an object based multi-formalism multi-solution modeling approach. In: Proceedings of the Second International Workshop on Modelling of Objects, Components, and Agents, pp. 47–66 (2002)
10. Goodfellow, I., et al.: Generative adversarial networks. Commun. ACM **63**(11), 139–144 (2020)
11. Han, R., Lu, X., Xu, J.: On big data benchmarking. In: Big Data Benchmarks, Performance Optimization, and Emerging Hardware, pp. 3–18 (2014)

12. Harchol-Balter, M.: Real-world workloads: high variability and heavy tails. Performance modeling and design of computer systems: Queueing theory in action, pp. 347–348 (2013)
13. Health, M.o.: National Minimum Dataset (hospital events) (2012)
14. Johnson, A.E., et al.: Mimic-iii, a freely accessible critical care database. Sci. Data 3(1), 1–9 (2016)
15. Lara, J.d., Vangheluwe, H.: Atom 3: A tool for multi-formalism and meta-modelling. In: Fundamental Approaches to Software Engineering: 5th International Conference, pp. 174–188 (2002)
16. Legato, M.J., Bilezikian, J.P.: Principles of Gender-specific Medicine. Gulf Professional Publishing (2004)
17. Moody, B., Moody, G., Villarroel, M., Clifford, G., Silva III, I.: Mimic-iii waveform database matched subset. PhysioNet (2017)
18. Rashidian, S., et al.: SMOOTH-GAN: towards sharp and smooth synthetic EHR data generation. In: Artificial Intelligence in Medicine, pp. 37–48 (2020)
19. Reinecke, P., Krauß, T., Wolter, K.: Phase-type fitting using hyperstar. In: Balsamo, M.S., Knottenbelt, W.J., Marin, A. (eds.) EPEW 2013. LNCS, vol. 8168, pp. 164–175. Springer, Heidelberg (2013). https://doi.org/10.1007/978-3-642-40725-3_13
20. Serazzi, G., Casale, G., Bertoli, M.: Java modelling tools: an open source suite for queueing network modelling and workload analysis. In: Third International Conference on the Quantitative Evaluation of Systems, pp. 119–120 (2006)
21. Tang, H.: Confronting ethnicity-specific disease risk. Nat. Genet. 38(1), 13–15 (2006)
22. Trivedi, K.S.: SHARPE 2002: symbolic hierarchical automated reliability and performance evaluator. In: Proceedings of the 2002 International Conference on Dependable Systems and Networks, p. 544 (2002)
23. Walonoski, J., et al.: Synthea: An approach, method, and software mechanism for generating synthetic patients and the synthetic electronic health care record. J. Am. Med. Inform. Assoc. 25(3), 230–238 (Mar2018)
24. Wang, X., Peng, Y., Lu, L., Lu, Z., Bagheri, M., Summers, R.M.: Chestx-ray8: Hospital-scale chest x-ray database and benchmarks on weakly-supervised classification and localization of common thorax diseases. In: Proceedings of the IEEE Conference on Computer Vision and Pattern Recognition, pp. 2097–2106 (2017)
25. Yale, A., Dash, S., Dutta, R., Guyon, I., Pavao, A., Bennett, K.P.: Generation and evaluation of privacy preserving synthetic health data. Neurocomputing 416, 244–255 (2020)

MONCHi: MONitoring for Cloud-native Hyperconnected Islands

Dulce N. de M. Artalejo$^{(\boxtimes)}$ (ID), Ivan Vidal (ID), Francisco Valera (ID), and Borja Nogales (ID)

Universidad Carlos III de Madrid, Av. de la Universidad, 30, 28911 Leganes, Spain
{dartalej,bdorado}@pa.uc3m.es, {ividal,fvalera}@it.uc3m.es

Abstract. Network performance monitoring is a crucial aspect in order to maintain reliable and efficient communications between different hosts and clusters. This is becoming more relevant as companies are progressively moving towards cloud-native environments, where hyperconnected islands are deployed. While monitoring for individual clusters and components is widely deployed, inter-domain metrics have not been yet added to present solutions. In this paper, we present MONCHi, an open-source based custom monitoring tool designed to collect and analyse traditional and custom metrics for network links within and among Kubernetes clusters. MONCHi consists on the flexible design and implementation on top of a conventional monitoring tool –Prometheus in this case– of several custom scripts that run in separate containers within a single pod and continuously collect and store metrics. These metrics are then exposed and visualised in an analysis tool, allowing for easy monitoring of network performance in clusters and for scalable multi-domain deployments based on multiple Prometheus instances. We also discuss the modification of Prometheus configuration to support the new endpoint for MONCHi. Finally, we present the results of several performance tests, focusing mainly on Round-Trip Time (RTT) and bandwidth, conducted to validate the effectiveness of MONCHi to accurately collect and analyse network performance metrics. Overall, MONCHi provides an effective and easy to customise solution for monitoring cloud-native multi-domain environments.

Keywords: Monitoring · Network metrics · Cloud-native · Inter-domain · Kubernetes · Prometheus · Grafana

1 Introduction

In recent years, virtualisation technologies have become increasingly popular as organisations seek to maximise efficiency, reduce costs, and enhance resiliency

This paper has partially been supported by the European H2020 FISHY Project (grant agreement 952644), and the TRUE5G project funded by the Spanish National Research Agency (PID2019-108713RB-C52/AEI/10.13039/501100011033).

[17]. This is very related to cloud-native applications, built and deployed on cloud computing infrastructures, that emphasize the use of virtualisation. Using containerization and microservices architectures, they allow for greater flexibility and agility making them the enablers of 5G networks [14]. This rise of cloud-native services –based in virtualisation–, with platforms such as Openstack [13], and the adoption of containers and container orchestration platforms, such as Kubernetes [10], has transformed the way applications are deployed and managed. As its use continues to grow, so does the need for effective monitoring solutions that provide real-time visibility into the performance, robustness, and security of these complex systems. Particularly in the context of 5G networks, assisting operators to execute appropriate networking decision, and to select the most efficient deployments.

While there are many monitoring tools available for individual Kubernetes clusters –such as SLATE [4] or OMNI [18]–, there is a lack of solutions that can measure the features of applications that communicate among different clusters, also called domains or islands. This is a significant limitation, given that cross-cluster communications are becoming increasingly common in 5G environments, where multi-partner setups are often used. This was sought to be addressed in this research.

The objective was to find the best solution for integrating a monitoring tool that could measure individual cluster and pod metrics while also providing the ability to add custom metrics related to the communication between applications in different places or any other type of required metric (e.g. security metrics). By addressing this gap, we aimed to provide an effective monitoring solution for 5G environments and other complex cloud-native architectures. Additionally, adherence to RFC 7594 [3] was sought, so that the design follows a predefined metric collector standard.

In this paper, we present a monitoring tool specifically designed for virtualisation environments: MONCHi, a MONitoring tool for Cloud-native Hyperconnected Islands. It has a focus on Kubernetes and is completely based on open-source tools and platforms, providing a comprehensive and integrated approach to monitoring, visualisation, and management, with the added value of including inter-workload metrics that have been forgotten in traditional methods hitherto developed. This solution provides real-time insights into the robustness and performance of the environments and enables organisations to quickly identify and resolve issues, improve system efficiency, and enhance the overall user experience by allowing the inclusion and customisation of scripts for the measurement of additionally desired metrics related to their inter-cluster communications.

The novelty of the presented solution lies in extending the conventional capabilities with the incorporation of new endpoints that can scrape custom metrics. Thus, providing a great versatility and flexibility, facilitating both the implementation phase and runtime operations. Moreover, the visualisation of all the components from different sources is concentrated into a single centralised platform, increasing the ease-of-use of this tool and providing great scalability. It is important to note that this monitoring solution goes beyond the sole provision of basic monitoring capabilities, since it is compatible with traditional metrics but also includes the inter-workload metrics collection. It provides organisations with the tools necessary to actively monitor their resources and ensure that their systems are functioning optimally across different platforms. By including all this, organisations can feel confident in the performance and reliability of their multi-island environment.

This work arose as part of the European FISHY project [2], which aims at delivering a coordinated cyber resilient platform towards establishing trusted supply chains of ICT systems. The project is led by a collaboration of experts from both industry and academia, where Universidad Carlos III de Madrid (UC3M) has allocated the resources to deploy the different components developed under the context of the project, the network infrastructure so as to facilitate the communication of them all, and the monitoring component that this document presents. This latter became crucial when dealing with inter-cluster workloads in a complex environment such as FISHY. MONCHi proved to be an indispensable component in facilitating the integration of various applications, by enabling the verification of their proper deployment, resource allocation, and monitoring of traffic flow across their respective interfaces in the different trials and demonstrations performed in FISHY.

2   Background

Monitoring tools are essential for cloud-native environments, providing real-time visibility and issue identification. Popular solutions like Zabbix, Nagios, and Prometheus have been tailored for Kubernetes. Grafana is often used alongside these tools, enabling custom dashboards and alerts.

In recent studies, similar research to ours was conducted to evaluate monitoring tools for cloud-native environments. In one of the first papers available on this area, a Kubernetes-based monitoring platform for dynamic cloud resource provisioning was introduced [5]. This solution required the deployment of Heapster for the monitoring, InfluxDB for the storage and Grafana for the visualisation, together with Apache JMeter for monitoring application performance metrics. This implied the deployment of many components and a very complicated management of the tool. SLATE was another monitoring solution proposed for distributed Kubernetes clusters [4]. Although the solution can be effective, its installation process was also more complicated, as Thanos was included, which implied additional expertise in several query languages and the insertion of more components in the system. Other solutions focused on the monitoring of edge devices [7], but in this case collecting very simple and basic metrics without taking into account the possibility of multi-cluster environments. In the case of MONCHi, a simpler solution for monitoring multi-cluster environments is presented, facilitating the introduction of custom metrics for inter-domain measurements.

Another interesting solution proposed, OMNI (Operations Monitoring and Notification Infrastructure), was based on a very similar architecture to ours, including Kubernetes with Prometheus and Grafana for facilitating the monitoring, viewing and alerting functionalities [18]. Although it has a similar core this solution does not include custom metrics between different applications, which was one of the motivations of MONCHi, so it was insufficient for the purpose of this investigation.

A well established tool that has been considered for the base of our system is Zabbix, an open source cloud monitoring tool that has been around since 2001 [11] and that uses agents to collect monitoring data. Despite being created before the rise of cloud computing, it is highly configurable and can be tailored to specific needs. However, its origin as a non-cloud tool means that it may not be easy to set up as a cloud monitoring platform, making it a potential disadvantage for cloud-native environments such as the one used for this research. Nonetheless, an additional limitation of this tool is its scalability, which is restricted to 1,000 nodes. Despite this, Zabbix could be a good option for small to medium-sized cloud setups that require a highly customised solution.

Nagios is another open-source cloud monitoring tool that has been around for almost two decades, making it a reliable solution [11]. One advantage is that it has a larger community and a wider variety of plugins than Zabbix, which can make it easier to monitor specific aspects of the cloud infrastructure. Anyway, it shares many similarities with Zabbix in terms of setup and data collection options. However, one significant difference is that it offers two options: Nagios Core, which is free, and Nagios XI, a commercial platform that provides more functionality and better usability. This could be a disadvantage in our case, as we require a solution that is both stable and free over time. It has some limitations when it comes to scalability, as it requires manual configuration and can become difficult to manage when monitoring a large number of hosts.

Furthermore, it is worth mentioning that Wireshark [20], a powerful network analysis tool, can also capture detailed network parameters, although without the displaying graphical capabilities that we required for our tool. While not covered extensively in this paper, Wireshark offers the ability to analyze network traffic in a highly granular manner, providing valuable insights into network performance and behavior.

Additionally, Prometheus was studied, an open-source cloud-native monitoring tool that was created in 2012 [16]. It is designed to collect time-series data, and is especially well-suited for microservices and containerised applications. One of its main advantages is its flexibility, as it can monitor a wide range of applications and systems. It also has a powerful query language and can generate alerts based on complex conditions. Prometheus is very scalable and can handle millions of time-series data points without any problems. However, one of its main disadvantages is that it requires a high level of expertise to set up and maintain, and it requires some familiarity with its query language, although good documentation can be found on this subject. Additionally, it does not have built-in support for distributed tracing, which can be a drawback for some cloud-native applications. Despite these drawbacks, Prometheus has been the most promising tool for our cloud-native approach, as it offers the best balance of functionality, flexibility, and scalability for monitoring applications in a multi-cluster environment.

To solve Prometheus' shortcomings, its combination with Grafana was considered. Grafana is a popular open-source visualisation tool that provides an efficient and powerful way of presenting data. It works seamlessly with multiple data sources, which makes it a great tool for the cloud-native environment. This feature enables the network administrator to monitor various islands on a single dashboard when needed, making it easier to compare and contrast data from different clusters or to use Grafana to monitor single clusters independently. Grafana can be highly customised, providing numerous graphing options and templates to create visually appealing and insightful dashboards. It also allows for alerting functions, making it easier to identify and respond to critical issues in the system. Overall, Grafana is a valuable addition to Prometheus, and it provides a user-friendly interface to monitor and understand the data generated by the system.

As a conclusion we can state that, while current platforms offer reliable monitoring capabilities, they have certain limitations that make them less ideal for our particular purposes. Moreover, while some papers have introduced additional components to address these limitations, the need for a more flexible and versatile tool that can incorporate custom metrics related to communication between applications across different domains has become increasingly clear. This is where MONCHi aims to fill the gap and provide a more comprehensive and effective monitoring solution for cloud-native environments.

3   Design of the Solution

This section presents the design of MONCHi, based on what has already been introduced, showing the main components and functionalities, the underlying architecture and the technologies used. Additionally, some examples of the types of metrics that can be collected and analyzed using MONCHi are provided.

Figure 1 presents the architecture in a compact and visual way, where several islands or clusters are included, although many more could be deployed anywhere over the Internet in the operating network (cluster A, cluster B, cluster X, etc.). The whole infrastructure is based on Kubernetes, implementing different clusters, each one deployed over one or several nodes, as preferred, and with a different number of pods and resources inside. Each component of the design will be implemented as a pod, to facilitate the management of the system.

In order to monitor each hyperconnected island, a monitoring system or collector is established to retrieve the relevant metrics. This chosen system deploys a database or repository to store the collected data, along with a Cluster Metric Collector (CMC) to gather default information from the cluster. An Inter-Cluster Metric Collector (ICMC) is also set up, which deploys different agents as separate containers to perform custom queries and collect new metrics. Additionally, a Graphical User Interface (GUI) –corresponding to the analysis tool–, is placed on each domain to provide a visual representation of the collected and processed

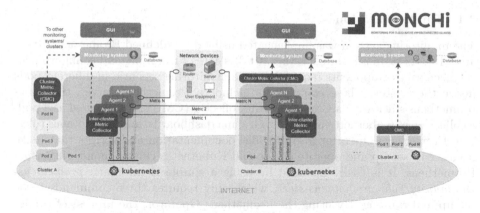

Fig. 1. MONCHi's architecture

data, making it easier to manage the hyperconnected environment. However, one Grafana may be chosen as central analysis tool for management purposes by grouping several domains and clusters into a single GUI. In this way, the administrator may choose to visualise all the metrics condensed or separated by clusters. This approach offers a significant advantage as it enables the addition of new clusters or endpoints to be monitored, making the solution multi-cluster capable.

It is worth noting that MONCHi complies with the terminology defined in RFC 7594 [3], used as the base for the design, clearly defining different agents with a collector, a repository and the analysis tool.

Among the network metrics that can be customised for their measurement and collection we may include the end-to-end delay, that can be calculated with tools such as *ping* or *traceroute*, or other metrics such as the bandwidth or throughput, percentage of lost packets or jitter.

In addition to the network metrics mentioned above, MONCHi also offers the ability to collect security metrics related to inter-site communications. This can be done by leveraging some IDS or software tools for security management such as Snort [6] or Wazuh [19] to collect security triggers and alerts. By incorporating security metrics into the monitoring solution, MONCHi can help ensure that inter-domain communications are not only efficient, but also secure and protected against potential threats. This is particularly important in today's landscape where cyber threats are becoming increasingly common and organisations need to take proactive measures to safeguard their networks and data.

Following this architecture, in the FISHY project several pods deploy different security functionalities as if it were a toolbox that environments can use. This test cases will be the ones used for the validation scenario. More information can be found in the project deliverables [8] –concretely in D6.1–, but due to space limitation no more details are included in this document.

3.1   Implementation

The monitoring system was implemented using a pre-defined Prometheus Operator [15]. Its main advantage is that it takes care of installing and configuring all the necessary components required for the monitoring scenario in different pods inside the cluster. These components include a Prometheus server, the kube-prometheus stack with a Cluster Metric Collector (CMC) implemented as a pod, a collection of Kubernetes manifests, Grafana dashboards, and Prometheus rules. The Prometheus Operator also provides documentation and scripts that make it easier to set up and operate end-to-end Kubernetes cluster monitoring using Prometheus. It is easily installed by using a simple Helm chart: *prometheus-community/kube-prometheus-stack*, which only requires three commands to be set up and running. By using the Prometheus Operator, the process of configuring Prometheus monitoring and Grafana visualisation for Kubernetes clusters becomes more streamlined and efficient.

For the implementation, Prometheus Operator was installed on three clusters inside 5TONIC labs [1]. Although for this investigation Prometheus has been used, and its setup is the one explained, the versatility of MONCHi architecture would allow the monitoring system to be any tool compatible with Grafana, such as InfluxDB, ElasticSearch or AWS CloudWatch.

In Fig. 1, the pod implementing MONCHi's functionality (named *Pod 1*) is highlighted inside two of the clusters, being the one that will generate, store, collect and expose the custom metrics for the inter-workload values. Inside this pod, different containers –one per task– will be deployed, including one or several agents, that will each run a script to measure network properties when communicating to: i) applications in a different cluster or ii) to other network devices in the environment (such as routers, servers or even user equipment). These agents will also parse the results to store them as metrics inside a custom file in a shared file space. This shared directory was created as a Kubernetes Persistent Volume and assigned to the pod, so that all containers can access it, either by storing the metrics in different files (agents) or by reading the different files and posting the metrics (metric collector). This inter-cluster metric collector is then deployed in another container and will access the created shared files and read them, posting their content to an internal HTTP server that the monitoring tool will be configured to scrape. In any case, monitoring of clusters implementing this pod is still compatible with simple islands that have no extended features.

The Prometheus instances, which are deployed as pods, have their own database, being the repository for the information in each cluster. Customisation of Prometheus monitoring tool is easily done through the *values.yaml* file available inside the Helm chart, by upgrading it with the new properties desired. Modifications to this file would allow a multitude of features to be set to the preferred values, for instance, Prometheus could be modified to scrape additional endpoints –such as the metric collector in this work–, and it would also allow to modify the retaining time of the metrics or change the scraping interval, among other possibilities. For MONCHi, the retaining time has been modified to store the metrics for 50 days, as the computation of monthly averages is of interest.

Finally, a central Grafana may be chosen for the management and visualisation of all the data. Although each cluster has its own Grafana instance and can be individually monitored, it may be interesting for the network administrator to show all the information in a centralised manner and use the analysis tool of MONCHi as a unique monitoring display point for the whole platform. In this case, the central Grafana has to be configured to include the remaining data sources; so, additionally to its own Prometheus instance, it should collect the values of the other domains' repositories. These external databases will be accessible thanks to a Kubernetes NodePort service that exposes a port through which the cluster's metrics are available for the analysis tool. Grafana extracts the information from the databases thanks to different PromQL queries, that can include mathematical operators to express the data as preferred and allow to query all Kubernetes components, up to selecting a specific container inside a pod. The inclusion of additional repositories from other islands would be as simple as creating their own NodePort service and adding that data source in Grafana's GUI by specifying their IP address and port.

In conclusion, MONCHi provides a highly scalable and easy-to-customise solution for cloud-native environments. Its ease of installation and configuration, combined with its coverage of previous limitations, makes it the ideal solution for monitoring hyperconnected islands in 5G and other cloud-native environments. The next section will delve into the practical experiments done, explaining the custom metrics that were implemented for the validation of the solution.

4   Practical Validation

For the purpose of practical validation, a series of tests were devised to verify the accuracy of network metrics recorded by MONCHi. The tests were carried out on three clusters inside the Fishy Reference Framework (FRF), namely *Fishy Control Services (FCS)*, *Domain 1* and *Domain 2*, as described in Fig. 2. We will not explain all the details of the given architecture as it falls out of the scope of our work, but it is essential to understand the complexity of the multi-cluster ecosystem in which MONCHi has been deployed. The Inter-Cluster Metric Collector (ICMC) was exclusively deployed in *Domain 2*. The key metrics that were of interest to us were receiving throughput, which was measured using *iperf3* commands, and Round-Trip Time (RTT) measurements, collected using *ping*. To facilitate the *iperf3* commands, it was necessary to deploy *iperf3* servers that would respond to the requests. Two identical, lightweight pods were thus deployed in *Domain 1* and *FCS*, respectively, and a Kubernetes NodePort service was attached to them so that they could be accessed from outside the cluster.

For the different tests done, Python programming language was utilised to create the custom metrics collection scripts. The RTT collection script was developed to capture a new measurement every 15 s, using 6 *ping* commands, discarding the initial result, and averaging the remaining responses. The specific number of responses to wait for was not specified, so as to take into account possible

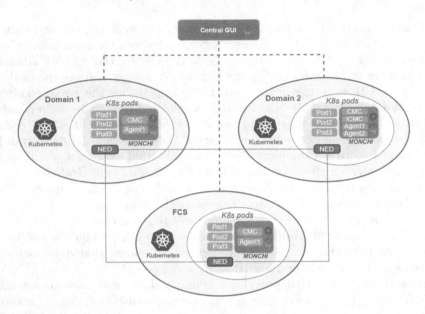

Fig. 2. FRF outline

packet losses. After the final average was computed, the result was transmitted in Prometheus format to a shared file named *rtt-metrics.prom*. The ICMC was programmed to extract these measurements from the shared file and submit them to the Prometheus monitoring system to be stored in the database.

Similarly, the bandwidth measurement script was programmed to send an *iperf3* command with a 1 min duration, specifying the custom port that was opened for the *iperf3* server with the NodePort service (any other less intrusive bandwidth estimation tool can be used, but *iperf3* was intentionally selected to clearly showcase the network activity in the monitoring tool). These commands were scheduled to be transmitted every 30 min to prevent network saturation. The result was stored in another shared file named *iperf-metrics.prom*. The collector was designed to scrape and post these metrics to the monitoring system to store them in the database.

In order to justify the sequences used to underpin Figs. 3 and 5, it is important to note that these values were randomly chosen with the aim of showcasing a diverse range of metrics. The selection was made deliberately to include scenarios with both significant bandwidth limitations and delays, as well as cases with smaller constraints. This approach allowed for the thorough validation of the tool under extreme conditions, ensuring its effectiveness across various challenging scenarios.

The monitoring system is finally designed to scrape information from the custom endpoint every 5 s. However, due to the nature of the endpoint being monitored in this particular case, metrics are not updated every instant, but rather the data remains the same until the personalised scripts generate new values in their corresponding files (every 15 s for RTT values and every 30 min for bandwidth metrics). As a result, when the monitoring system scrapes the endpoint, it will always retrieve the same data, which causes the visual representation of the metrics on graphs to appear as if they had constant values. This phenomenon is because the data displayed on the graphs does not reflect any changes or updates to the metrics until the personalised scripts generate new data and the monitoring system scrapes it again.

4.1 Performance Tests

The primary objective was to assess whether MONCHi accurately collects network bandwidth limitations, as reflected in its Graphical User Interface (GUI). To achieve this, the Linux traffic control (*tc*) tool was employed on the link connecting *Domain 2* with the other two clusters, along with the *token buffer filter (tbf)*, which restricted the maximum burst to 32 kbit and maximum limit of the queue to 1,500,000 bytes, which is the equivalent of 1,000 packets (Linux queue default size). The test spanned 24 h, during which the sustained maximum rate was modified every two hours in the following sequence: 350 - 600 - 400 - 250 - 500 - 650 - 100 - 300 - 150 - 450 - 500 - 700 Mbps. The results that this test originated can be seen in Fig. 3, where the image on top shows the outcome of the link between *Domain 2* and *Domain 1*, and the image below shows the results from *Domain 2* to the *FCS*.

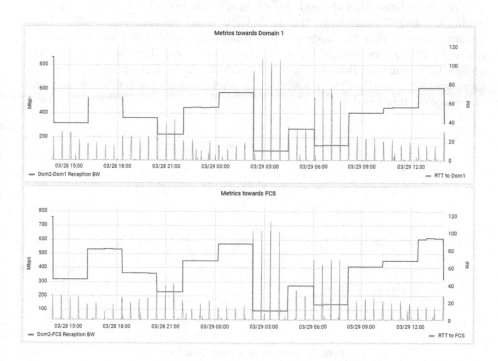

Fig. 3. Throughput and RTT results over 24 h

After analysing the graphs, it is evident that MONCHi is achieving its intended purpose as the variations in throughput conform to the predetermined limitations. However, the highest achieved throughput remains slightly below the maximum limit due to the functioning of *tc*. These observations confirm that the network is functioning correctly since more restricted maximum rates and new *iperf3* commands (the yellow peaks in the graphs) lead to considerably higher RTT. Conversely, a nearly unrestricted maximum rate leads to lower RTT during *iperf3* commands, indicating lower network congestion and more margin for *ping* commands to get through. When we limit the size of the queue and the maximum rate, we are essentially changing how long packets will wait in the queue. For example, if we have a queue dimension of 1,000 packets, each with a size of 1,500 bytes (8 bits per byte), then the total capacity of the queue is 12,000,000 bits. If we have very limited rates, then there will be more congestion and higher RTT values. On the other hand, if we have small limitations on the maximum rate, there will be less congestion and lower RTT values. To illustrate this, let us consider some numbers: when we divide 12,000,000 bits by 100 Mbps, we get 120 ms, whereas if we divide it by 600 Mbps, we get 20 ms.

To assess the impact of the custom metrics collection pod on the node where it is deployed, we monitored the CPU usage during the same 24 h test period, as shown in Fig. 4. The yellow line represents the whole node's CPU usage, while the blue line represents the pod's usage, which is significantly lower. The right axis expresses the pod's CPU consumption as a percentage of the 3 assigned cores. The results indicate that the node's CPU usage remains below 0.5 cores, and even during the *iperf3* commands, the pod's CPU usage remains below 15%. These findings suggest that the node could accommodate additional metrics without causing CPU saturation. Additionally, the graph shows that the increase in the node's CPU consumption during *iperf3* commands is slightly higher than the increase in the pod's CPU consumption, likely due to packet processing and other node-related tasks.

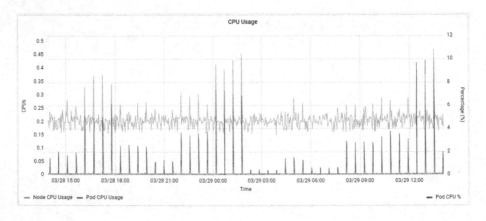

Fig. 4. CPU results over 24 h test

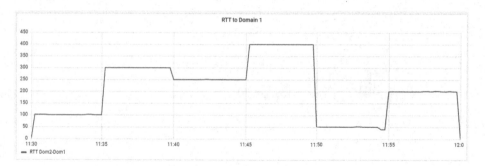

Fig. 5. RTT results over 30 min test

These results were highly encouraging, as they confirmed the efficacy of MONCHi in incorporating custom metrics across various clusters. Not only did MONCHi exhibit accurate performance, but it also demonstrated remarkable efficiency. Despite the network limitations, the custom-metrics pod had a minimal impact on the node's functioning, resulting in only a slight increase in resource consumption. In a regular environment without manual limitations, this consumption would be even lower, rendering the system even more efficient.

For the purpose of conducting Round-Trip Time (RTT) tests, we chose a specific time frame of approximately 30 min. Within this time frame, we incorporated a different delay every 5 min to evaluate the functionality of the custom-metrics exporter. To achieve this, we utilised the Linux traffic control tool, *tc*, to add varying delays to the queuing packets. Specifically, we introduced the following sequence of delays: 100 - 300 -250 - 400 - 50 - 200 ms.

The results of the RTT tests were represented graphically using Grafana, and the originating chart is displayed in Fig. 5. This diagram depicts the performance of the system under test by displaying the RTT values against time. By introducing delays at the different time intervals and verifying their correspondence in the graph, we were able to assess the responsiveness of the system to changing network conditions and evaluate the effectiveness of the custom-metrics exporter functionality.

4.2 Compatibility with Traditional Metrics

Finally, as we have mentioned, these custom metrics panels are compatible with the traditional integration of Prometheus and Grafana for monitoring different data sources. Figure 6 shows from the central analysis tool located in the *FCS* –the central Grafana–, some metrics collected by default by Prometheus. We have decided to display a pod located in *Domain 1*, called *Network Edge Device* or *NED*, which was shown in Fig. 2, that is in charge of performing networking between other pods in the system. For this pod, we wanted to measure its CPU, memory usage and traffic sent and received in general and through each of its interfaces. The results are quite visual and help managing the whole infrastructure quite easily from a central administrative view.

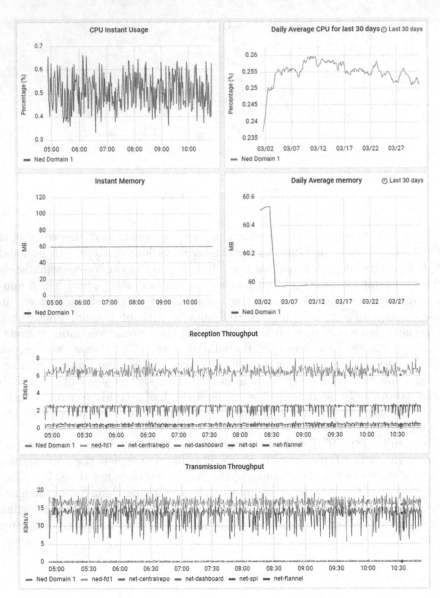

Fig. 6. *Domain 1* NED dashboard over 6 h

5 Conclusions

In conclusion, MONCHi monitoring tool has proven to be a crucial component in the FISHY project ecosystem, addressing the long-standing issue of inter-domain monitoring in a real-world case. Through rigorous testing in a realistic environment, where the metrics between different islands were collected and the links were deliberately tampered with, MONCHi has effectively detected these changes and presented them in an easy-to-understand graphical interface. The tool has demonstrated its ability to collect and display relevant metrics from different domains, which will aid, not only in this project but also all operators, in making informed decisions regarding the deployment and provisioning of components. This will significantly ease the provisioning and deployment operations, allowing a more efficient use of resources and a better overall performance in cloud-native environments.

Furthermore, this investigation highlights the importance of conducting non-invasive tests when integrating systems, to avoid resource consumption and performance impact on other components within the cluster. However, our tests aimed to push the network to its limits, yet surprisingly, the resource consumption was lower than anticipated, not only in terms of CPU but also bandwidth, despite conducting saturating tests every 30 min. It is worth noting that to ensure connectivity with the other islands, Kubernetes services will be required, particularly NodePort services in this case. Moreover, the usage of open-source tools has significantly reduced the cost of the project, making it a highly cost-effective solution. Overall, MONCHi monitoring tool has the potential to improve inter-domain monitoring and streamline the decision-making process for network stakeholders.

In addition to its current capabilities, MONCHi has the potential for further development and improvement in the future. One of the most important areas of future work is the implementation of an alerting system that can detect and report link disturbances. This will enable network managers to take timely and appropriate actions to address any issues that arise. Another interesting possibility for future development is the use of blockchain technology to secure and verify the integrity of critical alerts. This would provide an additional layer of security and assurance that the alerts have not been tampered with.

Moreover, MONCHi could benefit from the introduction of new metrics, such as security ones or different ways of implementing the existing ones. This would help to identify the most efficient metrics for measuring and analyzing network performance, and would enable MONCHi to expand its scope and capabilities. In addition, further research could be done to explore how MONCHi may be integrated with other tools and systems to enhance its overall functionality and usefulness. As part of future work, we aim at exploring the alignment of MONCHi with different network telemetry initiatives such as OTel [12] and RFC 9232 [9]. Ultimately, the potential for future work and development of MONCHi is significant, and the tool holds great promise for further improvement.

References

1. 5TONIC (2023). https://www.5tonic.org. Accessed May 2023
2. FISHY (2023). https://fishy-project.eu. Accessed May 2023
3. Abeele, N.G.V.D., Daigle, L.A., Kumari, W.A.: A Framework for Large-Scale Measurement of Broadband Performance (LMAP). RFC 7594, Internet Engineering Task Force, September 2015. https://doi.org/10.17487/RFC7594

4. Carcassi, G., Breen, J., Bryant, L., Gardner, R.W., McKee, S., Weaver, C.: SLATE: monitoring distributed Kubernetes clusters. In: Practice and Experience in Advanced Research Computing, pp. 19–25 (2020)
5. Chang, C.C., Yang, S.R., Yeh, E.H., Lin, P., Jeng, J.Y.: A Kubernetes-based monitoring platform for dynamic cloud resource provisioning. In: GLOBECOM 2017 IEEE Global Communications Conference, pp. 1–6 (2017). https://doi.org/10.1109/GLOCOM.2017.8254046
6. Cisco Systems: Snort: network intrusion detection and prevention system. https://www.snort.org. Accessed May 2023
7. Fathoni, H., Yen, H.Y., Yang, C.T., Huang, C.Y., Kristiani, E.: A container-based of edge device monitoring on Kubernetes. In: Chang, J.W., Yen, N., Hung, J.C. (eds.) Frontier Computing: Proceedings of FC 2020, Lecture Notes in Electrical Engineering, vol. 747, pp. 231–237. Springer, Singapore (2021). https://doi.org/10.1007/978-981-16-0115-6_22
8. FISHY: Use cases settings and demonstration strategy (IT-1). Deliverable D6.1, European Commission, October 2021. https://fishy-project.eu/library/deliverables. Accessed May 2023
9. Song, H., Qin, F., Martinez-Julia, P., Wang, L.A.: Network Telemetry Framework. RFC 9232. Internet Engineering Task Force (IETF), May 2022
10. Kubernetes: Documentation. https://kubernetes.io/docs. Accessed May 2023
11. MWTeam: 15 Best cloud monitoring tools & services in 2023 (Updated) (2023). https://middleware.io/blog/cloud-monitoring-tools. Accessed May 2023
12. Open Telemetry: Tools to instrument, generate, collect, and export telemetry data. https://opentelemetry.io. Accessed May 2023
13. OpenStack Foundation: Open source cloud computing infrastructure. https://www.openstack.org. Accessed May 2023
14. Oracle: What is cloud native? https://www.oracle.com/in/cloud/cloud-native/what-is-cloud-native. Accessed May 2023
15. Prometheus: kube-prometheus-stack: prometheus end-to-end K8s cluster monitoring. https://github.com/prometheus-community/helm-charts/tree/main/charts/kube-prometheus-stack. Accessed May 2023
16. Prometheus: An open-source systems monitoring and alerting toolkit (2023). https://prometheus.io/docs. Accessed May 2023
17. Shamir, J.: Virtualization is the foundation of cloud computing - key benefits it can bring to your organization? (2021). https://www.ibm.com/cloud/blog/5-benefits-of-virtualization. Accessed May 2023
18. Sukhija, N., Bautista, E.: Towards a framework for monitoring and analyzing high performance computing environments using Kubernetes and prometheus. In: 2019 IEEE SmartWorld, Ubiquitous Intelligence & Computing, Advanced & Trusted Computing, Scalable Computing & Communications, Cloud & Big Data Computing, Internet of People and Smart City Innovation, pp. 257–262. IEEE (2019)
19. Wazuh: Wazuh: Open source security platform. https://wazuh.com. Accessed May 2023
20. Wireshark: Network protocol analyzer. https://www.wireshark.org. Accessed May 2023

A Quantitative Approach to Coordinated Scaling of Resources in Complex Cloud Computing Workflows

Laura Carnevali[1], Marco Paolieri[2], Benedetta Picano[1], Riccardo Reali[1(✉)],
Leonardo Scommegna[1], and Enrico Vicario[1]

[1] Department of Information Engineering, University of Florence, Florence, Italy
{laura.carnevali,benedetta.picano,riccardo.reali,
leonardo.scommegna,enrico.vicario}@unifi.it
[2] Department of Computer Science, University of Southern California,
Los Angeles, USA
paolieri@usc.edu

Abstract. Resource scaling is widely employed in cloud computing to adapt system operation to internal (i.e., application) and external (i.e., environment) changes. We present a quantitative approach for coordinated vertical scaling of resources in cloud computing workflows, aimed at satisfying an agreed Service Level Objective (SLO) by improving the workflow end-to-end (e2e) response time distribution. Workflows consist of IaaS services running on dedicated clusters, statically reserved before execution. Services are composed through sequence, choice/merge, and balanced split/join blocks, and have generally distributed (i.e., non-Markovian) durations possibly over bounded supports, facilitating fitting of analytical distributions from observed data. Resource allocation is performed through an efficient heuristics guided by the mean makespans of sub-workflows. The heuristics performs a top-down visit of the hierarchy of services, and it exploits an efficient compositional method to derive the response time distribution and the mean makespan of each sub-workflow. Experimental results on a workflow with high concurrency degree appear promising for feasibility and effectiveness of the approach.

Keywords: Cloud computing · coordinated scaling · stochastic workflow · end-to-end response time distribution · complex workflow structure

1 Introduction

Cloud Computing (CC) systems [7,16] need to store, manage, and process enormous amounts of data continuously generated by a variety of sources within the Internet of Things (IoT) [28]. Excessive network traffic or heavy computational workload may lead to violations of Quality of Service (QoS) attributes granted through Service level Agreements (SLAs) [29]. Therefore, CC systems

M. Iacono et al. (Eds.): EPEW/ASMTA 2023, LNCS 14231, pp. 309–324, 2023.
https://doi.org/10.1007/978-3-031-43185-2_21

must autonomously adapt their operation in response to time-varying changes both in the software system itself and in its operating environment [26,32]. Adaptation can be achieved through *autoscaling* systems [13], which dynamically change software configurations and provision hardware resources on demand, with the goal of continuously satisfying cost objectives as well as non-functional Service Level Objectives (SLOs), i.e., specific measures agreed within an SLA. Scaling actions can be *horizontal*, if the system adds or removes containers or Virtual Machines (VMs) where services can be deployed, or *vertical*, if the system changes specifications of those containers or VMs, e.g., CPU cores or available memory.

Horizontal scaling can optimize resource provisioning for *individual* services orchestrated in larger applications [1,3,14,20,37], e.g., composite web services [8], Functions as a Service (FaaS) platforms [23,33], microservice architectures [2]. In [20], bottlenecks in a multi-tier application are automatically detected and resolved, minimizing the number of web servers and database instances while guaranteeing a maximum response time. Dynamic scaling of the number of VMs in cloud services is performed based on the number of active sessions of each web server instance [14], using queueing theory to estimate demand [1], and leveraging also time series analysis to forecast load intensity [3]. Few approaches exploit *coordinated* scaling of resources to avoid undesired effects of *local* scaling like bottleneck shifting and oscillations [37], e.g., by reconfiguring services of small web applications together [35], by exploiting time-series analysis and queueing theory to determine the number of VM instances that minimizes energy consumption without violating SLAs [4], or by collectively providing application tiers with a number of servers or VMs that guarantees meeting contracted [37] or average response times [6]. Though horizontal scaling has received more attention [13] and has better support from cloud vendors [15], being easier to implement and manage, it performs coarse-grained adaptation through static replication of VMs or containers with fixed-size configurations, and it suffers from non-negligible lags to instance and start VMs or containers [39], which, despite lag-mitigating actions like dynamic VM cloning [24], may negatively affect time-critical applications.

Vertical scaling performs fine-grained resource adaptation by modifying attributes of VMs or containers [15,21,34], thus limiting resource overprovisioning and resulting preferable for applications with time-critical requirements. In [34], CPU voltage and frequency of VMs in multi-tenant cloud systems are individually adapted to meet SLOs, supporting migration to new VMs in case of overloading. Optimization of CPU usage and memory allocated to a cloud application is performed in [15] to meet requirements on mean response time, exploiting a performance model based on an inverse relationship between the application mean response time and the number of allocated CPU cores [21]. In [39], CPU power tuning and hotplugging are performed to improve CPU usage efficiency in a web server, with minimum SLA violation rate. Few approaches perform vertical scaling in a *coordinated* manner. In [22], soft resources of web application servers (e.g., number of server threads and database connections) are allocated based on measured throughput and concurrency. A resource-management framework is defined in [25] to manage shared resources among microservices, exploiting machine learning methods both to localize microservice instances responsible for SLO violations and to define methods to mitigate resource contention. Horizontal and vertical scaling are combined in [19] to determine a load distribution policy for co-located distributed applications by exploiting multi-class queueing networks and model predictive control, and in [17] to adapt the number of replicas and the CPU capacity of each microservice by using layered queueing networks to assess potential performance improvement of scaling actions.

The few approaches that address coordinated resource scaling [26] mainly consider simple cloud applications consisting of few services orchestrated as sequential workflows. Notably, no approach takes into account the end-to-end (e2e) response time distribution in scaling decisions, which instead becomes relevant when SLAs are characterized by soft deadlines and penalty functions [27] defined as rewards calculated from such distribution.

In this paper, we present an efficient approach to perform *coordinated* vertical scaling of resources in complex stochastic workflows, aimed at satisfying a SLO by improving the workflow *e2e response time distribution*. Specifically, workflows compose IaaS services running on dedicated clusters whose size must be determined in advance, reserving and statically assigning resources to services before execution. Services are composed through sequence, fork-join, and choice-merge patterns [30], and have generally distributed (GEN) response times possibly with bounded supports, facilitating representation of real-time constraints and fitting of analytical distributions from observed data. Each service is characterized by a job size [5], representing its makespan (i.e. its expected response time) with a given amount of assigned resources; we assume the makespan to be inversely proportional to the amount of assigned resources [15,21,31]. The defined heuristics uses a structured workflow model [10] to perform a top-down visit of the hierarchy of services, assigning resources so as to minimize the makespan of the workflow e2e response time and to satisfy the agreed SLO. To this end, the heuristics exploits an efficient compositional analysis method [11] to derive the response time distribution of each sub-workflow and to compute its makespan. Feasibility and effectiveness of the approach are assessed on a non-trivial synthetic workflow stressing computational complexity. Results show that the heuristics is effective at improving the e2e response time distribution of the entire workflow, and very efficient, enabling its application at runtime in reaction to QoS changes.

In the framework of [9], our approach is defined by the following attributes: the *goal of resource adaptation* is to ensure that workflow execution fulfils non-functional requirements specified by percentiles of quality attributes, i.e., that the makespan of the workflow response time satisfies the agreed SLO; the *stage of system lifetime* at which resource adaptation is performed is the runtime stage with proactive mode (i.e., anticipating resource adaptation), though the approach can be applied also in reactive mode (i.e., after changes in quality attributes) as shown by the experimental results; the *composition level* at which resource adaptation is performed involves both services (i.e., abstract composition made of tasks orchestrated by some composition logic) and workflow (i.e., concrete composition where tasks of an abstract composition are mapped to implementations); the *scope of resource adaptation*, in terms of *number of systems* and *granularity* of adaptation, considers a single system and a single request; *adaptation actions* mainly consist of service tuning operations changing behavior of concrete services (e.g., reducing the makespan by increasing the amount of resources), though adaptation is performed in a coordinated manner; and, resource adaptation is performed by a *single authority*.

The rest of the paper is organized as follows. Section 2 recalls the hierarchical formalism for workflow modeling and the compositional method for evaluation of the workflow e2e response time distribution. Section 3 illustrates the proposed resource assignment method. Section 4 presents the experimental results achieved on a complex workflow. Finally, Sect. 5 draws conclusions and outlines possible extensions and improvements of the proposed approach.

2   Background: Workflow Modeling and Evaluation

We model workflows as recursive compositions of blocks specified by Stochastic Time Petri Nets (STPNs) [38]. Each STPN block has a single starting place, which receives a token when workflow execution starts, and a single final place, which receives a token when workflow execution eventually ends with probability 1 (w.p.1). As shown in Fig. 1, blocks model sequential, balanced split/join, and choice/merge workflow patterns [30,40], with the following EBNF syntax:

$$\text{BLOCK} := \text{ACT} \mid \text{SEQ}\{\text{BLOCK}_1, \ldots, \text{BLOCK}_n\} \tag{1}$$
$$\mid \text{AND}\{\text{BLOCK}_1, \ldots, \text{BLOCK}_n\} \mid \text{XOR}\{\text{BLOCK}_1, \ldots, \text{BLOCK}_n, p_1, \ldots, p_n\}$$

where ACT is an elementary activity with non-Markovian distribution possibly with bounded support (e.g., block A in Fig. 1b), $\text{SEQ}\{\text{BLOCK}_1, \ldots, \text{BLOCK}_n\}$ models n sequential blocks $\text{BLOCK}_1, \ldots, \text{BLOCK}_n$ (e.g., block S1 in Fig. 1b), $\text{AND}\{\text{BLOCK}_1, \ldots, \text{BLOCK}_n\}$ models n concurrent blocks $\text{BLOCK}_1, \ldots, \text{BLOCK}_n$ (e.g., block A1 in Fig. 1b), and $\text{XOR}\{\text{BLOCK}_1, \ldots, \text{BLOCK}_n, p_1, \ldots, p_n\}$ models n alternative blocks $\text{BLOCK}_1, \ldots, \text{BLOCK}_n$ with probability $p_1, \ldots, p_n$, respectively (e.g., block X1 in Fig. 1b). Note that associating activity durations with non-Markovian distributions possibly with bounded supports facilitates fitting of analytical distributions from data collected from real web applications and enables representation of firm constraints on execution times of activities.

This workflow model can be represented as a *structure tree* [10] $S = \langle N, n_0 \rangle$, where N is the set of nodes (i.e., blocks) and $n_0 \in N$ is the root node (i.e., the entire workflow). In turn, each node $n_i \in N$ is a tuple $\langle C_i, \text{type}_i \rangle$, where C_i is the set of the children nodes of n_i and $\text{type}_i \in \{\text{ACT}, \text{SEQ}, \text{AND}, \text{XOR}\}$ is the type of the block modeled by node n_i, e.g., in Fig. 1a, node A1 models an AND block composing nodes I and J.

For complex workflows made of several concurrent activities with duration characterized by GEN Cumulative Distribution Functions (CDFs) possibly with bounded supports, the e2e response time distribution cannot be evaluated by

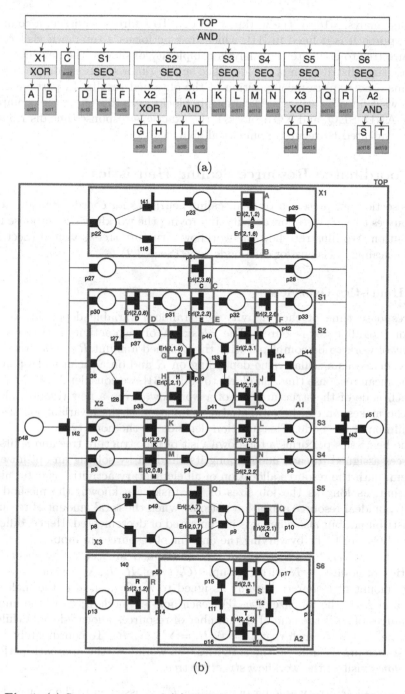

(a)

(b)

Fig. 1. (a) Structure tree and (b) STPN modeling a synthetic workflow.

transient analysis [18] of the workflow STPN. To address the issue, a compositional approach is defined in [11], which first performs a top-down visit of the structure tree to estimate the analysis complexity of blocks, then evaluates the response time distribution of the identified sub-workflows in isolation, and finally performs a bottom-up recomposition of the obtained results. In particular, for workflows defined by well-nested composite blocks as in this paper (i.e., composition of AND, SEQ, and XOR blocks), the exact e2e response time distribution can be evaluated by recursive numerical analysis [11].

3   Coordinated Resource Scaling Heuristics

In this section, we present a vertical scaling heuristics for coordinated allocation of resources to each workflow activity, improving the workflow e2e response time distribution. We illustrate how the workflow structure tree is visited (Sect. 3.1) and the scaling decisions for each node type (Sect. 3.2).

3.1   Heuristics Overview

The response time of each activity is a random variable depending on the amount R of allocated resources. The *job size* X of each activity is the invariant amount of work to be completed with the assigned amount R of resources, and it is evaluated as a scalar value depending on R and on a mean makespan T_R, i.e., the mean response time. Similarly to [15,21,31], we consider $X := R T_R \forall R$. For each node of the structure tree of a workflow, given a new resource allocation, the makespan can be estimated as a function of the invariant job sizes of the children of the node. This function depends on the node type.

The heuristics performs a top-down visit of the structure tree and splits the resources assigned to each node among its children by solving an optimization problem, until the resource allocation of all elementary activities is determined. Note that, as long as the job sizes of the tasks are known, the method can identify an ideal resource allocation not only when the total amount of resources is redistributed, but also when it is incremented or decremented. Hence, different scalar SLOs can be fit by varying the amount of resources in input.

Let $S = \langle N, n_0 \rangle$ be a structure tree as defined in Sect. 2. We extend the definition of node $n_i \in N$ with the tuple $\langle C_i, \text{type}_i, R_i, T_i, X_i \rangle$, where $R_i \in \mathbb{R}_{>0}$ is the amount of resources initially assigned to n_i, $T_i \in \mathbb{R}_{\geq 0}$ is the makespan of n_i, and X_i is the job size of n_i. For each non-leaf node $n_i \in N$, the number of resources of n_i is the sum of the number of resources allocated to its children nodes, i.e., $\forall n_i \in N$ such that $C_i \neq \emptyset$, $R_i = \sum_{n_j \in C_i} R_j$. To coordinately adapt the resource provisioning of the activities of a workflow, the approach performs a *top-down* visit of the workflow structure tree.

- Initially, an arbitrary amount of resources R_0^{in} is assigned to the root node n_0.
- For each non-leaf node n_k, the amount of input resources R_k^{in} is split by assigning an amount $R_j^\star$ to each child node n_j, i.e., $\sum_{n_j \in C_k} R_j^\star = R_k^{\text{in}}$; the assignment depends on the node type.

By induction, the sum of the amounts of resources allocated to the leaf nodes is equal to the amount of resources of the root node, i.e., $\sum_{n_k|C_k=\emptyset} R_k^\star = R_0^{in}$.

3.2 Scaling Decisions

We characterize the different resource scaling decisions based on the node types.

Sequential Activities. Let n_k be an activity with $type_k = SEQ$ and children $C_k = \{i, j\}$, and let R_k^{in} be the amount of resources to be split. The makespan of node n_k can be obtained as

$$T_k = T_i + T_j = \frac{X_i}{R_i} + \frac{X_j}{R_k^{in} - R_i} \tag{2}$$

which has a minimum when the following resources are allocated to node n_i:

$$R_i^\star = \frac{\sqrt{X_i}}{\sqrt{X_i} + \sqrt{X_j}} R_k^{in} . \tag{3}$$

The result is obtained by imposing $\frac{dT_k}{dR_i} = 0$, and it can be extended by induction to the ordered sequence $SEQ(n_1, \ldots, n_J)$ of $J > 2$ activities:

$$R_i^\star = \frac{\sqrt{X_i}}{\sum_{j=1}^{J} \sqrt{X_j}} R_k^{in} \quad \forall i \in \{1, \ldots, J\}. \tag{4}$$

Each node allocation is thus obtained considering the allocation evaluated for previous nodes, that is removed from the input resources R_k^{in}. Note that, since we reserve resources before service execution, the optimum does not exploit resources used by activities that have already been executed.

Concurrent Activities. Let n_k be an activity with $type_k = AND$, children $C_k = \{i, j\}$, and amount R_k^{in} of resources to be split. In this case, the makespan (mean response time) cannot be defined as an analytical function of R_i, precluding to evaluate the minimum by exploiting the Fermat theorem. Hence, we provide a heuristics evaluation of $R_i^\star$ depending on a parameter $\alpha \in \mathbb{R}^+$ modulating the weight of the job size in determining the solution. In particular, $R_i^\star$ is evaluated by imposing equality between the ratio of response times of node n_k children, and the $(\alpha + 1)$-th power of the resources provisioned to the children:

$$\frac{X_i}{R_i^{\alpha+1}} = \frac{X_j}{(R_k^{in} - R_i)^{\alpha+1}} \tag{5}$$

This leads to the allocation:

$$R_i^\star = \frac{\sqrt[\alpha+1]{X_i}}{\sqrt[\alpha+1]{X_i} + \sqrt[\alpha+1]{X_j}} R_k^{in} \tag{6}$$

The solution is extended to J activities by induction:

$$R_i^\star = \frac{\sqrt[\alpha+1]{X_i}}{\sum_{j=1}^{J} \sqrt[\alpha+1]{X_j}} R_k^{in} \quad \forall i \in \{1, \ldots, J\} \tag{7}$$

Note that, when $\alpha = 0$, the heuristics determines $R_i^\star$ by imposing equality between the response times of the considered node children.

Alternative Activities. Let n_k be an activity with $type_k = \text{XOR}$, children $C_k = \{i, j\}$ having probabilities p_i and $p_j = 1 - p_i$ to occur, and R_k^{in} the amount of resources to be split. The makespan of n_k is

$$T_k = p_i T_i + p_j T_j = p_i \frac{X_i}{R_i} + p_j \frac{X_j}{R_k^{\text{in}} - R_i} \tag{8}$$

which has the minimum

$$T_k^\star = \frac{\left(\sqrt{p_i X_i} + \sqrt{p_j X_j}\right)^2}{R_k^{\text{in}}} \tag{9}$$

when the following resources are allocated to node i:

$$R_i^\star = \frac{\sqrt{p_i X_i}}{\sqrt{p_i X_i} + \sqrt{p_j X_j}} R_k^{\text{in}} \tag{10}$$

which is in turn obtained by exploiting the Fermat theorem. As the solution shows, the optimal allocation of an XOR node is a generalization of the optimal allocation for a SEQ node. Hence, the extension to the case with more than 2 activities can be derived from the solution obtained for the SEQ node. Also note that the available resources R_k^{in} are split among the activities i and j: this assumption is useful to model workflows in microservice and service-oriented applications, where each service has a reserved amount of resources. In contrast, for microservices deployed with FaaS cloud solutions, resources are allocated on-demand for each service execution: in this case, cloud costs are accrued only for the resources of the selected service; the expected cost is in this case $p_i R_i + p_j R_j$ instead of $R_i + R_j$, resulting in a different optimal allocation.

Once resources are assigned to each simple node, the response time CDF of each activity can be determined by leveraging linearity of the job size invariance, e.g., if resources assigned to an activity with Erlang response time distribution are doubled, then the rate of the Erlang distribution doubles.

4   Experimental Evaluation

In this section, we assess feasibility and effectiveness of the proposed heuristics. First, we show how different values of parameter α produce different resource allocations, with consequent different improvement of the workflow e2e response time CDF (Sect. 4.1). The experiment is performed for two different initial resource allocations. Then, we test the ability of the heuristics to meet an agreed SLO while minimizing the amount of allocated resources (Sect. 4.2).

Both experiments are performed on the synthetic workflow of Fig. 1, which consists of 20 elementary blocks combined through well-nested patterns, yielding a model with up to 10 concurrent activities. For each elementary activity,

Table 1. Resources allocated to the activities of the workflow of Fig. 1, before (column r_{start}) and after heuristics execution with different values of α (columns r_α with $\alpha \in \{0, \frac{1}{4}, \frac{1}{2}, 1, 2, 4\}$): (a) balanced and (b) unbalanced initial allocation.

ACT	r_{start}	r_0	$r_{1/4}$	$r_{1/2}$	r_1	r_2	r_4
A	1.00	0.52	0.64	0.73	0.85	0.97	1.07
B	1.00	0.55	0.68	0.78	0.90	1.03	1.14
C	1.00	0.25	0.41	0.57	0.84	1.23	1.65
D	1.00	2.85	2.51	2.29	2.01	1.75	1.54
E	1.00	1.49	1.31	1.19	1.05	0.91	0.81
F	1.00	1.37	1.20	1.01	0.97	0.84	0.74
G	1.00	0.72	0.71	0.70	0.68	0.65	0.62
H	1.00	0.84	0.83	0.82	0.79	0.75	0.72
I	1.00	0.63	0.66	0.68	0.69	0.69	0.68
J	1.00	1.03	0.98	0.94	0.88	0.81	0.75
K	1.00	0.54	0.67	0.77	0.89	1.03	1.15
L	1.00	0.50	0.61	0.70	0.82	0.95	1.05
M	1.00	1.61	1.67	1.69	1.69	1.68	1.65
N	1.00	0.97	1.00	1.02	1.02	1.01	0.99
O	1.00	0.48	0.47	0.46	0.44	0.42	0.40
P	1.00	1.70	1.66	1.63	1.56	1.48	1.40
Q	1.00	1.19	1.16	1.14	1.09	1.04	0.98
R	1.00	1.49	1.52	1.52	1.52	1.49	1.45
S	1.00	0.71	0.71	0.70	0.68	0.66	0.63
T	1.00	0.53	0.56	0.57	0.59	0.59	0.59

(a) Balanced initial res. allocation.

ACT	r_{start}	r_0	$r_{1/4}$	$r_{1/2}$	r_1	r_2	r_4
A	0.375	0.21	0.32	0.43	0.60	0.80	0.99
B	0.375	0.22	0.34	0.45	0.63	0.85	1.05
C	0.375	0.01	0.21	0.33	0.59	1.01	1.52
D	0.375	1.13	1.27	1.34	1.41	1.44	1.42
E	0.375	0.59	0.66	0.70	0.74	0.75	0.74
F	0.375	0.54	0.61	0.65	0.68	0.69	0.68
G	3.75	2.49	2.22	2.02	1.76	1.48	1.26
H	0.375	0.92	0.82	0.75	0.65	0.55	0.46
I	3.75	3.14	2.63	2.28	1.83	1.40	1.08
J	0.375	0.51	0.62	0.68	0.74	0.77	0.75
K	0.375	0.21	0.34	0.45	0.63	0.85	1.05
L	0.375	0.20	0.31	0.41	0.58	0.78	0.97
M	0.375	0.64	0.84	0.99	1.19	1.38	1.51
N	0.375	0.39	0.51	0.60	0.72	0.83	0.91
O	3.75	2.05	1.93	1.83	1.67	1.47	1.29
P	0.375	2.29	2.16	2.04	1.86	1.64	1.44
Q	0.375	1.01	0.95	0.90	0.82	0.72	0.63
R	0.375	1.46	1.44	1.40	1.32	1.20	1.09
S	3.75	2.68	2.51	2.33	2.04	1.67	1.34
T	0.375	0.20	0.31	0.41	0.55	0.70	0.80

(b) Unbalanced initial res. allocation.

we assume that the response time CDF achieved with a given resource assignment is known, either from measurements of previous implementations or by contract. To easily manage the linear relation between response time and allocated resources assumed by the performance model of Sect. 3.1, and to facilitate the interpretation of the experimental results, without loss of generality, we consider Erlang CDFs for the response times of elementary activities. In particular, we consider Erlang CDFs with 5-phases and rates randomly selected in $[0, 5]$, so guaranteeing variability in expected response times of activities. We remark that any numerical CDF could be considered as well, or any analytical CDF in the class of expolynomial functions (also termed exponomials [36]) supported by the compositional analysis technique of [11] exploited by the proposed heuristics.

All the experiments reported in this section have been performed on a MacBook Pro 2021 equipped with an Apple M1 Pro octa-core processor with a rate clock up to 3.2 GHz and 16 GB RAM. Experiments have been performed using an extension of the Eulero Java library [12], currently under development.

4.1 Heuristics Sensitivity to Parameter α

We consider two arbitrary different initial allocations of resources to the activities of the workflow of Fig. 1. In the first (column r_{start} of Table 1a), we consider a

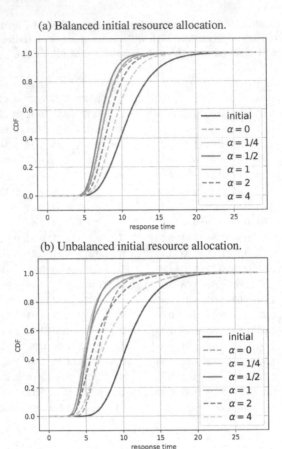

Fig. 2. CDF of the e2e response time of the workflow of Fig. 1, obtained through the execution of the proposed heuristics with different values of parameter α, by assuming: (a) the balanced initial resource allocation shown in Table 1a, and (b) the unbalanced initial resource allocation shown in Table 1b.

balanced allocation where each activity is assigned 1 resource, for a total amount of 20 resources (*balanced initial resource allocation*). In the second (shown in column r_{start} of Table 1b), we consider an unbalanced allocation, where activities G, I, O, and S have 10 times the resources of the other activities (*unbalanced initial resource allocation*). To make results comparable, we maintain a total of 20 resources, thus allocating 15 resources to G, I, O, and S (i.e., 3.75 resources each), and 5 resources to the remaining activities (i.e., 0.375 each). Tables 1a and 1b also show the resource allocation computed by the proposed heuristics.

We evaluate how the resource allocation provided by the heuristics improves the workflow e2e response time CDF for different values of $\alpha \in \{0, \frac{1}{4}, \frac{1}{2}, 1, 2, 4\}$. Results reported in Figs. 2a and 2b show an improvement for any considered value of α and for any of the two initial allocations of resources, with better results achieved for $\alpha \in [0, 1]$ and nearly the best result obtained for $\alpha = \frac{1}{4}$. This result suggest that, being $\alpha = \frac{1}{n}$, the e2e response time CDF improves as n increases, up to a certain value beyond which the CDF gets worse.

If the initial resource allocation is balanced, different values of $\alpha \in [0, 1]$ produce nearly comparable e2e response time CDFs. This result suggests that, in this case, balancing the makespan of the children of AND nodes determines a good resource allocation, significantly improving the workflow e2e response time CDF. Conversely, if the initial resource allocation is unbalanced, better results are obtained for values of $\alpha \in (0, 1]$, with a worse result obtained for $\alpha = 0$ with respect to the case of balanced initial allocation of resources.

It is worth noting that, for both variants of the experiment, the proposed heuristics runs in 0.2 s on average, proving to be efficient even for complex workflows, which is an essential requirement for modern microservice architectures where hundreds of services are orchestrated as workflows of activities.

4.2 Heuristics Ability to Achieve SLO Guarantees

For the workflow of Fig. 1 with the initial balanced resource allocation of Table 1a, we consider an SLO expressed in terms of e2e response time CDF. In particular, the SLO is obtained as the e2e response time CDF of a randomly generated well-nested workflow, having expected response time equal to $T_{\mathrm{SLO}} = 6.17$ s. As shown in Fig. 3, with this allocation of resources, the workflow e2e response time CDF (violet curve) does not meet the SLO (light blue curve). Rebalancing the existing resources by applying the proposed heuristics with $\alpha = \frac{1}{4}$ (which is the value yielding the best results in Sect. 4.1) produces an improvement of the e2e response time CDF, which however still violates the required SLO. Therefore, an additional amount of resources is needed not to violate the SLO.

To determine the new amount of resources, we consider two different strategies whose results are reported in Fig. 3. We perform an experiment (*reallocate resources first*) where we first compute a new resource allocation through our heuristics with $\alpha = \frac{1}{4}$ (green curve in Fig. 3a), by assuming that the total amount of allocated resources does not change, i.e., $R_0 = 20$. Then, we evaluate the additional amount of resources that is needed to meet the specified SLO, by exploiting the assumption of invariance of the job size of a workflow (as discussed in Sect. 3.1). In fact, by knowing the expected response time T_{SLO} of the SLO, the initial amount of allocated resources $R_0 = 20$, and the expected response time T_0 obtained after resource allocation (i.e., considering the resource allocation of column $r_{1/4}$ of Table 1a as initial resource allocation), the amount of resources required not to violate the SLO can be computed as $R^\star = \frac{R_0 T_0}{T_{\mathrm{SLO}}}$. In particular, the additional amount of resources turns out to be equal to 4.06. Finally, the amount $R^\star$ is allocated to the activities using the heuristics, and a new e2e response time CDF is computed (fuchsia curve in Fig. 3a).

Then, we perform a variant of the experiment (*add resources first*), where we directly evaluate the additional amount of resources needed to meet the specified SLO as $R^{\star\prime} = \frac{R_0 T_0'}{T_{\mathrm{SLO}}}$, where T_0' is the workflow expected response time obtained

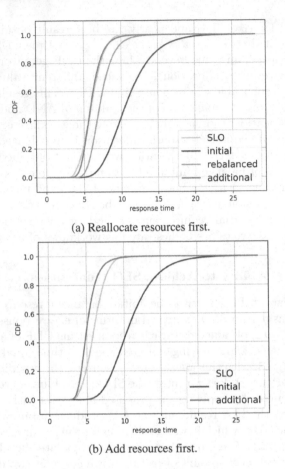

(a) Reallocate resources first.

(b) Add resources first.

Fig. 3. CDF of the e2e response time of the workflow of Fig. 1, obtained through two strategies: (a) reallocating resourced through the heuristics, determining the amount of resourced needed to satisfy the SLO, and allocating resources again through the heuristics; and, (b) determining the amount of resourced needed to satisfy the SLO and allocating resources through the heuristics.

by considering the initial resource allocation of column r_{start} of Table 1a. In particular, the additional amount of resources turns out to be equal to 16.15. The allocation of the increased amount of resources through our heuristics yields a new e2e response time CDF (fuchsia curve in Fig. 3b).

Figure 3 shows that the e2e response time CDF provided by the reallocate-resources-first strategy is stochastically larger than the one provided by the add-resources-first strategy, and is characterized by a larger expected response time. However, the add-resources-first strategy allocates a larger number of resources, which actually turn out to be over-provisioned, given that the obtained e2e response time CDF is stochastically lower than the SLO. Moreover, note that the time to calculate the new resource provisioning is 0.89 s for the reallocate-resources-first strategy and 0.51 s for the add-resources-first strategy, meaning that there is not a significant loss in performance when the heuristics is executed twice. Therefore, the reallocate-resources-first strategy is preferable.

5   Conclusions

We have presented a heuristics to perform coordinated scaling of resources in cloud computing workflows, with the aim of improving the e2e response time CDF. The heuristic is developed around the concept of job size of an activity, which is assumed to be invariant with respect to the amount of resources provisioned to the activity, and it is guided by the mean makespan indicators of sub-workflows. The method has been successfully tested on a complex workflow with a high degree of concurrency, and applied to the problem of identifying additional resources needed to guarantee a given SLO, proving to be not only effective at improving the workflow e2e response time CDF but also efficient. We are also planning to compare the approach with some state-of-the art method.

Though scaling actions considered in this paper are vertical, the heuristics could be easily extended to perform horizontal scaling actions. In fact, it is sufficient to intend the involved resources as discrete, i.e., as containers or VMs with fixed capacities. In this case, the approach should be adapted so as to round up or down the identified amounts of resources to be allocated. Moreover, the proposed heuristics can be extended to manage workflow blocks that break the structure of well-formed nesting of activities, requiring to compute a makespan indicator and a (sub-optimal) resource assignment for such blocks. The heuristics could also be extended to efficiently derive the value of $\alpha \in [0,1]$ that minimizes the makespan indicator of each block. Finally, the heuristics could also be improved by considering different performance models, so as to ensure the applicability of the method to contexts in which linearity between response time and amount of allocated resources may not be sufficient to properly characterize the behaviour of the system, e.g., due to the presence of not negligible VM start up times.

Acknowledgments. This work was partially supported by the European Union under the Italian National Recovery and Resilience Plan (NRRP) of NextGenerationEU, partnership on "Telecommunications of the Future" (PE00000001 - program "RESTART").

References

1. Ali-Eldin, A., Tordsson, J., Elmroth, E.: An adaptive hybrid elasticity controller for cloud infrastructures. In: IEEE Network Operations and Management Symposium, pp. 204–212. IEEE (2012)
2. Alshuqayran, N., Ali, N., Evans, R.: A systematic mapping study in microservice architecture. In: IEEE International Conference on SO Computing and Applications, pp. 44–51. IEEE (2016)
3. Bauer, A., Herbst, N., Spinner, S., Ali-Eldin, A., Kounev, S.: Chameleon: a hybrid, proactive auto-scaling mechanism on a level-playing field. IEEE Trans. Parallel Distrib. Syst. **30**(4), 800–813 (2018)

4. Bauer, A., Lesch, V., Versluis, L., Ilyushkin, A., Herbst, N., Kounev, S.: Chamulteon: coordinated auto-scaling of micro-services. In: IEEE International Conference on Distributed Computing Systems, pp. 2015–2025. IEEE (2019)

5. Berg, B., Dorsman, J.L., Harchol-Balter, M.: Towards optimality in parallel scheduling. Proc. ACM Meas. Anal. Comput. Syst. **1**(2), 40:1–40:30 (2017)

6. Bi, J., Zhu, Z., Tian, R., Wang, Q.: Dynamic provisioning modeling for virtualized multi-tier applications in cloud data center. In: IEEE International Conference on Cloud Computing, pp. 370–377. IEEE (2010)

7. Buyya, R., Yeo, C.S., Venugopal, S., Broberg, J., Brandic, I.: Cloud computing and emerging IT platforms: vision, hype, and reality for delivering computing as the 5th utility. Futur. Gener. Comput. Syst. **25**(6), 599–616 (2009)

8. Canfora, G., Di Penta, M., Esposito, R., Villani, M.L.: QoS-aware replanning of composite web services. In: IEEE International Conference on Web Ser, pp. 121–129. IEEE (2005)

9. Cardellini, V., Casalicchio, E., Grassi, V., Iannucci, S., Presti, F.L., Mirandola, R.: Moses: a framework for GOS driven runtime adaptation of service-oriented systems. IEEE Trans. on Softw. Eng. **38**(5), 1138–1159 (2011)

10. Carnevali, L., Paolieri, M., Reali, R., Vicario, E.: Compositional safe approximation of response time distribution of complex workflows. In: Abate, A., Marin, A. (eds.) Quantitative Evaluation of Systems: 18th International Conference, QEST 2021, Paris, France, August 23–27, 2021, Proceedings, pp. 83–104. Springer, Cham (2021). https://doi.org/10.1007/978-3-030-85172-9_5

11. Carnevali, L., Paolieri, M., Reali, R., Vicario, E.: Compositional safe approximation of response time probability density function of complex workflows. ACM Trans. Model. Comput. Simul. (2023)

12. Carnevali, L., Reali, R., Vicario, E.: Eulero: a tool for quantitative modeling and evaluation of complex workflows. In: Ábrahám, E., Paolieri, M. (eds.) Quantitative Evaluation of Systems: 19th International Conference, QEST 2022, Warsaw, Poland, September 12–16, 2022, Proceedings, pp. 255–272. Springer, Cham (2022). https://doi.org/10.1007/978-3-031-16336-4_13

13. Chen, T., Bahsoon, R., Yao, X.: A survey and taxonomy of self-aware and self-adaptive cloud autoscaling systems. arXiv preprint arXiv:1609.03590 (2016)

14. Chieu, T.C., Mohindra, A., Karve, A.A., Segal, A.: Dynamic scaling of web applications in a virtualized cloud computing environment. In: IEEE International Conference on e-Business Engineering, pp. 281–286. IEEE (2009)

15. Farokhi, S., Lakew, E.B., Klein, C., Brandic, I., Elmroth, E.: Coordinating CPU and memory elasticity controllers to meet service response time constraints. In: International Conference on Cloud and Autonomic Computing, pp. 69–80. IEEE (2015)

16. Fox, A., et al.: Above the Clouds: A Berkeley View of Cloud Computing. Dept. Electrical Eng. and Comput. Sci., University of California, Berkeley, Rep. UCB/EECS **28**(13), 2009 (2009)

17. Gias, A.U., Casale, G., Woodside, M.: Atom: Model-driven autoscaling for microservices. In: International Conference on Distributed Computing System, pp. 1994–2004. IEEE (2019)

18. Horváth, A., Paolieri, M., Ridi, L., Vicario, E.: Transient analysis of non-Markovian models using stochastic state classes. Perf. Eval. **69**(7–8), 315–335 (2012)

19. Incerto, E., Tribastone, M., Trubiani, C.: Combined vertical and horizontal autoscaling through model predictive control. In: Aldinucci, M., Padovani, L., Torquati, M. (eds.) Euro-Par 2018: Parallel Processing: 24th International Conference on Parallel and Distributed Computing, Turin, Italy, August 27 - 31, 2018, Proceedings, pp. 147–159. Springer, Cham (2018). https://doi.org/10.1007/978-3-319-96983-1_11

20. Iqbal, W., Dailey, M.N., Carrera, D., Janecek, P.: Adaptive resource provisioning for read intensive multi-tier applications in the cloud. Futur. Gener. Comput. Syst. **27**(6), 871–879 (2011)

21. Lakew, E.B., Klein, C., Hernandez-Rodriguez, F., Elmroth, E.: Towards faster response time models for vertical elasticity. In: IEEE/ACM International Conference on Utility and Cloud Computing, pp. 560–565. IEEE (2014)

22. Liu, J., Zhang, S., Wang, Q., Wei, J.: Coordinating fast concurrency adapting with autoscaling for SLO oriented web applications. IEEE Trans. Parallel Distrib. Syst. **33**(12), 3349–3362 (2022)

23. Lynn, T., Rosati, P., Lejeune, A., Emeakaroha, V.: A preliminary review of enterprise serverless cloud computing (function-as-a-service) platforms. In: IEEE International Conference on Cloud Computing Technology and Science, pp. 162–169. IEEE (2017)

24. Nguyen, H., Shen, Z., Gu, X., Subbiah, S., Wilkes, J.: Agile: Elastic distributed resource scaling for infrastructure-as-a-service (2013)

25. Qiu, H., Banerjee, S.S., Jha, S., Kalbarczyk, Z.T., Iyer, R.K.: Firm: An intelligent fine-grained resource management framework for SLO-oriented microservices. In: USENIX Symposium on Operating Systems Design and Implementation (2020)

26. Qu, C., Calheiros, R.N., Buyya, R.: Auto-scaling web applications in clouds: a taxonomy and survey. ACM Comput. Surv. **51**(4), 1–33 (2018)

27. Rahman, J., Lama, P.: Predicting the end-to-end tail latency of containerized microservices in the cloud. In: International Conference on Cloud Engineering, pp. 200–210. IEEE (2019)

28. Rose, K., Eldridge, S., Chapin, L.: The internet of things: an overview. The internet society (ISOC) **80**, 1–50 (2015)

29. Rosenberg, F., Leitner, P., Michlmayr, A., Celikovic, P., Dustdar, S.: Towards composition as a service-a quality of service driven approach. In: IEEE International Confernce on Data Engineering, pp. 1733–1740. IEEE (2009)

30. Russell, N., Ter Hofstede, A.H., Van Der Aalst, W.M., Mulyar, N.: Workflow control-flow patterns: A revised view. BPM Center Report BPM-06-22, BPMcenter. org 2006 (2006)

31. Salah, K., Elbadawi, K., Boutaba, R.: An analytical model for estimating cloud resources of elastic services. J. of Network and Sys. Manag. **24**, 285–308 (2016)

32. Salehie, M., Tahvildari, L.: Self-adaptive software: landscape and research challenges. ACM Trans. Auton. Adapt. Syst. **4**(2), 1–42 (2009)

33. Shahrad, M., Balkind, J., Wentzlaff, D.: Architectural implications of function-as-a-service computing. In: IEEE/ACM International Symposium on microarchitecture, pp. 1063–1075 (2019)

34. Shen, Z., Subbiah, S., Gu, X., Wilkes, J.: Cloudscale: elastic resource scaling for multi-tenant cloud systems. In: ACM Symposium on Cloud Computing, pp. 1–14 (2011)

35. Stieß, S., Becker, S., Ege, F., Höppner, S., Tichy, M.: Coordination and explanation of reconfigurations in self-adaptive high-performance systems. In: International Conference on Model Driven Engineering Languages and Systems: Companion Proc, pp. 486–490 (2022)

36. Trivedi, K.S., Sahner, R.: Sharpe at the age of twenty two. ACM SIGMETRICS Performance Eval. Rev. **36**(4), 52–57 (2009)
37. Urgaonkar, B., Shenoy, P., Chandra, A., Goyal, P., Wood, T.: Agile dynamic provisioning of multi-tier internet applications. ACM Trans. Auton. Adapt. Syst. (TAAS) **3**(1), 1–39 (2008)
38. Vicario, E., Sassoli, L., Carnevali, L.: Using stochastic state classes in quantitative evaluation of dense-time reactive systems. IEEE Trans. on Soft. Eng. **35**(5), 703–719 (2009)
39. Yazdanov, L., Fetzer, C.: Vertical scaling for prioritized VMs provisioning. In: International Conference on Cloud and Green Computing, pp. 118–125. IEEE (2012)
40. Zheng, Z., Trivedi, K.S., Qiu, K., Xia, R.: Semi-Markov models of composite web services for their performance, reliability and bottlenecks. IEEE Trans. Serv. Comput. **10**(3), 448–460 (2015)

A State-Size Inclusive Approach to Optimizing Stream Processing Applications

Paul Omoregbee$^{(\boxtimes)}$ ⓘ, Matthew Forshaw ⓘ, and Nigel Thomas

School of Computing, Newcastle University, Newcastle upon Tyne NE17RU, UK
{p.o.omoregbee2,matthew.forshaw,nigel.thomas}@newcastle.ac.uk

Abstract. In stream processing applications, accurately measuring a system's processing capacity is critical for ensuring optimal performance and meeting Service Level Objectives (SLOs). Traditionally, operator throughput has been used as a proxy for the application's state size, but this approach can be misleading when dealing with window-based applications. In this paper, we explore the impact of window selectivity on the performance of streaming applications, demonstrating how a growing application state can artificially decrease the operators' throughput, resulting in false positives that could trigger premature scaling-down decisions. To address this problem, we conduct empirical evaluations to assess the relationship between operators' throughput and state size, showcasing the state size pattern typically does not correspond to the operator's processing rate in window-based applications. Our findings highlight the importance of considering the state size of the application in performance monitoring and decision-making, particularly in the context of window-based applications.

Keywords: Auto-scaling · State Size · Offered Load

1 Introduction

Dataflow execution in streaming processors involves the encapsulation of distributed operator logic that is centred on records, with the aim of describing complex data pipelines. In Flink-based data processing pipelines, a consistent application state is a critical element and persisted using a modular state backbone. When required, the system orchestrates failure recovery and reconfiguration (scale-out/in) without having a significant negative effect on execution or violating consistency [1]. Apache Flink is an open-source framework that offers stateful stream processing capabilities that are scalable, distributed and fault-tolerant.

Typically, distributed systems are built to give the user the impression of a single entity while concealing from them any issues linked to their distributed

M. Iacono et al. (Eds.): EPEW/ASMTA 2023, LNCS 14231, pp. 325–339, 2023.
https://doi.org/10.1007/978-3-031-43185-2_22

structure. When using a distributed computing system, we frequently need to make assumptions about the state of a pipeline that is being used in real-world scenarios at any given moment.

The technique of snap-shotting in streaming systems has been utilised for a considerable duration [24]. It has been rigorously tested in production at some of the largest stream processing deployments on thousands of nodes handling hundreds of gigabytes of state [2].

Despite potential partial failures or reconfigurations, such as upgrades to the application code or changes in execution parallelism, Flink's runtime makes sure that consistency is guaranteed for any managed state stated [9]. The size of an application's state is the amount of memory or storage space needed to store the application's current state. Data like user preferences, current settings, and any other details required for the application to operate properly might be included in this. During the design and development process, the size of the application state is frequently a crucial factor to consider because it can significantly affect the program's performance and usability.

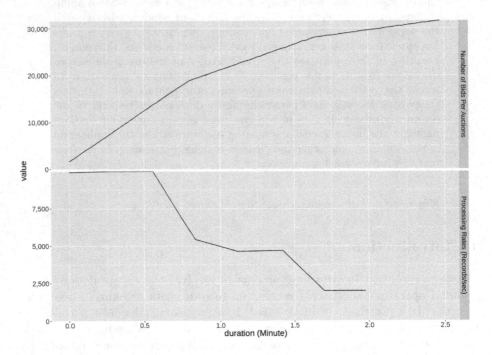

Fig. 1. State growth despite a declining operator throughput.

Offered load refers to the volume of data that the system needs to process at any given time. It is a measure of the amount of data that is being ingested or generated by streaming sources. The offered load is typically measured in terms of the number of data units (e.g. events, messages, records) that are being processed per second. The offered load can be affected by various factors, such as the number of users, backpressure, or the system resources available. In this paper, we consider the impact of state size on operator throughout. We demonstrate experimentally the impact on choice of window length and sliding period for winding operators, and their resulting impact on throughput. The objective is to evaluate the processing capacity of our streaming operators over bigger state sizes and to analyse the impact of window selectivity amongst various deployments. To achieve this, we adopted two workloads: Nexmark Query 3 and Query 5.

Figure 1 illustrates a non-alignment between the state size and the operator's throughput of a streaming application. Without sufficient knowledge of the relationship between these two metrics poses a potential challenge for auto-scalers. In a scenario where an operator, for example, has a window of 1 h, if the offered load (arrival rate) stops or reduces either because a sensor fails or due to a lack of customers, most auto-scalers will recommend downscaling, but the state size of that operator could still be maintained due to the window. This will mean scaling down too quickly while the state size is still large and creating additional time to recover. Our previous paper showed the impact of state size on auto-scalers' rescaling time [16].

In this paper, we make the following contributions: first, we demonstrate that offered load is an inadequate proxy for state size; second, we model the impact of growing state size on operator processing capacity. Finally, we show the importance of carefully selecting an appropriate window size and criteria to balance the selectivity and accuracy of the window operator, taking into consideration the performance requirements of the streaming application.

The remainder of this paper is organised as follows. Section 2 provides the background and related work. In Sect. 3, we provide details of the system design. Section 4 provides a summary of our experimental results and Sect. 5 provides concluding remarks and some future work.

2   Background and Related Work

A variety of strategies have been put forth to manage distributed stream processing systems' flexibility in virtualized environments. They differ in the type of data that is monitored, the quality-of-service objective that is addressed, the deployment environment (cluster, cloud, fog), and the optimisation technique that is used [4,17].

To store and process the application state during normal operation and quickly restore it after a failure, checkpointing is a crucial and practical crash-tolerant strategy. However, choosing the checkpoint frequency to reduce a suitable cost function would be the primary interest question for different research use cases. Knowing the conditions where checkpoints are useful and where they are not will help optimize the performance of a streaming application [6,24]. In each of our experiments, we chose a checkpointing interval that our system resource could handle. Our interest is in the amount of state produced and collecting the state size for each deployment through an API call.

Prior research has introduced automatic rescaling controllers that consider various metrics in arriving at a scaling decision [7, 22, 23], and to understand the performance and robustness of these systems [10, 11, 15]. This paper seeks to extend the DS2 project, which offers an autonomous scaling system that aims to balance resource over-provisioning and on-demand scaling [12]. DS2 measures each operator's true and observed processing ability. The number of records that may be processed by an operator instance in one unit of useful time is known as the true processing rate. This logically determines the operator's hardware capacity. The number of records a particular operator instance processes in a certain amount of time is known as the observed processing rate.

Most auto-scaling systems do not consider the impact of the application state and rely on offered load as a proxy for application state size, resulting in a untimely scale-down decision when the operator's throughput declines. This assumption can mislead the auto-scaler to make a wrong scaling decision.

Using machine learning approaches, a different class of solutions uses measured or profiled data to identify patterns [3, 5, 14, 19]. Typically, patterns are improved at runtime to increase precision. To make precise scaling decisions, these systems must first undergo extensive training, which could take a very long time.

2.1 Evaluate the Impact of System Resources on Operator Throughput

Most public and private cloud systems require users to specify resource needs for running their workloads efficiently. For example, users must select the type and quantity of virtual machines (VMs) they will rent on public cloud platforms; in a Kubernetes cluster, users must specify the number of pod replicas and the resource limits for individual pods; Google requires users to identify the number of containers they require as well as the resource limits for each [18]. These restrictions allow the cloud architecture to provide appropriate resource utilization estimates, which makes cloud computing possible.

However, because many end-user serving activities have daily or weekly load patterns, and traffic changes over longer time scales as a service gets popular, these estimates will become outdated as resource needs vary over time. Finally, as the underlying software or hardware stack gets updated, optimised, or added new capabilities, the resources required to handle a particular load change accordingly. When the requested resources are exceeded, the performance may suffer if the CPU is capped, or the process may be terminated because the memory is exhausted [18].

2.2 Evaluating the Impact of Window Selectivity Across Various Deployments

Window selectivity is an important concept in stream processing applications, where data is processed as it arrives in real time. Selectivity refers to the ability to filter and process only the relevant data within a specific time window. However, in practice, the length of the window and the sliding window period can vary widely between deployments, leading to variations in selectivity. These variations can impact the effectiveness and efficiency of stream processing applications [8]. Therefore, understanding the likely performance implications of the window size and slide period configuration choice is crucial to meet optimal business needs of streaming applications that rely on window-based processing.

3   System Design

Figure 2 shows our experimental pipeline setup and the relationship among the various unit of our setup. These units are self-contained, which makes it easy to scale the experiment. This setup architecture has four major areas: the data source (Nexmark Query 3 & Query 5), Streaming framework (Flink and DS2 scaling controller), Visualisation (we interrogate the offered load and state size and show their relationship and impact), and Datastore (this is stored in local file). Our experimental aim is to simulate a growing application state size of a streaming environment and measure the impact on the operator's observed processing rate. We also collect the system resource utilisation metrics to know the highest resource-consuming system users during the experiment period.

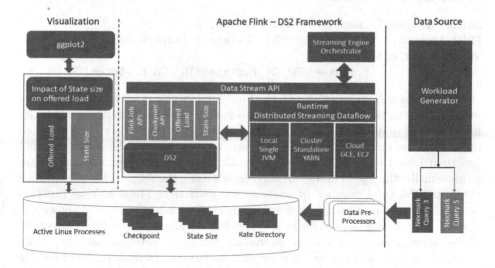

Fig. 2. Experimental System Architecture

Hardware and Software Configuration. This test was carried out using a single instance of Flink 1.4.1 running on Ubuntu 4.15.0-74-generic. Intel® Core^{TM} i5-8500 CPU @3.00GHz (six cores), 16GB of memory, and 1TB of storage make up the hardware setup.

The file system state backend was our choice (`FsStateBackend`). The different effects of alternate state backends, such as `MemoryStateBackend` or `RocksDB-StateBackend`, require further study.

3.1 Nexmark Workload

We adopt two Nexmark [21] queries for this experimental research, Query 3, and Query 5. We shall explain in more detail how each of these queries works.

3.2 Nexmark Query 3

Nexmark Query 3. is a complex data processing query that involves filtering, joining, and aggregation operations [21]. The data source generates a stream of auction and person events. The person event contains information about the person, such as their name, email, credit card, state, and city, while the auction event contains information about the auction, such as the auction id, the seller id, the initial bid, and the description. Other components include the Filter, Join, Aggregation, Key-value store and the Sink.

The architecture of Nexmark Query 3 is designed to process a large volume of events in real time. The choice of implementation and configuration of the system will depend on the specific requirements of the use case. In this experiment, our objective is to instrument the workload to generate a bigger state size that is required to test our hypothesis. The summary of the workload operation is summarized in the equation below:

```
SELECT Istream(P.name, P.city, P.state, A.id)
FROM Auction A [ROWS UNBOUNDED], Person P [ROWS UNBOUNDED]
WHERE A.seller=P.id AND
            (P.state='OR' OR P.state='ID' OR P.state='CA')}
```

In the case of Nexmark Query 3, the FlatMap function is used to apply a series of operations to the stream of auction and person events. The function starts by filtering the events using the `PersonFilter` and `AuctionFilter` classes, which only allow events that match certain criteria to pass through. The filtered events are then joined using the Join class, resulting in a stream of auction-person pairs. Finally, the joined events are aggregated using the Aggregator class, which computes the total number of auctions for each city. The FlatMap function is an important part of the Nexmark Query 3 architecture, as it allows the system to process a large volume of events in real-time and produce the results of the aggregation operation. To enable my workload to retain more state, we disabled the default operation of removing the `person.id` from the `auctionMap` collection. This measure prevents the workload from searching and removing the `person.id` from the map in the instance where there is a match between a person and the auction state. This action has the potential to decrease the number of elements contained within the map, which would not align with our objective of accumulating state.

3.3 Nexmark Query 5

In Nexmark Query 5, the state size management of a windowing operation is a crucial factor that can impact the query's performance. It allows the query to process events in a sliding time window, which can be used to aggregate events over a certain period [21]. The size of the window determines the amount of time for which events are aggregated and stored in state. When a sliding window is used, the state size can increase over time as new events are added to the window and older events are discarded [20]. State size management involves controlling the size of the state by managing the window size, the number of windows, and the number of events stored in each window. The objective is to keep the state size within acceptable limits while ensuring that the query maintains an optimal latency and throughput. The selection of a parameterization method for a sliding window has an impact on the state size.

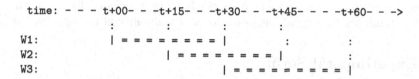

Fig. 3. Three sliding windows, with overlapping elements across windows.

A sliding window can overlap, as shown in Fig. 3. For instance, each window might record data for 60 s, but a new window might start every 30 s. The period refers to how frequently sliding windows open. As a result, our example's window and period would both be 60 s long. In our experiment, we update the following Query 5 parameters to generate more state.

```
.timeWindow(Time.minutes(60),Time.minutes(1))
```

The aforementioned parameter pertains to the configuration of the time window for a stream processing application in Apache Flink. Time.minutes(60) sets the sliding window size to 60 min. Data in the stream will be grouped and processed every 1 min, across a past horizon of 60 min. Time.minutes(1) defines the sliding window period. Each processing step advances the window by 1 min. Thus, each processing step creates a new window that spans data from the previous minute to the present minute and discards the prior window. Other common windowing approaches are tumbling or fixed windows, as shown in Fig. 4 below, where the size of the window is fixed, and events are processed in batches. Tumbling windows have the advantage of being simple to implement and easy to manage, but they can result in high latency and may not be suitable for processing real-time events.

Increasing the size of the sliding window period can result in a larger state size and a higher memory footprint, even when the window period is not completed, due to the increased amount of data that needs to be stored in memory as intermediate results [13]. This trade-off must be carefully considered when configuring the sliding window period to ensure that the application can handle the increased state size while still delivering the desired performance and scalability. State size management is a crucial aspect of designing a scalable and

```
time: - - - t+00- - -t+15- - -t+30- - -t+45- - - -t+60- - ->
               :           :           :           :
    W1:        | = = = = = = = = |           :           :
    W3:                    | = = = = = = = = = = |
    W2:                                            | = =
```

Fig. 4. Tumbling windows, with non-overlapping elements across windows

performance streaming application, and it is important to choose the appropriate windowing operation and state size management strategy based on the requirements of the application [1]. Additionally, larger windows may result in longer processing times, as more data needs to be processed for each window, which can impact the overall performance and scalability of the application [13].

4   Experimental Setup

We ran three experiments using two different workloads. The first and second sets of experiments were run using Nexmark Query 5, and the third experiment with Nexmark Query 3.

4.1   Experiment One (State Size Impact on Capacity Using NexMark Query 5)

Modelling impact of offered load on state size. The aim of this study is to demonstrate the impact of a growing state size on the processing capabilities of streaming operators. We keep a uniform arrival rate of 20,000 records/seconds with a transient window width size (20, 40, 60, 80, 100, 120, 140, 160, 180, 200, 400 and 600 min). The study involved conducting 12 deployments, each with a consistent checkpoint interval of 10 min. The total deployment circles were carried out during a single experimental period. Upon completion of each experiment, the observed and true processing rates are collected for each sliding window configuration as well as the state size metrics for each deployment.

4.2   Experiment Two (State Size Impact on Capacity Using NexMark Query 3)

The preceding experiment were conducted utilising Nexmark Query 5, which serves as an exemplar of sliding window and combiner. Consequently, we proceed to evaluate the state size impact on operator capacity using Nexmark Query 3, which serves as an exemplar of incremental join and filter. Query 3 comprises five operators. (Auction, Person, Filter, Incremental Join and Sink). The value of each operator's parallelism has been set to one since we are running on a single node cluster installation.

The workload comprises two source operators, namely the Auction operator and the Person operator. The Auction source has a data processing rate of

1,000,000 records per second, whereas the Persons source has a rate of 500,000 records per second. The arrival rate specified is deemed adequate for producing the necessary state size as per the requirements of this experiment. The duration of deployment is 15 min, while the frequency of checkpoint intervals is 2 min. During the experimental deployment, the system underwent seven checkpoints and state metrics were collected for each of them.

4.3   Experiment Three (Relationship Between Parameters of Streaming Window and State Accumulation)

This experiment involves a thorough examination of the constituent data elements that constitute the Query 5 workload state, with the aim of enhancing our understanding of the internal workings of the application state. The aim of this experiment is to assess and contrast the effects of diverse window size configurations in a stream processing system. A transient offered load is constructed, consisting of three distinct tasks. Each task exhibits an arrival rate that initially increases and subsequently decreases, following a predetermined sequence (100,000, 200,000 and 50,000). Each task has a duration of one minute.

The available window sliding periods are 5, 10, and 20 s. A sliding window size of 30 s is upheld for all deployments, while the overall duration of the experiment is estimated to be around 3 min. Following this, the auctions, bids, and date time values are collected, and subsequently, the quantity of bids produced in each window configuration is assessed. The absence of back pressure in this experiment allows for a focused examination of the effects of windowing in a narrow window size, contingent upon the chosen window size configuration.

5   Results

After concluding 12 executions of the first experiment measuring the effect of state size on capacity using NexMark Query 5, the synchronised results depicted in Fig. 5 illustrate the processing capability of the true and observed processing rates in Query 5. As indicated previously, we maintained a constant arrival rate of 20,000 records per second with a window width of 20, 40, 60, 80, 100, 120, 140, 160, 180, 200, 400, and 600 min. During the initial hour, the true processing capacity of the operator surpasses the observed processing rate to a significant extent. The reason for this is that the true processing rate reflects the maximum possible capacity of the operator, which is typically greater at the outset of the experimental deployment when system resources are less utilised. A consistent decrease in the processing rate is observed with an increase in state size caused by the increase in window size as demonstrated by the data spanning from 1:30 min to 2 h.

The observed processing rate and state size growth are measured through the utilisation of Query 3 and Query 5 workloads. The findings depicted in Fig. 6 and Fig. 7 demonstrate the relationship between the processing rate and the growth of state size. A decrease in the processing capacity of operators is noted with

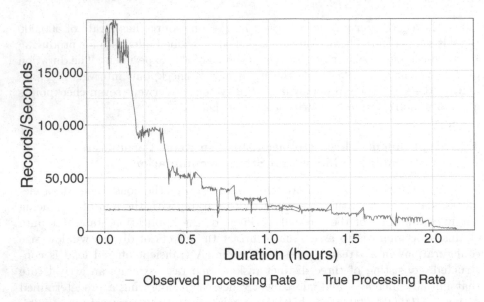

Fig. 5. Evolution of true and observed processing rates for Nexmark Query 5.

the increase in state size over time. The findings of this study provide evidence in favor of our hypothesis that a higher state could impact the throughput of streaming operators.

Building upon the findings of the second experiment, a comprehensive analysis is conducted on the individual data components that comprise the Query 5

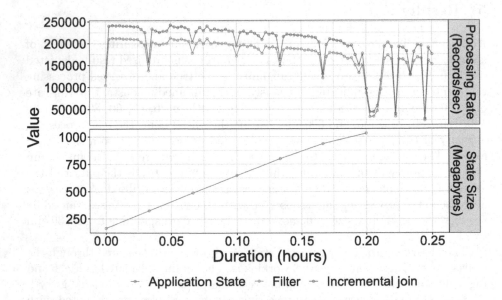

Fig. 6. For Nexmark Query 3, the evolution of processing rates for Filter and Incremental Join operators, and the interplay with application state.

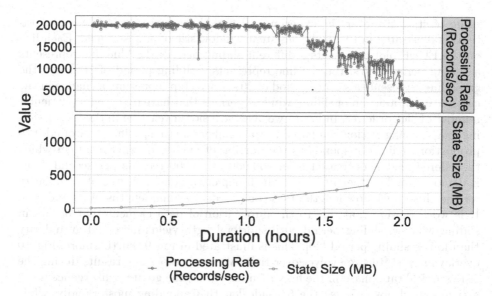

Fig. 7. For Nexmark Query 5, the evolution of processing rates and the interplay with application state.

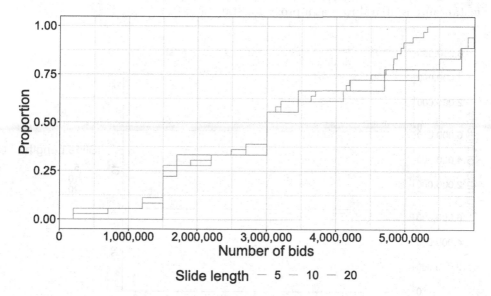

Fig. 8. ECDF plot of instantaneous state size for a sliding window operator of 30 s, with slide lengths of 5, 10 and 20 s.

workload state. The objective is to augment our comprehension of the internal mechanisms of the application state and to evaluate and compare the impact of various window size configuration on a stream processing system. The percentile value of state size over the experimental period for various window configurations is illustrated in Fig. 8. A consistent window size of 30 s is upheld. Each line on the ECDF plot represents a different sliding period: the blue line represents a sliding period of 5 s, the red line represents a sliding period of 10 s, and the green line represents a sliding period of 20 s. Each step represents a change in the cumulative proportion of sliding window periods that have observed a particular number of bids or fewer. In Fig. 8, we observe a slower reaction in the longer sliding length configuration compared to the smaller length, specifically around 0.75 proportion. This corresponds to the portion of our third experiment, in which the offered load is decreased from 200,000 to 50,000 records per second. This is demonstrated further in Fig. 9, which depicts a greater degree of granularity for the shorter window length of 5 s compared to sliding lengths of 10 and 20 s. It is worth noting that due to the duplication of data element that occurs in sliding window, sliding period with shorter length yields increased granularity than longer sliding period [20]. This is illustrated in Fig. 9. Furthermore, Fig. 10 overlays how state size evolves over time. We see comparable results during the rising arrival rate phase in the first 1.5 min. We see the greatest divergence when arrival rate drops. We see the 5 s slide length responding most rapidly, while longer slide lengths retain elevated state sizes for a longer period. This elevated state size represents a gap not currently captured by approaches which leverage instantaneous arrival rate as a proxy for state size.

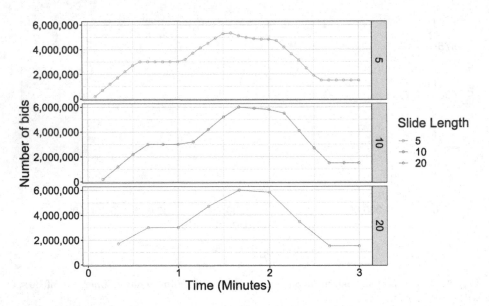

Fig. 9. Visualisation of state size (number of bids) for a window operator of length 30 s, for different slide lengths of 5, 10 and 20 s.

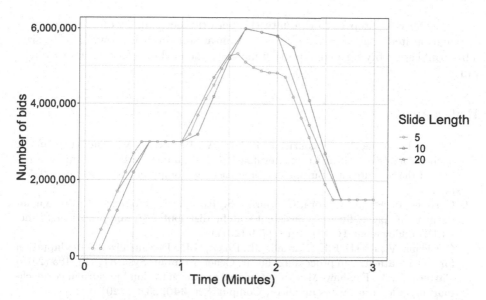

Fig. 10. Visualisation of state size (number of bids) for a window operator of length 30 s, for different slide lengths of 5, 10 and 20 s.

6   Conclusion and Future Work

The experimental results obtained provide evidence in favour of the assertion that the size of the state has the potential to present a challenge for auto-scalers in accurately assessing the presented offered load. Streaming applications with a larger state or longer window size may result in fluctuations in the convergence between the application state and the offered load. Like most auto-scalers, which utilises the true and observed processing rate to determine rescaling actions, most auto-scalers do not consider the influence of state size when making a scaling decision. Our empirical findings indicate that the growth of the state can compel an auto-scaling system to downscale rapidly, owing to a decrease in the operator processing rate. Nonetheless, the decrease in question is contrived and does not accurately depict the arrival rate. Such decisions by the auto-scaler may result in difficulties in rescaling, leading to reduced efficiency.

To evaluate the effects of window selectivity options in streaming applications, an experiment was done. We demonstrate the impact of different sliding window length configuration on instantaneous state size. Therefore, our results lend credence to the claim that window size selectivity is essential for streaming programmes to run as efficiently as possible in a windowed environment.

An interesting future work will be to optimise stream processors' data initialization period and the ability for the system to reach its optimal processing capacity within a shorter period. This could become a distinctive factor amongst streaming systems, especially when real-time data for analytics is required.

A useful next step is to evaluate the performance impact of the streaming application iterating over a larger collection arising from a growing state size. This could arguably be a contributing factor to the declining operator's throughput.

References

1. Asyabi, E., Wang, Y., Liagouris, J., Kalavri, V., Bestavros, A.: A new benchmark harness for systematic and robust evaluation of streaming state stores. In: Proceedings of the Seventeenth European Conference on Computer Systems, pp. 559–574 (2022)
2. Carbone, P., Ewen, S., Fóra, G., Haridi, S., Richter, S., Tzoumas, K.: State management in apache flink®: consistent stateful distributed stream processing. Proc. VLDB Endowment **10**(12), 1718–1729 (2017)
3. Cardellini, V., Presti, F.L., Nardelli, M., Russo, G.R.: Decentralized self-adaptation for elastic data stream processing. Futur. Gener. Comput. Syst. **87**, 171–185 (2018)
4. Cattermole, A., Forshaw, M.: An automated approach to cloud performance benchmarking. Electron. Notes in Theor. Comput. Sci. **340**, 23–39 (2018)
5. Cengiz, M., Forshaw, M., Atapour-Abarghouei, A., McGough, A.S.: Predicting the performance of a computing system with deep networks. In: Proceedings of the 2023 ACM/SPEC International Conference on Performance Engineering, pp. 91–98 (2023)
6. Ezhilchelvan, P., Mitrani, I.: Checkpointing models for tasks with widely different processing times. In: Gilly, K., Thomas, N. (eds.) Computer Performance Engineering: 18th European Workshop, EPEW 2022, Santa Pola, Spain, September 21–23, 2022, Proceedings, pp. 100–114. Springer, Cham (2023). https://doi.org/10.1007/978-3-031-25049-1_7
7. Floratou, A., Agrawal, A., Graham, B., Rao, S., Ramasamy, K.: Dhalion: self-regulating stream processing in heron. Proc. VLDB Endowment **10**(12), 1825–1836 (2017)
8. Gou, X., et al.: Sliding sketches: a framework using time zones for data stream processing in sliding windows. In: Proceedings of the 26th ACM SIGKDD International Conference on Knowledge Discovery and Data Mining, pp. 1015–1025 (2020)
9. Hueske, F., Kalavri, V.: Stream processing with Apache Flink: fundamentals, implementation, and operation of streaming applications. O'Reilly Media (2019)
10. Jamieson, S.: Dynamic scaling of distributed data-flows under uncertainty. In: Proceedings of the 14th ACM International Conference on Distributed and Event-based Systems, pp. 230–233 (2020)
11. Jamieson, S., Forshaw, M.: Measuring streaming system robustness using non-parametric goodness-of-fit tests. In: Gilly, K., Thomas, N. (eds.) Computer Performance Engineering: 18th European Workshop, EPEW 2022, Santa Pola, Spain, September 21–23, 2022, Proceedings, pp. 3–18. Springer, Cham (2023). https://doi.org/10.1007/978-3-031-25049-1_1
12. Kalavri, V., Liagouris, J., Hoffmann, M., Dimitrova, D., Forshaw, M., Roscoe, T.: Three steps is all you need: fast, accurate, automatic scaling decisions for distributed streaming dataflows. In: 13th USENIX Symposium on Operating Systems Design and Implementation (OSDI 2018), pp. 783–798 (2018)

13. Li, S., Gerver, P., MacMillan, J., Debrunner, D., Marshall, W., Wu, K.L.: Challenges and experiences in building an efficient apache beam runner for IBM streams. Proc. VLDB Endowment **11**(12), 1742–1754 (2018)
14. Li, T., Xu, Z., Tang, J., Wang, Y.: Model-free control for distributed stream data processing using deep reinforcement learning. arXiv preprint arXiv:1803.01016 (2018)
15. Mohamed, S., Forshaw, M., Thomas, N., Dinn, A.: Performance and dependability evaluation of distributed event-based systems: a dynamic code-injection approach. In: Proceedings of the 8th ACM/SPEC on International Conference on Performance Engineering, pp. 349–352 (2017)
16. Omoregbee, P., Forshaw, M.: Performability requirements in making a rescaling decision for streaming applications. In: Gilly, K., Thomas, N. (eds.) Computer Performance Engineering: 18th European Workshop, EPEW 2022, Santa Pola, Spain, September 21–23, 2022, Proceedings, pp. 133–147. Springer, Cham (2023). https://doi.org/10.1007/978-3-031-25049-1_9
17. Röger, H., Mayer, R.: A comprehensive survey on parallelization and elasticity in stream processing. ACM Comput. Surv. (CSUR) **52**(2), 1–37 (2019)
18. Rzadca, K., et al.: Autopilot: workload autoscaling at google. In: Proceedings of the Fifteenth European Conference on Computer Systems, pp. 1–16 (2020)
19. da Silva Veith, A., de Assunçao, M.D., Lefevre, L.: Monte-carlo tree search and reinforcement learning for reconfiguring data stream processing on edge computing. In: 2019 31st International Symposium on Computer Architecture and High Performance Computing (SBAC-PAD), pp. 48–55. IEEE (2019)
20. Tangwongsan, K., Hirzel, M., Schneider, S.: Sliding-window aggregation algorithms (2019)
21. Tucker, P., Tufte, K., Papadimos, V., Maier, D.: Nexmark-a benchmark for queries over data streams (draft). Tech. rep., Technical report, OGI School of Science and Engineering at (2008)
22. Van Dongen, G., Van Den Poel, D.: Influencing factors in the scalability of distributed stream processing jobs. IEEE Access **9**, 109413–109431 (2021). https://doi.org/10.1109/ACCESS.2021.3102645
23. Vogel, A., Griebler, D., Danelutto, M., Fernandes, L.G.: Self-adaptation on parallel stream processing: a systematic review. Concurrency Comput.: Pract. Experience **34**(6), e6759 (2022)
24. Zhang, Z., Li, W., Qing, X., Liu, X., Liu, H.: Research on optimal checkpointing-interval for flink stream processing applications. Mobile Networks Appl. **26**(5), 1950–1959 (2021). https://doi.org/10.1007/s11036-020-01729-7

Execution Time Prediction Model that Considers Dynamic Allocation of Spark Executors

Hina Tariq[(✉)][iD] and Olivia Das[iD]

Electrical, Computer and Biomedical Engineering, Toronto Metropolitan University
(Formerly Ryerson University), Toronto, Canada
hina1.tariq@torontomu.ca

Abstract. We propose a deterministic analytical model that considers dynamic allocation of spark executors while predicting execution time of spark applications. Our new model uses idle time and backlog time metrics to determine whether to add or remove executors. Following the update of executors, this model traverses every stage of a direct acyclic graph using a graph traversal algorithm. We repeat this process until the total execution time of the spark application is calculated. We validate our model against the measured execution time for Query-52 and K-Means workloads that reveal error rates of 4.96% and 4.74%, respectively. A comparison of our model to four classic machine learning models indicates that it is more effective than linear regression, neural networks, decision trees, and random forest. To the best of our knowledge, this is the first deterministic analytical model that accounts for dynamic allocation of executors.

Keywords: Dynamic Allocation · Backlog Time · Idle Time ·
Analytical Model · Deterministic Model · Execution Time · Spark
Application

1 Introduction

Apache Spark is a powerful framework for processing and analyzing large-scale datasets in parallel, making it an essential tool for big data applications. Spark's *dynamic allocation* feature allows applications to adjust the number of executor instances based on workload, at runtime. As a result, resource utilization can be optimized and the overall cost of running the spark application can be reduced in comparison to running it on a fixed number of executors (i.e. *static allocation* of executors) [2].

In Apache Spark, executors are worker processes that run on worker nodes in a cluster. Each executor has a specified number of cores (i.e. physical threads). Executors are responsible for executing tasks (each task being executed through a core) that are assigned to them by the Spark driver program (that runs on the master node of the cluster). Idle time and backlog time are important metrics that can be used to monitor the performance of executors. Idle time refers to

M. Iacono et al. (Eds.): EPEW/ASMTA 2023, LNCS 14231, pp. 340–352, 2023.
https://doi.org/10.1007/978-3-031-43185-2_23

the amount of time an executor spends waiting for a new task to be assigned to it. When an executor completes a task, it becomes idle and waits for the driver program to assign it a new task. The idle time for an executor should be minimal as it indicates that the executor is fully utilized and is not wasting any resources. On the other hand, backlog time refers to the amount of time a task spends waiting to be executed by an executor. This occurs when there are no available executors to run the task. Backlog time can occur when there are not enough resources available in the cluster, or when tasks are being launched faster than they can be completed. High backlog time can lead to slower application performance and longer job completion times [2].

Multiple works have used machine learning models for predicting performance of spark applications [5,6,9]. These models entail reasonably large amount of previously measured data to make successful predictions. Collecting such data is a tremendously time-consuming and resource-intensive task that often comes at a prohibitively high financial cost—it is time-consuming since one has to run a spark application repeatedly while varying the values of several different parameters and record the measurements for every run; it is resource-intensive since one has to acquire multiple nodes for substantially long period of time to run the experiments; and, this entire process of experimentation incurs huge monetary expenditure to pay for the acquired cloud resources, and human intervention to conduct such experiments. Analytical models [1,12–14], on the other hand, can predict performance of a spark application in less time. Developing such models needs considerably fewer runs to collect measurement data for model parameter estimation, compared to the machine learning models.

Performance modeling of Spark applications with dynamic allocation can be challenging as it involves multiple factors that can affect the application's performance. Before modeling the performance, it is important to understand the resource requirements of the application. This includes memory and CPU requirements of the application, as well as the number of tasks that need to be executed in parallel.

In this work, we incorporate the *dynamic allocation* capability to our prior modeling work that was limited to modeling the static allocation of executors [1]. To the best of our knowledge, this work is the first deterministic analytical model that accounts for dynamic allocation of executors. We utilize the Spark policy of adding and removing executors by considering the backlog time and idle time respectively in our algorithm.

We validate our model against the measured execution time obtained by running a popular big data query—Query-52—of TPC-DS industry benchmark [3], and a popular machine learning algorithm—K-means clustering algorithm—of SparkBench industry benchmark (developed at IBM Research) [4,8]. All experiments are performed on the spark cluster of the Google Cloud. Our model results in 4.96% error in comparison to the measured execution time of Query-52 and in 4.74% error in comparison to the measured execution time of K-Means algorithm.

We also test the results of our novel model results against four classic machine learning models—linear regression (LR), neural network (NN), decision tree (DT), and random forest (RF). Each machine learning model is run as a blackbox as well as a whitebox. The blackbox method refers to making predictions using features that capture knowledge available prior to the execution of the applications, e.g. configuration parameters of the execution. On the other hand, whitebox method refers to making predictions using features capturing detailed structure of the application in terms of its stages and tasks.

The rest of the paper is organized as follows. In Sect. 2, we discuss the Spark Architecture with dynamic allocation within the context of this work. In Sect. 3, we provide the background of this work. Here, we briefly discuss an existing model that we utilize in this work. In Sect. 4, we explain our proposed model. In Sect. 5, we present the results and analyse them. Section 6 enumerates some related work. Finally, Sect. 7 concludes the paper with a suggestion for future research.

2   Apache Spark Architecture

Apache Spark is a distributed computing system that allows for the processing of large datasets in a distributed and parallel manner. The architecture of Spark consists of several components that work together to provide efficient and fault-tolerant data processing capabilities.

Apache Spark executes a job by dividing it into stages, and each stage is further divided into tasks. The tasks are then executed in parallel across the cluster. The Spark driver program submits the job to the cluster manager, which allocates resources for the job. The cluster manager then launches the Spark executors on the worker nodes, which run the tasks assigned to them. The executors communicate with the driver program to exchange data and status updates. Once all the tasks in a stage are completed, the results are sent back to the driver program, which then submits the next stage. This process continues until all stages are completed, and the job is finished [7].

2.1   Dynamic Allocation

The dynamic allocation mechanism in Apache Spark works by monitoring the cluster workload and by adding or removing executors based on the workload. In dynamic allocation, the Spark application can request additional executors when it needs more resources to process data, and release executors when it no longer needs them.

Dynamic allocation in Spark can be enabled by setting the following configuration properties [2]:

- spark.dynamicAllocation.enabled: Set this property to true to enable dynamic allocation.
- spark.shuffle.service.enabled: Set this property to true to enable the external shuffle service, which is required for dynamic allocation.
- spark.dynamicAllocation.minExecutors: Set this property to the minimum number of executors that should be running at all times.

– spark.dynamicAllocation.maxExecutors: Set this property to the maximum number of executors that can be added dynamically.

These properties can be set either in the Spark configuration file or by using the SparkConf object in the application code. Once dynamic allocation is enabled, Spark will automatically adjust the number of Executors based on the workload of the cluster.

Adding Executors: The process of adding executors is as follows:

1. The Spark Driver (the master process) requests a new executor from the cluster manager (e.g., YARN, Kubernetes).
2. The cluster manager creates an executor process in its own container and offers it to Spark Driver.
3. After the new executors are joined to the existing cluster, Spark Driver reschedules jobs to take the new executor into production.

Removing Executors: The process of removing executors is as follows:

1. The dynamic allocation mechanism detects that some executors are idle and no longer needed.
2. The Spark Driver sends a message to the executor to stop running tasks and release its resources.
3. The executor stops running tasks and releases its resources back to the cluster manager.
4. The cluster manager removes the executor process from its container.

Two crucial indicators for Spark applications dynamic resource allocation are backlog time and idle time. Backlog time refers to the time that a Spark application spends waiting for resources to become available in the cluster. This happens when the application has submitted more tasks than the cluster can currently handle. The tasks are then queued up in a backlog, waiting for resources to become available. Backlog time can have a negative impact on the performance of the application, as it can lead to increased latency and reduced throughput. Therefore, backlog time can be exploited to allocate executors. Idle time, on the other hand, refers to the time that resources in the cluster are not being used by any Spark application. This occurs when there are more resources available than are needed to process the tasks submitted by all the applications. Idle time can be seen as a wasted opportunity to improve the overall efficiency of the cluster. Hence, idle time can be exploited to release executors.

To optimize the performance of a dynamic resource allocation, it is important to minimize both backlog time and idle time. One way to achieve this is by fine-tuning the parameters of the allocation system, such as the minimum and maximum number of resources that can be allocated to an application. Another approach is to use workload prediction techniques to anticipate the resource needs of the applications and proactively allocate resources accordingly.

3   Background – An Existing Static Allocation Model (SAM)

The work in [1] developed a graph-based, deterministic analytical model that predicts the execution time of a spark application in presence of static allocation of executors—that is, allocation of a fixed number of executors prior to running the application. Henceforth, we call this model SAM. The SAM represent the structure of the application as a flat monolithic Directed Acyclic Graph (DAG) of stages that captures the precedence relationship among the stages of the application. A graph traversal algorithm—henceforth called SAM algorithm, coupled with a fixed scheduling policy of the cloud (hosting the spark cluster) is used to process the stages of the DAG.

The input parameters for the SAM algorithm include a list of executor cores available to process the tasks in parallel; the warm-up time of the spark application—it includes the time needed for starting the driver and the executors on the master and the worker nodes; the DAG of the spark application; the number of tasks in each stage of the DAG; and, the average execution time per task for each stage of the DAG.

By exploiting two reference executions of the spark application on a cloud for two different input data sizes, SAM is able to estimate the input parameters of the SAM algorithm, and subsequently predict the execution time of the application for an arbitrary big input data size.

In order to maximize parallelism, SAM identifies the stages of the DAG that can be executed concurrently. Based on an analysis of the dependencies between stages, SAM is able to determine which stages can be executed concurrently and which ones need to be executed sequentially. Furthermore, the SAM allots free executor cores to succeeding stages that are ready to begin instead of waiting for the preceding parallel stage, with the largest execution time, to finish.

We next propose a model that accounts for dynamic allocation of executors leading to better utilization of resources and lower cost of running a spark application, compared to SAM.

4   Proposed Model – The Dynamic Allocation Model (DAM)

In this section we propose our novel model for prediction of execution time of a spark application while accounting for dynamic allocation (i.e. support for variable number of executors during runtime). We call our new model DAM. Our model DAM utilizes SAM, adds and removes executor cores dynamically at runtime, and subsequently predicts the execution time for the application. Our model uses the following parameters as inputs:

- backlogTime - refers to the amount of time an application waits for a resource allocation (executor) request to be fulfilled by the cluster manager.

- idleTime - refers to the amount of time an executor remains idle before it is released and returned to the resource pool.
- Static Allocation Model (SAM) - The model developed in [1].

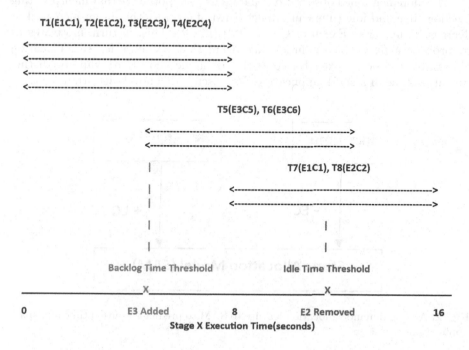

Fig. 1. Dynamic Addition and Removal of Executors Example

In our experiments on Google Cloud, we observed that when there are pending tasks with all executors occupied, new executors are added once a backlog time threshold is reached. In a similar manner, if an executor has to wait more than the idle time for the tasks to be assigned, that executor will be removed from the stage. To explain this observation, let us consider an example mentioned in Fig. 1. Consider a Stage X with eight tasks, idle time of 4 s, backlog time of 5 s, and average task time of 8 s. In the beginning, there are two executors, each of which has two cores. The presence of two cores means that one executor can handle two tasks at the same time. We can therefore say that E1C1 and E1C2 represent executor number one with core one and core two. In the same way, Task1 can be referred to as T1. In the beginning of stage, tasks are assigned as follows: T1 (E1C1), T2(E1C2), T3(E2C3), T4(E2C4). Each of the four tasks assigned will take eight seconds to complete. As of now, there are four pending tasks in the queue, and after 5 s, the backlog threshold will be reached, and a new executor E3 will be added to execute tasks T5 (E3C5) and T6 (E3C6). Tasks T5 and T6 will be completed after 13 s of stage execution time. Once the timeline reaches 8 s, T7 (E1C1) and T8 (E1C2) will be assigned to executor one, which will eventually complete execution at stage execution time of 16 s. E1 and E3 are both occupied until 13 and 16 s, respectively. However, E2 remains free since stage execution became 8 s. As a result, E2 will be removed when the stage execution time exceeds 12 s since idle time is set at 4 s.

The aforementioned observation enables us to hypothesize that incorporating backlog time and idle time can impact SAM of [1], resulting in an increased or decreased number of Executor Cores (EC) (Fig. 2). This, in turn, motivates us to propose a new model solution algorithm that accounts for the allocation of variable number of executor cores (i.e. dynamic allocation) while predicting execution time of a spark application. We discuss this novel algorithm next.

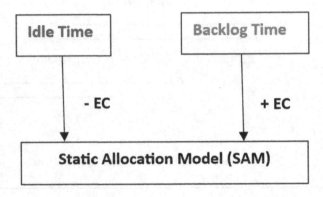

Fig. 2. Adding dynamic allocation feature to SAM to predict execution time of a spark application.

Algorithm 1 presents our new model solution algorithm that accounts for varying number of executors during runtime of a spark application. We refer to this new algorithm as DAM algorithm. The DAM algorithm takes backlog time, idle time, and the Static Allocation Model (SAM) of [1] as its input. The output of DAM is the predicted execution time of the spark application.

In the description of DAM, the notation $Y(SAM)$ implies that the data Y is obtained from SAM. Thus Line 2–7 of the algorithm expresses that the six items—the list of stages S, the list of available executor cores EC, the time elapsed per stage $Telapse1$, the tasks per stage $tasks$, the average task execution time per stage $AvgTaskExecTime$ and, the time that elapsed since all executors were assigned tasks $Telapse2$—they are all extracted from SAM.

We begin the processing of each stage one by one starting from line number 9 to 16. In the course of executing the tasks for each stage, we check to see—Have

Algorithm 1. Dynamic Allocation Model (DAM) Algorithm

Input: $backlogTime$, $idleTime$, $StaticAllocationModel(SAM)$[1]
Output: $AppExecutionTime$

```
 1: +//Pseudocode below initializes paramaters+
 2: S ← Stages(SAM)
 3: EC ← Executor_Cores(SAM)
 4: Telapse1 ← Time_Elapsed_Per_Stage(SAM)
 5: tasks ← Tasks_Per_Stage(SAM)
 6: AvgTaskExecTime ← Average_Task_Execution_Time(SAM)
 7: Telapse2 ← Time_Elapsed_After_Commencing_Task_Execution(SAM)

 8: +//Pseudocode below adds and removes executors dynamically+
 9: for stage s ∈ S do
10:     if tasks ≥ 1 and Telapse > backlogTime and EC == occupied then
11:         EC ← Add_ExecutorCores_to_SAM
12:     end if
13:     if AvgTaskExecTime + idleTime < Telapse2 then
14:         EC ← Remove_ExecutorCores_from_SAM
15:     end if
16: end for
17: AppExecutionTime ← Computed_Through_SAM
```

---

we ensured that all executor cores are engaged? Has the backlog time threshold been reached? Are there any tasks that still need to be completed? If answer to these questions is yes, then more executor cores are allocated and SAM begins to process the tasks with the newly increased number of executor cores.

Otherwise, we check to see—Is there are any free executor cores left after all tasks have been assigned executor cores last time? While keeping in mind the average task execution time, has the idle time threshold been crossed? If answer to these questions is yes, we remove executor cores from SAM, which in turn starts processing tasks using the newly decreased number of executors.

The aforementioned processing steps continues until all the stages have completed their tasks. In the final step (line 17), the spark application's execution time $AppExecutionTime$ is computed through SAM.

5   Result and Analysis

In this section, we discuss our results and analyze our model. Our model is validated using the measurement results from Query-52, a big data query from TPC-DS benchmark, and K Means from Sparkbench. Models are all Python 3 programs that are run using Jupyter Notebook on a machine having the following specifications: Intel(R) Core(TM) i7-8550U CPU @ 1.80 GHz, 2.00 GHz Processor, and 16 GB of RAM.

The TPC Decision Support (TPC-DS) benchmark suite is a standard tool for evaluating the performance of decision support systems [3]. It consists of a set of

SQL queries that simulate real-world decision support workloads, including ad hoc queries, reports, and data mining. This benchmark is designed to assess the performance of database systems in terms of query execution time, data loading time, and data storage requirements. To evaluate the performance of Spark applications, SparkBench provides a set of workloads for use with Apache Spark [4,8]. These workloads include machine learning, graph processing, and SQL queries, which are representative of real-world use cases. Using SparkBench, one can evaluate the performance of various Spark configurations, including cluster sizes, memory settings, and tuning parameters. Furthermore, it provides tools for collecting and analyzing performance metrics, such as execution time and resource utilization.

5.1 Model Validation Using a Big Data Query Query-52 and K-Means

In this section, we validate our model using the measured execution time of Query-52 and K-Means on Google's Spark cluster.We aim here to determine the execution time for Query-52, starting with a minimum of four executor cores, for input data sizes of 100 GB, 200 GB, and 500 GB. In addition, we intend to determine the execution time of K-Means based on input row sizes of 10,000,000, 15,000,000, and 20,000,000, starting with a minimum of four executor cores. For each case, we run the spark application ten times by fluctuating backlog time and idle time for each stage of the DAG of Query-52 and K-Means, and then, for each stage, we take the average of the task execution times obtained from those ten runs.According to the 100 GB and 200 GB datasets for Query-52, we predict execution time for input size 500 GB. With 10,000,000 and 15,000,000 rows of input for K-Means, we predicted the execution time for 20,000,000 rows. In comparison with the measured execution times for Query-52 and K-Means workloads, the error rates for our model came out 4.96% and 4.74%, respectively.

Our K-Means experiment is conducted in Google Cloud, using its provided services to create a cluster of three n1-standard-4 workers, where each worker node has four virtual CPUs, totaling 12 virtual CPUs with 15 GB of RAM. The Query-52 experiment is carried out on Google Cloud with four n1-standard-4 worker nodes rather than three. Each node has four virtual CPUs. As a result, there were now 16 virtual CPUs available for processing and 15 GB of memory available.

Comparison with Machine Learning Models. The results of our model were also compared with those of machine learning models. The data was collected by running Query-52 and K-Means with dynamic allocation enabled. As a result of these data, 16 machine learning models were trained (eight for each of the two benchmarks) to predict the execution time of the benchmarks. We selected linear regression, random forest, decision tree, and neural network models for analysis.

To be able to learn and evaluate the prediction models, the data collected from the experiments were split into training and test sets, where the training set was used to learn the models and the test set was used to evaluate the models. In this experiment, we used the following machine learning models:

1. Linear Regression (LR): A linear regression analysis is a statistical method for analyzing the relationship between two or more variables. This technique is commonly used in data analysis, machine learning, and predictive modeling.As a result of linear regression, the best-fit line is determined that describes the relationship between the independent variable (also known as the predictor variable or feature) and the dependent variable (also known as the response variable or target).
2. Neural Network(NN): A neural network is a type of machine learning model that is based on the structure and function of the human brain. A neural network consists of multiple interconnected nodes or neurons, arranged in layers, and is able to learn and predict based on the data that it receives.
3. Decision Tree(DT) A decision tree is a machine learning algorithm that is used for classification and regression tasks. It works by recursively partitioning the input data into subsets based on the values of their features. This is done until each subset contains only instances of a single class or has reached maximum depth.
4. Random Forest (RF): A random forest is a machine learning algorithm that is commonly used for classification, regression, and other tasks. It is an ensemble learning technique in which multiple decision trees are combined in order to create a more accurate and robust model.

For all four of the aforementioned machine learning models, we construct both blackbox and whitebox models for Query-52 and K-Means. A blackbox model and a whitebox model are two approaches to building predictive models in machine learning. Blackbox models are models that are trained on input-output pairs without knowledge of the underlying process that generates the data. Whitebox models, on the other hand, incorporate knowledge of the underlying process into their design. It employs domain-specific knowledge to guide the selection of features and construction of models [5]. In our blackbox models, data size, backlog time, and idle time were considered. During the development of our whitebox model, we also considered additional features such as tasks per stage, the average task time, and the maximum task time.

From Table 1 it can be seen that the blackbox models perform better as opposed to the whitebox models, indicating that the features extracted posteriori are not necessarily better to have in the dataset. This is the case for Query-52 benchmark queries. However, for K-Means, it is quite the opposite excluding Neural Networks. In fact, the models performed better with features available for extraction priori and posteriori. Considering, different configurations and techniques were used on each benchmark, it shows that there isn't a single technique or benchmark to outperform another, therefore choosing the best model can also vary accordingly. However, our dynamic model outperforms all machine learning models with a minimum error of 4.96 for Query52 and 4.74 for K-Means.For error percentages, we took the absolute error value of $(|(measuredTime - predictedTime)|/measuredTime) \times 100\%$.

Table 1. Error Percentages for Predicted Executed Times using Dynamic Allocation

Models	Query-52 Workload	K-Means Workload
Black-Box DT	9.4	18.7
Black-Box RF	6.6	14.5
Black-Box LR	7.8	15.6
Black-Box NN	5.9	13.9
White-Box DT	12.4	15.7
White-Box RF	7.5	12.6
White-Box LR	20.5	15
White-Box NN	19.2	37.3
Dynamic Model	**4.96**	**4.74**

6   Related Work

The works in [5,6,9] modelled spark applications using machine learning techniques. For example, the work in [5] uses supervised machine learning models to predict the performance of Spark cloud applications. These models are trained on historical data collected from the application and its environment, and then used to make predictions about future performance. On the other hand, the works in [10,11] modelled such applications using queuing networks and simulation modelling techniques. Besides, several works reported deterministic analytical models [1,12–14] that can provide insight into the performance of a spark application in comparatively less time. Yet, all these works are *limited* to fixed allocation of executors—*none* accounts for dynamic allocation (addition and removal of executors at runtime) although it could provide better resource utilization and decreased financial cost in running a spark application.

In contrast, our novel modeling technique incorporates dynamic allocation of executors in the process of predicting execution time of spark applications [1]. Besides, our technique enables more accurate prediction compared to four classic machine learning models.

7   Conclusion and Future Work

We have presented a graph-based, deterministic analytical model that accounts for dynamic allocation of executors for a spark application—that is, where the executors are allocated on demand for the application at runtime. It predicts the execution time of spark applications, using idle time and backlog time metrics to measure the performance of the executors. An evaluation of our model against measured execution time for a Query-52 and a K-Means workload revealed an error rate of 4.96% and 4.74%, respectively. Our model also outperformed four classic machine learning models, including linear regression, neural networks, decision trees, and random forest. In future, we intend to investigate the possibility of combining our model with an optimization framework. On doing so, we will be able to make informed decisions regarding the number of executors that should be allocated for running the application.

Acknowledgements. We acknowledge the assistance of undergraduate students Grahi Desai, Yiran Chen, Marc Lima, and Asma Fawzia Kawser Maisha in collecting the results of machine learning models. We would also like to thank NSERC Canada for financial support.

References

1. Tariq, H., Das, O.: A deterministic model to predict execution time of spark applications. In: Gilly, K., Thomas, N. (eds.) Computer Performance Engineering. EPEW 2022. LNCS, vol 13659, pp. 167–181. Springer, Cham (2023). https://doi.org/10.1007/978-3-031-25049-1_11
2. Spark Job Scheduling. https://spark.apache.org/docs/latest/job-scheduling.html. Accessed 28 Mar 2023
3. TPC-DS decision support benchmark. https://www.tpc.org/tpcds/
4. SparkBench. https://codait.github.io/spark-bench/workloads/. Accessed 11 Apr 2022
5. Maros, A., et al.: Machine learning for performance prediction of spark cloud applications. In: 2019 IEEE 12th International Conference on Cloud Computing (CLOUD), Milan, Italy, pp. 99–106 (2019). https://doi.org/10.1109/CLOUD.2019.00028
6. Didona, D., Quaglia, F., Romano, P., Torre, E.: Enhancing performance prediction robustness by combining analytical modeling and machine learning. In: ACM/SPEC 6th International Conference on Performance Engineering (ICPE), pp. 145–156 (2015)
7. Wang, K., Khan, M.M.H., Nguyen, N., Gokhale, S.: A model driven approach towards improving the performance of apache spark applications. In: IEEE International Symposium on Performance Analysis of Systems and Software (ISPASS), Madison, WI, USA, pp. 233–242 (2019). https://doi.org/10.1109/ISPASS.2019.00036
8. Li, M., Tan, J., Wang, Y., et al.: SparkBench: a spark benchmarking suite characterizing large-scale in-memory data analytics. Cluster Comput. **20**, 2575–2589 (2017). https://doi.org/10.1007/s10586-016-0723-1
9. Venkataraman, S., Yang, Z., Franklin, M., Recht, B., Stoica, I.: Ernest: efficient performance prediction for large-scale advanced analytics. In: 13th USENIX Symposium on Networked Systems Design and Implementation NSDI 2016, pp. 363–378 (2016)
10. Ardagna, D., et al.: Performance prediction of cloud-based big data applications. In: 2018 ACM/SPEC 9th International Conference on Performance Engineering (ICPE), pp. 192–199 (2018)
11. Ardagna, D., et al.: Predicting the performance of big data applications on the cloud. J. Supercomput. **77**, 1321–1353 (2021)
12. Shah, S., Amannejad, Y., Krishnamurthy, D., Wang, M.: Quick execution time predictions for spark applications. In: IEEE 15th International Conference on Network and Service Management (CNSM), pp. 1–9 (2019)

13. Shah, S., Amannejad, Y., Krishnamurthy, D., Wang, M.: PERIDOT: modeling execution time of spark applications. IEEE Open J. Comput. Soc. **2**, 346–359 (2021)
14. Shah, S., Amannejad, Y., Krishnamurthy, D.: Diaspore: diagnosing performance interference in Apache Spark. IEEE Access **9**, 103230–103243 (2021)

Exploiting Structural Dependency Relations for Efficient Agent Based Model Simulation

Marzio Pennisi[1]([✉])(iD), Elvio G. Amparore[2]([✉])(iD),
and Giuliana Franceschinis[1]([✉])(iD)

[1] Università del Piemonte Orientale, Alessandria, Italy
{marzio.pennisi,giuliana.franceschinis}@uniupo.it
[2] Università di Torino, Torino, Italy
elviogilberto.amparore@unito.it

Abstract. In the last few years Agent Based Models (ABMs) have attracted growing interest in the field of computational simulation thanks to their applicability in very heterogeneous landscapes, usability for fine-grained descriptions and comprehensibility for application domain experts. However, the lack of a well-defined semantics for specifying how agents behave and how they get coupled and scheduled may lead to inconsistent results. To fill this gap we proposed a well defined ABMs semantics that, using Extended Stochastic Symmetric Nets for model description, allows the modeller to automatically derive the corresponding ABM simulator that is directly executable in the NetLogo ABM framework. In the present paper we propose an improvement that exploits locality of state change effects to avoid recomputing the rates of the enabled events at each state change. This is achieved by exploiting structural properties of the ESSN model to generate optimized NetLogo code (semi)automatically. The results obtained for an example case-study demonstrate a relevant improvement in terms of execution time when structural optimizations are employed to reduce rates calculations.

Keywords: ABM Simulation · Symmetric Nets · Structural analysis

1 Introduction

Mechanistic modeling and computer simulation remain two fundamental approaches for the study of many real-world phenomena. Such approaches allow in fact to predict how a system evolves over time, and then to execute what-if analyses even on unseen conditions, such as, for example, the testing of novel pharmaceutical compounds for in Silico Trials, or for forecasting the effects of contingency polices during pandemic spreads, allowing some sort of results extrapolation that it is not usually easily achievable with other kinds of models.

Giuliana Franceschinis and Marzio Pennisi are members of the CNIT (Consorzio Nazionale Interuniversitario per le Telecomunicazioni) Research Unit of the Universitá del Piemonte Orientale.

There are two main approaches for the mechanistic modeling of real-world phenomena. The first approach is to look at the phenomena from the top, by defining rules and equations that describe the behavior of the system as a whole. This is typically achieved by means of complex systems of differential equations on which there exists a solid mathematical theory. The second approach is to look at the phenomena under study from the bottom, by looking at the single behaviors and interactions of the entities composing it. This is achieved exploiting Agent-Based Models (ABMs), a more recent computational modeling paradigm that describes a system as the sum of the behaviors of autonomous decision-making entities called agents, allowing to observe "emergent phenomena", i.e. predictable complex dynamics of the global system that arise from the simple actions and interactions among the agents.

While there is a growing interest on the use of ABM models, also thanks to their applicability in very heterogeneous landscapes, usability for fine-grained descriptions and comprehensibility for application domain experts, they lack well-defined semantics specifying how the agents and environment behaviors get coupled and scheduled, leading to different results for the same domain model not only with respect to other computational approaches, but even among different ABM programming environments and simulation engines. There is indeed no consensus on the definition of the concept of "agent" and its behavior, on time advance mechanism [10] that drives the system evolution from its initial conditions up to a final state/time and on how conflicts and concurrency among the agents have to be handled.

To fill this gap we proposed the use of Extended Stochastic Symmetric Nets (ESSN) [11], a generalization of Symmetric Nets [7], as starting point to provide a well-defined semantics for the description of ABMs. We then defined a translation algorithm to automatically derive the corresponding ABMs [4], and implemented it into GreatSPN [2]: given a properly annotated ESSN, it produces an ABM directly executable on the NetLogo framework [3]. The simulation algorithm makes use of a Next Event Time Advance (NETA) approach. The current implementation recomputes the rates of all enabled events after each state change resulting rather inefficient in many scenarios.

In the literature it is possible to find methods that propose to exploit locality to speed up the simulation. In [12] this approach was applied to agent based ABMs expressed in the ML3 language [13]: the dependencies can be detected by analysing the model structure, that in ML3 takes the form of rules including an enabling condition (guard) on the state, a rate used to establish when an enabled rule can be applied, and a command to be executed to update the state when the rule is applied. In [9] the optimization has been introduced in the agent based simulation framework MASON in a more dynamic way and using Aspect Oriented Programming to achieve separation of concerns, i.e. keeping the embedded code, used to trace the dependencies and exploit them, separated from the code expressing the actual model: experiments have shown that this very flexible approach achieves the goal to exploit locality, but it introduces an overhead that penalizes performance if compared to a manually optimized code.

In the present paper we propose to exploit the event "locality" by following an approach that statically detects the dependencies on the ESSN model used to define the ABM, so it is more similar to the method proposed in [12]; the novelty is due to the exploitation of existing algorithms and a tool for structural analysis of SNs allowing to derive from the model some expressions denoting the dependency relations that can then be used for the (semi)automatic generation of optimized (NetLogo) code from the SN model.

The paper is organized as follows: In Sect. 2 we recall the ESSN formalism. In Sect. 3 we briefly sketch the ESSN to ABM translation algorithm and as novel contribution we show how to use structural analysis to identify structural dependencies by means of an example case-study. Section 4 presents the results of structural analysis on the case-study and the optimized simulation algorithm, also discussing the feasibility of its implementation, while Sect. 5 shows numerical results obtained on the case-study. Finally, in Sect. 6 conclusions and future work will be drawn.

2   Extended Stochastic Symmetric Net Formalism

Petri Nets (PN) and extensions are effective formalisms used to model many contexts thanks to their ability to simply and clearly represent both quantitative and qualitative aspects of systems. One extension of PN is the ESSNs [11], which constitute our starting point, as the idea is to use such graphical formalism to define the models, and to automatically derive from them an equivalent ABM that can be directly executed exploiting the NetLogo framework. In the following we briefly review the ESSN formalism.

Like PNs, ESSNs are bipartite graphs whose nodes are *places*, and *transitions*. Places are graphically denoted as circles and are used to describe a system local state, while transitions, graphically denoted as boxes, represent the system events. Places and transitions are connected by directed and annotated *arcs*, which express the connection between states and events. While places and transitions and the arcs connecting them define the network structure, tokens (that are contained into places) define the state of the PN, namely the *marking*. More specifically, let a color domain be defined as the Cartesian product of elementary types called *color classes* $C = \{C_1, \ldots, C_n\}$ that represent finite and disjoint sets, and that can be further partitioned into (static) subclasses: the marking of place p is a multiset of tokens belonging to a given color domain $cd(p)$.

A color domain is also associated with transitions, assigning to each transition $t \in T$ a set of variables, each with values in a given $C_i \in C$ (the variable's type). The color domain of t, $cd(t)$, is then defined as the Cartesian product of its variables' types. Thus, given a transition t, an instance of t (denoted as $\langle t, c \rangle$) is defined by the binding c of the transition variables to a color of proper type.

Arcs are annotated by the functions $I(p, t)$, if the arc connects a place p to a transition t, and $O(p, t)$ for the opposite direction. Given a legal binding of t, the evaluation of $I(p, t)$ and $O(p, t)$ returns the multiset of colored tokens that will be withdrawn from, or added to, the place connected to the arc when the transition instance fires. The set of input/output places of transition t is denoted $^\bullet t$ / $t^\bullet$. Guards are logical expressions defined on the color domain of the transition, and can be used to define constraints on its legal instances. A transition instance $\langle t, c \rangle$ is enabled and can fire in a marking m, if the evaluation of its guard on c is true, and $\forall p \in {}^\bullet t : I(p, t)(c) \leq m(p)$, where $\leq$ is the comparison operator between multisets. The firing of an enabled transition removes a multiset of tokens from the input places and adds a multiset of tokens on the output places (based on the input/output arc expressions). The set of all instances of t enabled in marking m is denoted by $E(t, m)$. Finally, each transition defines a mean velocity characterizing its random firing times, represented as the parameter λ of an exponential distribution. ESSN add the possibility to define firing rates using generalized real functions. The set of transitions T is partitioned into two subsets T_{ma} and T_g depending on the associated velocity. The former contains all transitions firing with a velocity defined by the Mass Action (MA) law [14]. The latter includes all transitions whose rates are defined as general real functions. Further details on ESSNs can be found in [11].

By using stochastic firing delays that are sampled from negative exponential distributions it is possible to automatically derive the underlying CTMC in order to quantitatively evaluate the system behaviour [1]. More specifically, the CTMC state space $\mathbb{S}$ corresponds to the reachability set of the corresponding ESSN, i.e. all possible markings that can be reached from the initial marking. The Master equations (MEs) for the CTMC are:

$$\frac{d\pi(m_i, \nu)}{d\nu} = \sum_{m_k} \pi(m_k, \nu) q_{m_k, m_i} \qquad m_i, m_k \in \mathbb{S} \qquad (1)$$

where $\pi(m_i, \nu)$ represents the probability to be in marking m_i at time ν, and q_{m_k, m_i} the element of the *infinitesimal generator* (i.e., the velocity to reach the marking m_i from m_k), which is defined as follows:

$$q_{m_k, m_i} = \sum_{\substack{t \in T \wedge c' \in cd(t) \wedge \\ \langle t, c' \rangle \in E(t, m_k)_{|m_i}}} F(m_k, t, c', \nu). \qquad (2)$$

where $E(t, m_k)_{|m_i}$ is the subset of $E(t, m_k)$ whose firing leads to marking m_i. For complex systems the equations (1) are often computationally intractable, thus various techniques can be exploited to study the system. As an example, a technique is represented by the Stochastic Simulation Algorithm (SSA) [8], a method that can be used for the simulation of stochastic systems whose behaviour can be described by the Master Equations.

3   Optimizing ABMs Simulation Times by Exploiting Net Structure Properties

One of the major concerns regarding the use of the Next Event Time Advance (NETA) algorithm for well-defined ABMs introduced in [4] refers to the fact that all the agents have to recalculate the rates of the actions that they may carry on after each event. This has a detrimental effect on the simulation times. Under the hypothesis that all the action times follow exponential distributions, it is possible to exploit some characteristics of the model to speed up the simulation; for instance the spatial representation that most ABMs have induces a locality of state changes. In this scenario, actions have localized effects, and then rate recalculations can be executed only for agents that are in a given neighbourhood of the position of the agent that performed the last event. This concept of rule "locality" can be extended and considered in a broader sense with the goal of selecting, for rate recalculations, only the agents that may be effectively influenced by the last event: this can be achieved by analysing the ESSN from which we derive the ABM model. It is in fact possible, by looking at the net structure, to determine the mutual dependencies between pairs of transitions.

In general, two types of structural properties are of interest for our purposes: *Structural Conflict* (SC) and *Structural Causal Connection* (SCC). SC between two transitions A and B holds when the transitions have in common one or more input places, so the firing of an instance of A may disable some instances of B (note that A and B may be two instances of the same transition). SCC between two transitions A and B occurs instead when one or more output places of A are input places of B, and then the firing of A modifies the marking of such places, possibly entailing the enabling of some instances of B. Such structural properties can be automatically determined by using SNexpression [6], a tool that performs the symbolic structural analysis of (ES)SNs.

To better clarify how such structural properties can be computed and used to reduce the ABM simulation execution time we refer to the SEIRS example in Fig. 1, already presented in [4]. This classical epidemiological model is an example of how infections spread among individuals. We considered one agent type that can be in four different states: Susceptible, Exposed, Infected and Recovered. States are represented using places. Furthermore, each agent (represented by a token) has three attributes: unique id, age and position, encoded as colors A, B and Z, respectively. Agent ids are restricted to be unique across the whole net. Agents in SUS (state susceptible) can move towards a different random position (transition MoveS) or can move towards place EXP (state exposed) after interacting (transition GetInfected) with an agent in the same position and in place INF (state infected). Agents in the exposed state can move towards a different random position (transition MoveE) and can become infected (transition EndIncubation). Agents in INF can move towards a different random position (transition MoveI), can die with different rates according to their age (transition Death), and can recover from the infection and become Recovered (transition Recovery). Finally, agents in REC (state Recovered) can become susceptible again with transition EndImmunity.

Before proceeding, we intuitively recall the main concepts of the ESSN-to-ABM translation approach (for the details see [3,4]). Places represent the possible states of the agents, and each type of agents has its (disjoint) set of states (and thus places). Each agent type is associated with a color class (representing the agents unique ids) that appears in the color domain of all places associated

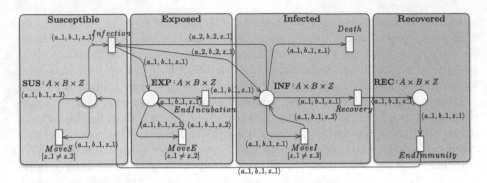

Fig. 1. The SEIRS infection model in the ESSN formalism.

with the corresponding agent type. The SEIRS example has only one agent type. Transitions represent the actions that the agents can carry on; those involving only one agent have only one variable whose type is an agent id color class in their color domain and model *internal actions* that the involved agent can directly execute without interacting with other agents. Transitions with more than one variable corresponding to an agent id, represent an *interaction among different agents*. As actions and interactions are executed by agents, for each action/interaction a specific agent (identified a priori) takes the role of leader and will be in charge to calculate the action rate and to carry on the action when selected. The choice of the leader for an interaction involving more than one agent is arbitrary, however structural dependencies may suggest the most appropriate choice.

Since the SNExpression language for expressing structural dependencies is not trivial, in order to explain its notation and application to minimize the number of rate re-computations (avoiding the net unfolding), we refer to an application example. The basic idea is to analyse the ESSN structure to determine transition instances that are in SC or SCC and may thus be enabled/disabled by the firing of a given transition. Let us consider an interesting case in the SEIRS model i.e., what happens to the Infection transition instances when an instance of MoveS fires. The input and output arc expressions of Infection can be seen in Fig. 1 and are reported in Table 1, the interpretation is the following: two different agents a_1 and a_2 of a given age b_1 and b_2 and residing in the same position $z_1 = z_2$ are involved; agent a_2 is infected and is not affected by the occurrence of the event represented by the transition, while agent a_1 is susceptible and will change its state to exposed. Transition *MoveS* represents the movement of agent a_1, age b_1 and position z_1, to a new position z_2 (with the constraint $[z_1 \neq z_2]$, expressed by the guard).

Let us consider the SC' relation (the transposed of the SC relation as defined in [5]) between MoveS and Infection, denoted as SC'(MoveS, Infection). This function has domain cd(MoveS) and co-domain $2^{cd(\text{Infection})}$:

$$SC'(\text{MoveS}, \text{Infection}) = \texttt{<a\_1,S\_A,b\_1,S\_B,z\_1>[z\_1 != z\_2]}$$

The interpretation of the formula is explained in the sequel; since the variables' scope is local to each transition, let's prefix the variable names with the name of the corresponding transition. The instances of Infection that may be disabled by the occurrence of an instance of MoveS are those with:

Infection.a_1 = MoveS.a_1, Infection.a_2 in A,
Infection.b_1 = MoveS.b_1, Infection.b_2 in B, Infection.z_1 = MoveS.z_1

that is, all the instances of Infection that involve a_1 and any other agent $(a_2 \in A)$ of any age $(b_2 \in B)$ and taking place in zone z_1 may be disabled by the occurrence of MoveS.

The remaining aspect to consider is how to assign an agent responsible for enumerating the disabled/newly enabled transition instances, every time a transition that may affect their enabling fires. This is the role of the leader agent, that can be any agent participating in the transition. In the example above, suppose that the leader is the infected agent (from place INF): The set of agents that need to recompute their rate is SC'(MoveS, Infection) composed[1] with the input function I(INF,Infection), resulting in

I(INF,Infection).SC'(MoveS,Infection) = <S_A,S_B,z_1>

So, if the leader agent of Infection is the infected one (agent a_2 from place INF) when MoveS fires on a_1 causing the susceptible agent to move to a different position, all the infected agents $(a_2 \in A)$ of any age $(b_2 \in B)$ in position z_1 will have to recalculate their rates for the Infection transition.

If instead the leader is the susceptible agent, the set of agents that will recompute their rates is SC'(MoveS, Infection) composed with the input function I(SUS,Infection), resulting in

I(SUS,Infection).SC'(MoveS,Infection) = <a_1,b_1,z_1>[z_1 != z_2]

so the susceptible agent a_1 must recompute the rate of the Infection transition.

MoveS and Infection are also in SCC relation. Intuitively, when a susceptible agent moves from a position to another, it could get the infection from the infected agents in the new position, so new instances of Infection may be enabled. To this end, we will take a look to the transposed of SCC(MoveS, Infection):

SCC'(MoveS,Infection) = <a_1,S_A,b_1,S_B,z_2>[z_1 != z_2]

The formula means that when a given instance of MoveS fires causing the agent with id a_1 to move to z_2, all the Infection instances that involve a_1 (with age b_1) and any infected agent $(a_2 \in A)$ of any age $(b_2 \in B)$ in position z_2 can become enabled. Hence, when the leader agent is the infected one, the set of agents that have to recompute their rates will therefore be

I(Infection,INF).SCC'(MoveS,Infection) = <S_A,S_B,z_2>[z_1 != z_2]

while if the leader agent is the susceptible one, only that agent will have to recompute the rates:

I(SUS,Infection).SCC'(MoveS,Infection) = <a_1,b_1,z_2>[z_1 != z_2]

This type of structural analysis not only determines which agents have to recalculate their rates, but gives also suggestions on how to minimize rate recalculations for a given transition by choosing the most convenient leader.

As further example we will show how `MoveS` transition has a self-SC and SCC. For such reason we will use `SNexpression` on `MoveS` obtaining[2]:

`SC'(MoveS,MoveS)=  <a_1,b_1,z_1,(S-z_2 * S-z_1)>[z_1 != z_2]`

The function means that all the possible instances of `MoveS` with same agent and same starting position are in conflict with those with same agent and initial position but different destination. If we compose the above function with `I(S,MoveS)` we obtain the susceptible agents that need to recalculate their rate for the `MoveS` action. In this specific case the result will be:

`I(MoveS,SUS).SC'(MoveS) =  <a_1,b_1,z_1>[z_1 != z_2]`

Table 1. Summary of structural relations between transitions in Fig. 1

Transition, cd(), In/Out expressions	Agent, State change	SC	SCC
Infection(a1,a2,b1,b2,z1) In: INF($<a_2,b_2,z_1>$),SUS($<a_1,b_1,z_1>$) Out: INF($<a_2,b_2,z_1>$),E($<a_1,b_1,z_1>$)	a1, place: SUS→ EXP A2, unchanged	Infection MoveS	EndIncubation MoveE
EndIncubation(a1,b1,z1) In: EXP($<a_1,b_1,z_1>$) Out: INF($<a_1,b_1,z_1>$)	a1, place: EXP → INF	EndIncubation MoveE	Infection Recovery Death, MoveI
Recovery(a1,b1,z1) In: INF($<a_1,b_1,z_1>$) Out: REC($<a_1,b_1,z_1>$)	a1, place: INF → REC	Infection Recovery Death, MoveI	EndImmunity
Death (a1,b1,z1) In: INF($<a_1,b_1,z_1>$)	a1, place: INF→	Infection Recovery Death, MoveI	
EndImmunity(a1,b1,z1) In: REC($<a_1,b_1,z_1>$) Out: SUS($<a_1,b_1,z_1>$)	a1, place: REC→ SUS	EndImmunity	Infection MoveI
MoveS(a1,b1,z1,z2) Input: SUS($<a_1,b_1,z_1>$) Output:SUS($<a_1,b_1,z_2>$)	a1, attribute: $z_1 \to z_2$	Infection MoveS	Infection MoveS
MoveE(a1,b1,z1,z2) Input: EXP($<a_1,b_1,z_1>$) Output: EXP($<a_1,b_1,z_2>$)	a1, attribute: $z_1 \to z_2$	EndIncubation MoveE	EndIncubation MoveE
MoveI(a1,b1,z1,z2) Input: INF($<a_1,b_1,z_1>$) Output: INF($<a_1,b_1,z_2>$)	a1, attribute: $z_1 \to z_2$	Infection Death, MoveI Recover	Infection Death, MoveI Recover

[1] The notation $f_1.f_2$ used in `SNexpression` corresponds to the composition of functions: $f_1 \circ f_2$; S_A is a constant function returning the whole set A.

[2] The symbol $*$ is used in `SNexpression` to denote the intersection operator.

meaning that only the agent a_1 will need to recalculate the rates (which will be 0 in this case, for position z_1). Similarly, MoveS features also a self-SCC, due to the fact that the agent has moved to a new position and the result is that the same agent a_1 in position z_2 needs to recalculate the rate of MoveS (see Table 2).

4   Structural Relations Derived from the SSN Model

Table 1 reports, for each transition, its variables (color domain), its input/output places, the impact on one or more agents' states, and the transitions that may be in SC or SCC relation with it. The impact on the involved agent(s) state could be a change of one of its attributes: the place (where the unique token with the agent's identity is) or the value of an attribute (other values included in the tuple representing the agent: in our example the age and the position).

The change of place and the change of attributes can be derived directly from the net graph and by algebraic manipulation of the arc expressions and in particular of the formulae defining the SC and SCC relations. For each place p connected to transition t, $I(p,t) - O(p,t)$ allows to derive the agent(s) removed from p while $O(p,t) - I(p,t)$ allows to derive those added into p.

It is important to highlight the fact that the ESSN representation of an ABM is subject to a few constraints which are a consequence of the specific type of model, meant to represent an ABM behavior: the first constraint concerns the uniqueness of agent ids and the fact that in all reachable markings each agent can be present (as a token) in only one place. Structural analysis can also be used (through verification of marking invariants) to check the following property: if the constraint is satisfied in the initial marking, then it is also satisfied in any reachable marking. The second constraint concerns the type of state change associated with transitions and as a consequence of the expressions that may appear on the arcs: in our case the terms of the expression represent the agents involved in that action (note that one of them is the leader) and it is easy to identify those who change state (move to a new place or update some attribute), those who die and the newborn. So based on the source and destination places of such token-agents (input/output arc expressions) it is possible to identify which actions are surely disabled for the agent(s) involved in the action, and which could be potentially newly enabled and need to recompute their rates.

The SC and SCC relations identify the agents that are not directly involved in the action (they do not change state) but nevertheless could be affected by the state change induced by the current action (e.g. when a susceptible moves to a new zone, all infected in the old zone have one less potential target to infect, while all the infected in the new zone have one more potential target to infect).

Table 2 shows the results of dependency computation on the model of Fig. 1 from which the information can be extracted to be then (semi)automatically incorporated into the optimized scheduling algorithm. The table shows only a few examples, since the others have a very similar structure. The first column indicates the pair of transitions and the structural relation (SC or SCC) considered (selected from the third and fourth column of Table 1). The second column contains the expression indicating the affected agents, obtained as explained in

Table 2. Structural locality of effect. In case transition `Infection` is involved, the leader agent can be (a) the agent in `INF`, or (b) the agent in `SUS`.

Transitions, Relation	Affected Agents	Description
`MoveS, Infection,` `SC` Agent a_1 moves from z_1 to z_2	A) $<S_A, S_B, z_1>$ b) $<a_1, b_1, z_1>[z_1!=z_2]$	a) all agents in zone z_1 and place `INF` must recompute their rates; b) only the agent that just moved must recompute the rates;
`MoveS, Infection,` `SCC` Agent a_1 moves from z_1 to z_2	a) $<S_A, S_B, z_2>[z_1!=z_2]$ b) $<a_1, b_1, z_2>[z_1!=z_2]$	a) all agents in zone z_2 and place `INF` must recompute their rates; b) only the agent that just moved must recompute the rates;
`MoveS, MoveS, SC` Agent a_1 moves from z_1 to z_2	$<a_1, b_1, z_1>[z_1!=z_2]$	These two relations express the fact that the single agent involved in MoveS will not be in z_1 any longer (so after executing this instance of MoveS it will be no more enabled) but it'll be in z_2
`MoveS, MoveS, SCC`	$<a_1, b_1, z_2>[z_1!=z_2]$	so the rate of the new instance will have to be computed.
`Infection,` `Infection, SC`	a) $<S_A, S_B, z_1>$ b) $<a_1, b_1, z_1>$	a) all agents in zone z_1 and place `INF` must recompute their rates; b) only the agent that just moved to place E must recompute its rates;
`Infection,` `EndIncubation,` `SCC`	$<a_1, b_1, z_1>$	the agent that just moved in `EXP` must recompute its rates
`Infection, MoveE,` `SCC`	$<a_1, b_1, z_1>$	same as above
`Recovery,` `Infection, SC` Agent a_1 moves from place `REC` to `INF`	a) $<a_1, b_1, z_1>$ b) $<S_A, S_B, z_1>$	a) only the agent that just moved from `INF` to `REC` must recompute its rate; b) all agents in place `SUS` and zone z1 must recompute the rates of transition Infection

Sect. 3, where the variables refer to the first transition in the pair: this is the information used to generate the code in the last "foreach" loop in Algorithm 1. Column 3 explains the interpretation of the expression in column 2. Whenever transition `Infection` is involved (the only one involving two agents) two cases are considered, depending on which is the leader agent. Observe that all transitions except `MoveE` have an impact on transition `Infection` (since they change the marking of `INF` or of `SUS`) and depending on which is the leader agent for `Infection`, the state change will require more or less computational effort for updating the rates: when `MoveS` fires, for instance, it is more convenient that the leader agent be the one in `SUS`, when `Recovery` fires, on the contrary, it is more convenient that the leader agent be that in `INF`.

Now that we have understood how to calculate possible structural relations and determined which agents will have to recalculate their rates after a given action is performed, we propose in Algorithm 1 an improved version of the scheduler function defined in [4] for continuous-time ABM simulation that extends the concept of "rule locality" from the spatial description only to a more general definition capable of exploiting structural dependencies (in blue new/updated sections). The idea is to keep inside each agent an array to save the rates for any action it can carry on as active agent. The rate variable for a given action is set to -1 if it should be recalculated as the rule matches a transition instance that is newly enabled. Instead, it can be set to 0 if the rule matches a surely disabled transition instance.

The information about the fact that a rule for a given agent corresponds to a transition that is enabled or disabled comes from the net structural properties as well as from the input and output functions. Since we assume that each token is uniquely associated with an agent through its id, we cannot have multiple tokens in the net in different places representing different states of the same agent (i.e., states are mutually exclusive). When a token moves from a place to another

---

Algorithm 1: Optimized scheduler function for continuous-time ABM simulation

---

// *Initialization*
Create initial set of agents and setup their initial state
CurrentTime $\leftarrow$ 0
foreach agent e_j **do**
  // *rules correspond to transition instances*
  foreach rule I_h where agent e_j is leader **do**
    if The rule is enabled in the current state **then**
      $\gamma_{(e_j,I_h)} \leftarrow$ -1 // *The rate of rule I_h for agent e_j needs to be recalculated*
    else
      $\gamma_{(e_j,I_h)} \leftarrow$ 0 // *rule not enabled for the agent status/configuration*

// *Algorithm execution*
while CurrentTime $\leq$ FinalTime **do**
  // *Calculation of agents' individual rates*
  foreach agent e_j with $\gamma_{(e_j,I_h)} = -1$ for some I_h **do**
    foreach $\gamma_{(e_j,I_h)} = -1$ **do**
      Calculate, for the agent e_j, the rule rate $\gamma_{(e_j,I_h)}$
  Set the agent cumulative rate $\gamma_{e_j} \leftarrow \sum \gamma_{(e_j,I_h)}$
  $\gamma_{tot} \leftarrow \sum \gamma_{e_j}$ // *Calculation of the global rate*
  // *Selection of the next agent and rule according to their rates*
  $e^* \leftarrow$ roulette_wheel_selection(e_j, γ_{e_j})
  $I^* \leftarrow$ roulette_wheel_selection($I_h, \gamma_{(e^*,I_h)}$)
  Execute the rule I^* on agent e^* (and on other involved passive agents)
  foreach agent e_i and rule I_k having a structural relation with I^* **do**
    if The rule I_k is now enabled for the agent e_i **then**
      $\gamma_{(e_i,I_k)} \leftarrow$ -1 // *The rate of I_k for agent e_i needs to be recalculated*
    else
      $\gamma_{(e_i,I_k)} \leftarrow$ 0 // *rule not enabled for the agent status/configuration*
  // *Time update*
  CurrentTime $\leftarrow$ CurrentTime $+ \frac{1}{\gamma_{tot}} \cdot \ln\left(\frac{1}{random\_float(0,\,1)}\right)$

one, all the actions that could occur only in the source state cannot be executed anymore as referring to transition instances that are now disabled. For example, we have seen that when the Infection action takes place an agent moves from the SUS to EXP. If that happens, the agent will no more be entitled to participate in the MoveS action nor in Infection as those actions require that the agent be in place SUS.

Currently the method has not been fully automated due to the need to adapt some of the involved components that are illustrated in Fig. 2. The steps to complete automation require: the adaptation of the SNexpression tool to be integrated in the pipeline, the choice of the exchange format for the structural information required by the new translation algorithm, and the automatic generation of the additional variables and code. In the sequel we briefly discuss the feasibility of each task. The integration of the SNexpression tool (see Fig. 2 of [6] for more details) is a matter of wrapping the two libraries (SN framework and the Symbolic Calculus Library) in a new artifact that performs the preliminary structural analysis, i.e. it should select all possible pairs of transitions sharing some place and perform the computation of the SC or SCC transposed and composition with the input arc function for the leader agent. This generates the functions shown in the second column of Table 2. Observe that at this point the leader agent for each interaction transition should already be defined. We assume that appropriate data structures associating the transitions that *may be / are surely not* enabled when an agent of given type is in a given place are also derived from the structural analysis phase. Concerning the translation algorithm, it should produce code for creating the rule-indexed vectors (for each agent type) that are used to filter the subset of rules that need to update the rates at each iteration. In the code implementing a rule application it is necessary to store the value of the variables identifying the specific rule (and hence transition) instance just applied: this allows to produce a concrete implementation of the statement:

for each agent e_i and rule I_k having a structural relation with I^*

The rules having a structural relation with I^* are obtained in the first step together with the functions that can now be evaluated on the specific instance of I^* to provide the set of potentially affected agents and rules to be marked as targets of rate recomputation. Since the functions to be evaluated have the same form of the arc functions, the methods for their evaluation are already

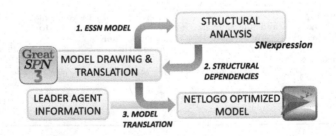

Fig. 2. The pipeline for optimized translation

implemented in GreatSPN; the result of the evaluation allows one to select the subset $\{e_i\}$ of agents that must set $\gamma(e_i, I_k) = -1$. For all the agents that have changed state as a consequence of the application of I^* the surely disabled rules can be identified by checking if the "place" variable has changed.

For what regards the structural analysis complexity, it is based on an expression rewriting system, and for specific instances the calculation may be very expensive. However, in the case of AB-ESSN models with the restrictions described above (unique id-colored tokens, places encoding state machines of agents), as class functions appearing on the arcs are just projections that may move tokens from one place to another and/or update some of their color elements, the expected complexity for computing the necessary structural relations is not critical. Furthermore, as the structural analysis does not depend on the initial marking, it is executed only once during the model translation. Translation from the Petri net model into the ABM is linear in the net dimensions (places, transitions, colors). ABM simulation time complexity is dominated by the rate recomputation, which is at most linear in the number of events. Of course, one does not want to recompute all rates of every event all the time, and recomputing only a minimal subset (the goal of this paper) reduces the computation time.

5   Results

To assess the potential advantages brought by the use of structural information we started by automatically generating the NetLogo [15] model of the SEIRS network in Fig. 1 using GreatSPN [2]. The generated model is already built to use the NETA simulation algorithm we presented in [4]. We note here that at the present stage the choice of the leader agent for every interaction is done at random by the algorithm, as structural information coming from SNexpression is not yet included in its current implementation. In the example model the leader agent for the Infection transition was the susceptible one.

Thus, we manually modified the NetLogo model to introduce the structural optimizations discussed in Sect. 4 to avoid unneeded rate recalculations.

We then run the two NetLogo models by increasing the initial number of *Susceptible* and *Infected* agents. We considered 100, 200, 400, and 800 agents, with a ratio (*Susceptible* vs *Infected*) of 9:1. The initial number of *Exposed* and *Recovered* agents was set to 0. For the initial displacement of agents we considered two different scenarios. In the first scenario (Scenario A) we supposed that all the agents were placed in the same initial position. So, in this first scenario the total number of enabled Infection instances should be at its maximum right at the beginning of the simulations.

In the second scenario (Scenario B) we supposed to partition our *Susceptible* agents in groups of 90 agents placed within the same position, and groups of 10 *Infected* agents placed in the same position, but with no overlap among the positions of *Susceptible* and *Infected* agents. So, for 100 agents we had 90 *Susceptible* in one position and 10 *Infected* agents in another position, whereas for 800 agents we had 8 groups of 90 agents in 8 different positions, and 8 groups of *Infected* agents in other 8 positions that were different from the positions of *Susceptible* agents. In this way, agents need to diffuse a little bit before we can observe the infection spread. The total number of positions was set to 30.

For each case-study we executed 100 runs with different random seeds. The experiments were executed by using 8 parallel threads exploiting the Netlogo BehaviorSpace feature on an Intel Core i9-9880H CPU (8 cores - 16 threads). Median execution times, interquartile ranges, max and min times (not outliers), and outliers for the two case-studies are presented in the boxplots in Fig. 3. A substantial reduction of the required computation time is observed in the optimized version, compared to the version without structural optimizations, in both scenarios. Absolute differences are more and more pronounced as the initial number of agents increases.

We then calculated the performance improvement in terms of time reduction using the formula $100 \cdot \frac{T_{old}-T_{new}}{T_{old}}$, where T_{old} represents the mean execution time without optimizations, and T_{new} represents the mean execution time in presence of the optimizations. The results in Fig. 4 show that the performance improvement tends to stabilize for cases ranging from 200 agents and upwards, with an advantage of about 75% and 83% for scenarios A and B, respectively.

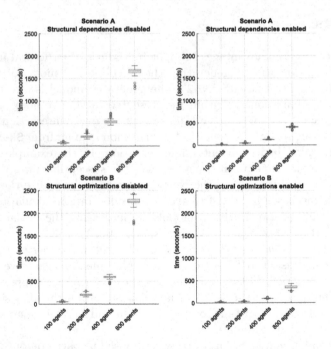

Fig. 3. Simulation times for the two tested scenarios

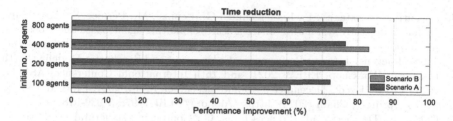

Fig. 4. Mean performance improvement (%) obtained for the two tested scenarios.

6 Conclusions

ABMs are increasingly adopted to model complex phenomena thanks to their expressive power, but the lack of clarity in their definition and formalization may lead toward results that are not consistent with other stochastic approaches, or even within different ABMs platforms.

We tried to fill this gap through a well-defined semantics for ABMs and supporting automatic generation of NetLogo code enforcing it; to improve the efficiency of the simulation algorithm in this paper we proposed to analyze the structure of the ESSN describing the domain model with SNexpression, a tool for structural analysis of (ES)SNs, that allowed us to detect structural relations between events and to substantially improve the execution times of an example model. The obtainable speed-ups may be influenced by the model type, the value of the rate parameters, the initial conditions, and the choice of the leader agents. However, the exploitation of this "locality" is beneficial for ABMs, as often in this context actions and interactions have strongly localized effects.

Structural optimizations have been implemented by hand here, so we plan as future work to fully automatize the generation of optimized Netlogo models by integrating SNexpression in the GreatSPN model generation workflow.

References

1. Ajmone Marsan, M., Balbo, G., Conte, G., Donatelli, S., Franceschinis, G.: Modelling with Generalized Stochastic Petri Nets. John Wiley & Sons Ltd. (1995)
2. Amparore, E.G., Balbo, G., Beccuti, M., Donatelli, S., Franceschinis, G.: 30 years of GreatSPN. In: Fiondella, L., Puliafito, A. (eds.) Principles of Performance and Reliability Modeling and Evaluation, pp. 227–254. Springer, Cham (2016). https://doi.org/10.1007/978-3-319-30599-8_9
3. Amparore, E.G., Beccuti, M., Castagno, P., Franceschinis, G., Pennisi, M., Pernice, S.: Multiformalism modeling and simulation of immune system mechanisms. In: Proceedings - 2021 IEEE International Conference on Bioinformatics and Biomedicine, BIBM 2021, pp. 3259–3266 (2021)

4. Beccuti, M., Castagno, P., Franceschinis, G., Pennisi, M., Pernice, S.: A petri net formalism to study systems at different scales exploiting agent-based and stochastic simulations. In: Ballarini, P., Castel, H., Dimitriou, I., Iacono, M., Phung-Duc, T., Walraevens, J. (eds.) Performance Engineering and Stochastic Modeling: 17th European Workshop, EPEW 2021, and 26th International Conference, ASMTA 2021, Virtual Event, December 9–10 and December 13–14, 2021, Proceedings, pp. 22–43. Springer, Cham (2021). https://doi.org/10.1007/978-3-030-91825-5_2

5. Capra, L., De Pierro, M., Franceschinis, G.: Computing structural properties of symmetric nets. In: Campos, J., Haverkort, B.R. (eds.) Quantitative Evaluation of Systems: 12th International Conference, QEST 2015, Madrid, Spain, September 1-3, 2015, Proceedings, pp. 125–140. Springer, Cham (2015). https://doi.org/10.1007/978-3-319-22264-6_9

6. Capra, L., De Pierro, M., Franceschinis, G.: SNexpression: a symbolic calculator for symmetric net expressions. In: Janicki, R., Sidorova, N., Chatain, T. (eds.) Application and Theory of Petri Nets and Concurrency: 41st International Conference, PETRI NETS 2020, Paris, France, June 24–25, 2020, Proceedings, pp. 381–391. Springer, Cham (2020). https://doi.org/10.1007/978-3-030-51831-8_19

7. Chiola, G., Dutheillet, C., Franceschinis, G., Haddad, S.: Stochastic well-formed coloured nets for symmetric modelling applications. IEEE Tran. Comput. **42**(11), 1343–1360 (1993)

8. Gillespie, D.T.: Exact stochastic simulation of coupled chemical reactions. J. Phys. Chem. **81**(25), 2340–2361 (1977)

9. Kreikemeyer, J.N., Köster, T., Uhrmacher, A.M., Warnke, T.: Inferring dependency graphs for agent-based models using aspect-oriented programming. In: 2021 Winter Simulation Conference (WSC), pp. 1–12 (2021)

10. Law, A.M.: Simulation Modeling and Analysis, Fifth Edition. McGraw-Hill (2015)

11. Pernice, S., et al.: A computational approach based on the colored Petri net formalism for studying multiple sclerosis. BMC Bioinform. **20**(6), 1–17 (2019)

12. Reinhardt, O., Uhrmacher, A.M.: An efficient simulation algorithm for continuous-time agent-based linked lives models. In: Proceedings of the 50th Annual Simulation Symposium, pp. 9:1–9:12. SCS International / ACM (2017)

13. Reinhardt, O., Warnke, T., Uhrmacher, A.M.: A language for agent-based discrete-event modeling and simulation of linked lives. ACM Trans. Model. Comput. Simul. (TOMACS) **32**(1), 1–26 (2022)

14. Voit, E.O., Martens, H.A., Omholt, S.W.: 150 years of the mass action law. PLoS Comput. Biol. **11**(1), e1004012 (2015)

15. Wilensky, U.: NetLogo (1999). www.ccl.northwestern.edu/netlogo/

Stability Analysis and Simulation of a Cellular Network with Retransmissions Policy

Taisiia Morozova$^{(\boxtimes)}$ (iD)

Uppsala University, Uppsala, Sweden
`taisiia.morozova@math.uu.se`

Abstract. We consider an uplink cellular network with static users and unlimited retransmissions. The users are assigned to the base stations (BSs) using the shortest distance association policy. The network cells are formed according to the Voronoi tessellation, and we study stability of this model with focus on a single cell. In particular, we consider a model with non-homogeneous users where the buffer size of each user depends on the number and locations of the active users at each time slot. We obtain a basic relation between input and output rate (coverage probability) of each user in steady-state regime. Moreover, we use stochastic simulation to verify sufficient stability conditions (obtained in the paper [8] for a more general system) which are reformulated in terms of the model under consideration. In particular, we find that these conditions turn out to be quite close to stability criteria in the most realistic case of the heavily loaded cell. In this regard and because of analytical unavailability of some metrics, we empirically study the convergence of the stability zone of the lightly-loaded cell to the zone defined by the sufficient stability conditions, when the load increases.

Keywords: cellular network · performance evaluation · stability analysis · Markov process · stochastic simulation

1 Introduction

We consider a no-loss uplink cellular network model in slotted time. One of the main features of the model is the signal retransmissions, when the signals are automatically re-sent through the same channel in case of damage or loss. This feature usually presents in protocols and serves as a tool to increase reliability and reassure that no data is lost. (The most common example of such system is the TCP protocol [16].) The stored signals thus form a buffer of an unlimited capacity (we do not limit the number of retransmissions). In this work we examine the stability conditions of such a model where the transmission rates are influenced by the interference. Although in general, under stability, a large buffer size may happen sometimes, nevertheless the fulfillment of the stability conditions prevents its dramatic increase. The problem of seeking of stability conditions is one of the most required and often the most challenging problem in the analysis of the modern communication systems [7,8]. It is therefore of our interest to analyze the buffer behaviour and derive conditions for the system's stability. We consider the general setting of the network model with the *non-homogeneous* users, where the buffer size of a single user depends on the states of other users (whether they are active or not). Such a setting is well-motivated (for example, in [7,17]) and explained in details in Sect. 2. We then focus on a fixed user i in a single cell and analyze the behaviour of its buffer.

The contribution of this research includes the two main parts. First, we derive a basic relation connecting the input and output rates of user i in steady-state regime, implying the necessary stability condition of each user (Lemma 1). Second, based on work [8], we adapt and then verify by simulation the sufficient stability condition for the i-th user. In particular, we demonstrate by simulation that the sufficient stability conditions obtained in [8] turn out to be very close to stability criteria when the traffic load of each user is close to one. We also discuss briefly a problem of underestimating the stability zone when not all users are fully loaded. In this regard, in the last section, we discuss how to apply the approach obtained in [8] to find the stability zone of the system in a general case.

It is worth mentioning that in general a direct application of the sufficient stability condition from [8] to our model is impossible because a required monotonicity property of the service allocation (the set of the state-dependent departure rates) is violated. In this regard, in Sect. 5 we outline the method from [8] and mention that it can be modified to overcome a non-monotonicity of the service allocation in our model. On the other hand, we show how the sufficient stability condition (5) (from paper [8]) may be obtained by comparison with the corresponding classical queuing model, which is also a contribution of this research.

1.1 Related Work

Similar systems have been investigated in many works, including [1,3,4,11,12,19], where the authors focus mainly on the coverage probability and coverage rate, as well as the network performance analysis. These works consider consecutive transmissions where each transmitter is allocated a certain time slot, and therefore the interference is only caused outside the cell and not inside. Moreover, in most of the mentioned research the networks allow for signal losses. That is, if the signal is not transmitted, it is lost with no possibility to resend. Networks with retransmissions are considered in [2,15,18] where the authors perform the analysis of the coverage probability, as well as outage (failure) probability and the average symbol error probability. The key difference with the model we observe is that the authors consider a downlink system with the limited number of retransmissions implying possible losses of the data. In our work, we do not allow for the signal loss. The dependence between the transmitters in general should be taken into account, since the transmission probability is different for each transmitter. This has been done in some recent works including [6–8,17].

In [17] the authors consider an infinite grid with a pair transmitter-receiver with no distance between them, and use some approximations in order to establish stability criteria. In [7] and [8] the authors consider abstract interfering queues with state-dependent departure rates and derive the stability criteria for each of the queues. As we will show the approach from [8] turns out to be the most suitable to analyze stability of the model under consideration.

The paper is organized as follows. In Sect. 2 we give definition of uplink network model and the interference and comment briefly a dependence between the users. In Sect. 3, we derive a basic relation between the input rate and the coverage probability and then deduce the necessary stability conditions. In Sect. 4 we present some simulation results in order to support the theoretical findings. In particular, we show that the sufficient stability condition we study are very close to be stability criteria when the network is heavily loaded. Finally in Sect. 5 we discuss a potential for future research and briefly describe the approach for obtaining more accurate sufficient stability conditions for the system with dependent users based on the work [8].

2   System Model

In this section we describe the mathematical model of a cellular network to be used in the further analysis. We consider a network in $R \times R =: R^2, R = [0, \infty)$, with retransmissions and unlimited buffer in slotted discrete time. Both users and base stations (BSs) are randomly and independently distributed on a R^2 plane according to Poisson point processes, with respective intensities λ_u and λ_b (this is a standard setting, see [4]). The network area is split into cells according to Voronoi tessellation, with one BS per cell. The signals are transmitted from the users as senders to the BSs as receivers, according to an uplink policy. The number of newly arrived signals for the user i per time slot n, $A_n^{(i)}$, is assumed to follow Binomial distribution with parameters 1 and p_i. Thus p_i is referred to as *input rate* or a data rate of the i-th user. We consider a Rayleigh fading policy where the signal power S_i of a user i follows exponential distribution with mean $1/\mu_i$ and is assumed to be independent from the other users. If the signal strength is not enough to get the signal through (due to the interference or a long travel distance), the transmission fails. Each user i has a buffer of infinite capacity where the non-transmitted signals are stored to be re-sent in the next time slot. Each user may attempt to re-transmit only once during a single time slot and the total number of retransmissions is not limited.

The buffer size in slot n is determined by the number of failed transmissions which is in turn determined by the transmission probability (also referred to as *coverage probability*). Let N be the total number of users in the cell. Let also $J_j(n)$ be a (random) *configuration* of j active users (or *active configuration*) in slot n and, respectively, $C_j(n)$ be the set of all possible configurations. (For instance, configurations $\{1, 2\}$ and $\{3, 5\}$ are different and belong to the set $C_2(\cdot)$.) Note that these configurations take into account different locations of the active users as well. Let also $J_j^{(k)}(n)$ be fixed k-th configuration from the (ordered) set $\mathcal{C}_j(n)$, that is

$$\mathcal{C}_j(n) = \{J_j^{(1)}(n),\ J_j^{(2)}(n),\dots,J_j^{(C_N^j)}(n)\}$$

(C_N^j is the number of j-combinations from N users). Let also $P_c^{(i)}(\mathbf{0})$ and $P_c^{(i)}(\mathbf{N})$ be the configurations where only user i is active and, respectively, all N users are active in the stationary regime. We will use these notation below in Sect. 3.

Now we define the interference $I^{(i)}(n)$ of user i in a single cell seen by a typical BS in the n-th time slot as

$$I^{(i)}(n) = \sum_{l \in J(n)} S_l a(x_l),\quad l \neq i,$$

where $J(n)$ is a current (random) configuration of active users in the n-th slot and

$$a(x_l) = \frac{1}{|x_l/r_0|^\beta}, \tag{1}$$

is a pathloss function, $x_l = \{x_1^{(l)}, x_2^{(l)}\}$ is the l-th user's location on a R^2 plane and $\beta > 2$ is a pathloss exponent [4]. For simplicity and without loss of generality we assume a typical BS to be located at the origin. As can be seen from (1), we allow a scaling of an attenuation function with a circle centered in $z = 0$ of a radius r_0. (For details see [10,13].) The coverage probability of the i-th user in slot n is then defined as

$$P_c^{(i)}(n) = \mathsf{P}(\mathrm{SINR}^{(i)}(n) > T) = \mathsf{P}\left(\frac{S_i a(x_i)}{I^{(i)}(n) + \sigma^2} > T\right)$$

$$= \mathsf{E}\left[\exp\{-\mu_i T |x_i/r_0|^\beta(\sigma^2 + \sum_{l \in J(n)} S_l a(x_l))\}\right]$$

$$= e^{-T\sigma^2 \mu_i |x_i/r_0|^\beta} \prod_{l \in J(n)} \frac{1}{1 + T|x_l/x_r|^\beta}, \tag{2}$$

where $\mathrm{SINR}^{(i)}(n)$ is a classic signal-to-noise ratio of user i in slot n, T is a pre-defined deterministic critical threshold, σ^2 is a noise constant. We stress that the interference and hence the coverage probability depends on the current configuration but we omit this notation for simplicity.

Remark 1. Note that when all users (except i) are idle the interference is zero and thus $P_c^{(i)}(\mathbf{0})$ can be expressed only in terms of x_i and μ_i as

$$P_c^{(i)}(\mathbf{0}) = \mathsf{P}\left(S_i > \frac{T\sigma^2}{a(x_i)}\right) = e^{-T\sigma^2 \mu_i/a(x_i)}.$$

In this work we focus on the stability conditions of the described network. We consider a general model where the buffers of users are dependent through the influence of the inter-cell interference generated by currently active users. We describe the dynamics of the (dependent) buffers as a multi-dimensional Markov chain (MC). The approach we use to analyze this MC is mainly about defining a balance equation for the input/output load in steady-state.

Remark 2. In some works (see for example [3,9]), the homogeneity assumption between users is often made to reach a better tractability. However in real systems the transmissions of the signals indeed correlate and depend on the states of other users (active or idle), see [17]. Later in Sect. 4 we will also show empirically that despite the locations of users being independently and identically distributed, their respective buffers differ significantly.

3   Stability Analysis

In this section we focus mainly on the behaviour of buffer size of the user i and perform the stability analysis of the users buffers. We denote by $B_n^{(i)}$ the number of stored signals in the buffer of user i in the n-th slot. The dynamic of the buffer process of the i-th user is

$$B_n^{(i)} = B_{n-1}^{(i)} + A_n^{(i)} - 1_{\{A_n^{(i)} + B_{n-1}^{(i)} \geq 1\}} 1_{\{S_i > T(\sigma^2/\mu + I^{(i)}(n))/a(x_i)\}}, \qquad (3)$$

where $1_{\{.\}}$ denotes the indicator function. Note that the use of indicators is motivated by the model property where only one transmission per slot is allowed. Therefore, $B_n^{(i)}$ is a jump process with a unite size of a jump. It is easy to see that the process

$$\mathcal{B}_n = \{B_n^{(1)}, \ldots, B_n^{(N)}\}, \; n \geq 1, \qquad (4)$$

is a multi-dimensional Markov chain (MC). Note that MC (4) is aperiodic and irreducible since both arrival probability p_i and departure (i.e. coverage) probability are positive at every slot. As a result, each state of the process can be reached from any other state with a positive probability implying irreducibility, while the aperiodicity of any state (some state is enough) follows since both p_i and transmission probability belong to the open interval $(0, 1)$. By the arguments above, the two indicators terms on the r.h.s. of (3) are dependent. Let us also assume that the transmission rate is insensitive to the buffer sizes of other users. Thus the buffer size itself does not contribute to the interference but indicates the state of the user, that is if the user is active or not. The transmission rate in the n-th slot is then affected by the interference coming from all users in the cell that are active in that slot.

Remark 3. It is evident and follows from the construction that the coverage probability is partially decreasing with the number of active users and it achieves its maximum when all users (except given user i) are idle, and minimum when all users are busy.

Now we formulate the sufficient stability conditions for our model adapted from the work [8] and then prove a fundamental relation connecting input and departure rates for each user in the stationary state which in turn implies the necessary stability condition. In terms of our notation, the sufficient condition from [8] for user i is

$$p_i < P_c^{(i)}(\mathbf{N}), \ i = 1, \ldots, N, \tag{5}$$

where the r.h.s. corresponds to the transmission probability of a user i in a *fully busy* system (that is, when the number of active uses coincides with the total number of users N). To explain conditions (5), we may consider a classic system with fixed departure probability $P_c^{(i)}(\mathbf{N})$. Then this new system is stable and hence its buffer size is a *tight process*. At the same time, it easily follows from a monotonicity property that the buffer size (of each user) in the new system dominates the buffer size $B_n^{(i)}$ in the original system. Then it is immediate that $B^{(i)}$ is a tight process, which is the definition of a stable system in [8]. Later in Sect. 4 we will verify condition (5) and analyze its precision using stochastic simulation.

It is worth mentioning that both necessary and sufficient conditions are important for stability analysis, but we will mainly use the latter in order to verify stability in simulation section. This is motivated by the fact that verification of the necessary condition can only show an instability zone, whereas the sufficient one can be used to establish stability. Therefore, the use of the sufficient condition in simulations appears to give more informative results.

Lemma 1. *Consider the system in steady-state. Then the input rate of user i is upper bounded as*

$$p_i < \max_j \max_{k \in \mathcal{C}_j} P_c^{(i)}(J_j^{(k)}) = P_c^{(i)}(\mathbf{0}), \ i = 1, \ldots, N. \tag{6}$$

Proof. Let $A^{(i)}(n)$ denote the total number of signals arrived to the (arbitrary fixed) user i in interval $[1, n]$. Let also $D^{(i)}(n)$ be the total number of departures from user i in the same interval. (In the further analysis we will address successfully transmitted signals as *departed* from the system.) Then the balance equation for the arrived traffic yields

$$A^{(i)}(n) = D^{(i)}(n) + B_n^{(i)}, \ n \geq 1. \tag{7}$$

Let $1_j^{(i)}(k, n)$ be the indicator function such that

$$1_j^{(i)}(k, n) = \begin{cases} 1, & \text{if } \text{SINR}^i(n) > T \text{ and } J_j(n) = J_j^{(k)}(n); \\ 0, & \text{otherwise.} \end{cases} \tag{8}$$

Using the strong law of large numbers (SLLN), we obtain

$$\frac{A^{(i)}(n)}{n} \to p_i, \ n \to \infty. \tag{9}$$

Now, for the total number of departures, we can write

$$D^{(i)}(n) = \sum_{l=1}^{n} \sum_{j=1}^{N} \sum_{k \in \mathcal{C}_j(l)} 1_j^{(i)}(k, l) 1_{\{A_l^{(i)} + B_{l-1}^{(i)} \geq 1\}}, \ n \geq 1. \tag{10}$$

It is important to note that we can no longer apply the SLLN to the sum above, as the two indicator terms are in general dependent. To show this, assume that in the first time slot a user obtains a signal but fails to transmit it, and the signal is stored in the buffer. In the next time slot, the same user does not obtain a new signal but remains active and therefore contributes to the interference. Thus, the indicator of transmission depends on the indicator of the user's activity. Because the basic MC $\{\mathcal{B}_n\}$ is aperiodic and irreducible, then in steady state the following limit, as $n \to \infty$,

$$\frac{D^{(i)}(n)}{n} \to \sum_{j=1}^{N} \mathsf{P}\left(\mathrm{SINR}^i(n) > T, J_j = J_j^{(k)}, A^{(i)} + B^{(i)} \geq 1\right)$$

$$= \sum_{j=1}^{N} \mathsf{P}\left(\mathrm{SINR}^i(n) > T | J_j = J_j^{(k)}, A^{(i)} + B^{(i)} \geq 1\right) \mathsf{P}(J_j = J_j^{(k)}, A^{(i)} + B^{(i)} \geq 1)$$

$$= \sum_{j=1}^{N} \sum_{k \in \mathcal{C}_j} P_c^{(i)}(J_j^{(k)}) P_a^{(i)}(J_j^{(k)}). \tag{11}$$

exists, where $P_a^{(i)}(J_j^{(k)})$ denotes the stationary probability that j active users compose configuration $J_j^{(k)}$, and $P_c^{(i)}(J_j^{(k)})$ is the stationary coverage probability of the user i under this configuration. Recall that we consider the system in steady state, which implies *positive recurrence* of the process $B^{(i)}$, and in particular it implies $B_n^{(i)} = o(n)$ [5,14]. It then follows from (7) that

$$\lim_{n\to\infty} \frac{A^{(i)}(n)}{n} = \lim_{n\to\infty} \frac{D^{(i)}(n)}{n},$$

which, together with (7), (9) and (11), gives

$$p_i = \sum_{j=1}^{N} \sum_{k \in \mathcal{C}_j} P_c^{(i)}(J_j^{(k)}) P_a^{(i)}(J_j^{(k)}). \tag{12}$$

Let J_0 denote the *zero configuration* when all users are idle. Because

$$\sum_{j=1}^{N} \sum_{k \in \mathcal{C}_j} P_a^{(i)}(J_j^{(k)}) + P(J_0) = 1 \tag{13}$$

and, by the assumed steady state, $\mathsf{P}(J_0) > 0$, then relation (6) follows directly from (12) and (13), and the proof is completed.

Remark 4. Note that in (12) the stationary distributions $\{P_a^{(i)}(J_j^{(k)}), k \in \mathcal{C}_j\}$ for each $j = 1, \ldots, N$, are analytically unavailable, while the explicit condition (6) is quite intuitive. Moreover, if we assume $p_i > P_c^{(i)}(0)$ then it is easy to show that the buffer size $B_n^{(i)} \to \infty$ a.s., implying the necessary stability condition in the form $p_i \leq P_c^{(i)}(0)$.

It is worth mentioning that, as we show in Sect. 5, the stability zone of the system is quite sensitive to configuration of the active users, which explains the presence of the configurations in the stationary distributions $\{P_a^{(i)}(J_j^{(k)}), k \in C_j\}$ in our general setting. However, in some types of networks it may happen that such dependence is weak enough to be neglected, in which case representation (12) is simplified as follows:

$$p_i = \sum_{j=1}^{N} P_c^{(i)}(j) P_a(j), \quad i = 1, \ldots, N, \tag{14}$$

where, evidently, $P_a(j)$ denotes the stationary probability of j active users (regardless of the configuration), and $P_c^{(i)}(j)$ denotes the stationary coverage probability of user i given that j users are active. We assume that relation (14) might work in some types of dense systems where the difference in distances between the users becomes more smooth as the number of transmitters increases. It is clear that the insensitivity of stability to the concrete configuration considerably reduces computation efforts to analyze the system.

4   Simulation Results

In this section we present some numerical and simulation results to verify the accuracy of the stability conditions (5) under different input rates of the users. In (6) we have shown the necessary stability condition for user i to be its highest departure rate. It is however more practical to verify the results related to sufficient conditions, and by this reason we verify by simulation the accuracy of condition (5). We will use discrete-time stochastic simulation to model the behaviour of the buffers in a described system and verify condition (5). For the first experiment, we set the network parameters as follows

$$\lambda_u = 450, \ \lambda_b = 50, \ \sigma^2 = 1, T = 0.5, \beta = 4, n = 1000, \mu = 0.5, \tag{15}$$

which gives on average 50 cells with expected number of users in the cell $\mathrm{E}N = \lambda_u/\lambda_b = 9$. Finally, we set the input rates p_i to be equal to the r.h.s. of (5), that is

$$p_i = \{0.002, 0.002, 0.11, 0.58, 0.002, 0.02, 0.044, 0.002, 0.21\},$$
$$i = 1, 2, \ldots, 9. \tag{16}$$

In order to perform the simulation, we use the classical discrete-time Monte-Carlo method with the statistical software R. We run the simulation and compute the users buffers in a fixed cell. For more informative results, we construct the plots averaged over 50 buffer size realizations (for each user), see Fig. 1.

Note that 'stability' of user i in [8] means that the buffer size process $\{B_n^{(i)}\}$ is *tight* (as a component of the positive recurrent N-dimensional MC $\mathcal{B}$). Therefore, a visualization of the (averaged) buffers dynamics can be used to confirm stability/instability of the user. On Fig. 1, the buffer size of each user in the cell is plotted as a function of time n. We see on the plots that all buffers seem to be stable, and no long-term decrease of the dynamics is visible. While we present on Fig. 1 the results for a fixed cell with $N = 9$ users, the experiment was conducted for over a hundred cells with different number of users, and for each cell the same effect as on Fig. 1 was observed. We thus conclude that condition (5) is indeed sufficient for the stability and works well for the observed model.

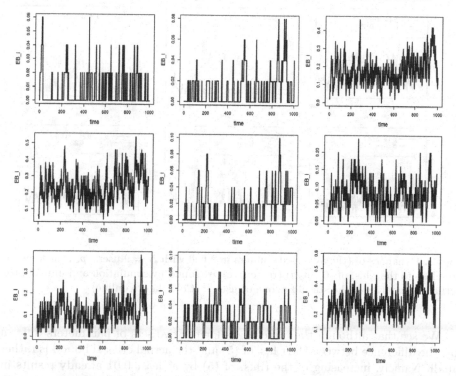

Fig. 1. Averaged buffer size for each user $i = 1, \ldots, N$ in a fixed cell (with $N = 9$ users) vs. time (x-axis).

The authors of [8] mention that condition (5) most probably underestimates the stability region. That is, setting the input rates $\{p_i\}$ in accordance with (5), might lead to losing a significant part of the stability region and thus result in a decrease in network capacity efficiency. Therefore, an important question is how close the r.h.s. of (5) is to the actual stability threshold. In this regard, we use stochastic simulation again to evaluate the 'real' stability bound of each user. For the next experiment, denote as $p_c^{(i)}$ the empirical critical threshold of user i and compare it with p_i satisfying (5). We run the simulation, keeping the network setting as in (15), and compute the buffers dynamics. The results are shown on Fig. 2, where the buffer dynamics is plotted as a function of time. Note that for stability analysis it is important to establish the non-increasing dynamics of the buffer while the buffer size itself does not play a significant role.

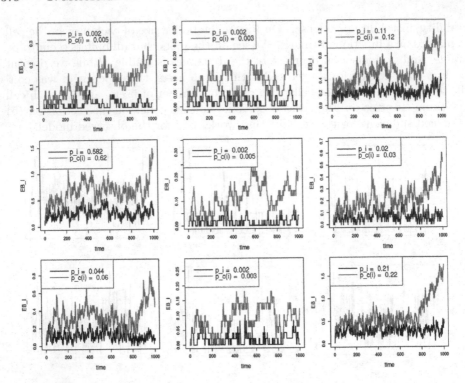

Fig. 2. Dynamics of (the averaged) buffers in a cell with $N = 9$ users: p_i (black curve) are close to the r.h.s. of (6); $p_c(i)$ (red curve) are found by simulation and indicate the actual boundaries of the stability zone of user i. (Color figure online)

We see that although the buffer sizes for the two sets of parameters differ a lot, the difference between the r.h.s. of (5) and the actual critical bound is rather small. Namely, increasing of the r.h.s. of (5) by at least 0.01 already results in violation of stability. This indicates that the condition (5) appears to be very close to the stability criteria.

At first sight, the observed effect seems unexpected because, as follows from Sect. 3 (as well as Theorem 4 in [8]), condition (5) is not a stability criteria. As we demonstrate below, the actual reason for this effect is that the coverage probability of each user is very sensitive to the states of other users (active/idle). Thus, if we choose the values of p_i for each user to be very close to its stability bound (5), the corresponding users appear to be almost at the threshold of stability.

Before we get to the experiment setting, let us clarify the notation used later. We address the users, for which the input rate is set close to $P_c^{(i)}(\mathbf{N})$ as *highly loaded* users. On the other side, if we set p_i to be much less than the r.h.s. of (5), for example, $p_i = P_c^{(i)}(\mathbf{N})/2$, it will result in a lower load, and respectively, lower activity time fraction of user i. Further in this section we address the users with such setting as *lightly loaded*.

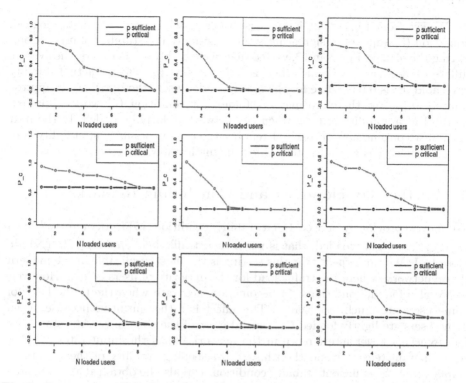

Fig. 3. Convergence of the boundary of real stability zone for each user obtained by simulation (red line) to the boundary of stability zone defined by the inequalities (5) (black line), as the number of 'highly loaded' users increases. (Color figure online)

For the next experiments, we set the parameters as follows. First, we fix the user i and set its input rate to be very close to the r.h.s. of (5) (unlike previous experiments where we used this setting for all users). For the rest of the users we set

$$p_j = P_c^{(j)}(\mathbf{N})/2, \ j \neq i, \tag{17}$$

that is, all users except for user i are set to be lightly loaded. Let $p_c^{(i)}(k)$ be the critical threshold of the stability zone for user i given that k users are highly loaded. (Note that, for setting (17), $k = 1$.) Note that in the further experiments k loaded users means that users $1, 2, \ldots, k$ are loaded, that is, we always use an identical permutation. We then compute the average buffer of user i and empirically find the critical stability threshold $p_c^{(i)}(k)$. We then repeat the experiment with increased k and, again, find the new $p_c^{(i)}(k)$. Thus, for 9 users we have a vector $p_c^{(i)} = \{p_c^{(i)}(1), p_c^{(i)}(2), \ldots, p_c^{(i)}(9)\}$. Repeating the procedure for each user in the cell we obtain 9 such vectors, the results are given on Fig. 3.

We see from Fig. 3 that in the case where not all users are highly loaded, condition (5) appears to significantly underestimate the stability of users. Comparing the plots in Fig. 2 and 3, we see that, with highly loaded users, the average difference between the bounds (between the r.h.s. of (5) and the actual stability threshold p_c^i) is 0.009 while for the lightly loaded users the difference is approximately 0.14. We therefore may conclude that condition (5) works (approximately) as a stability criteria when the network is heavily loaded. In the next section we outline an approach of refining the obtained sufficient stability conditions, which has a good potential for the further work.

5   Further Development and Concluding Remarks

We showed previously in Sect. 4 that stability conditions (5) work quite well when all users are highly loaded, that is when their traffic loads $\rho_i =: p_i/P_c^{(i)}(\mathbf{N})$ are close to 1. We can expect that this setting is the most applicable in the modern dense networks where the traffic load for the majority of users is high. However, as we also found, conditions (5) become less accurate when the traffic loads of some users are much less than 1. This effect is rather intuitive because, when some users are lightly loaded (i.e., spend most of the time in the idle state), we can expect a lower interference in the network. Thus the input rates of other users may increase beyond the r.h.s. of (5) keeping stability of these users. It implies that the sufficient stability conditions can also be obtained as a weighted sum of the coverage probabilities similar to (12) using the stationary distribution of the configurations. However, this distribution and in particular activity and busy probabilities of users are in general unknown, which complicates the stability analysis. For this case, the authors of [8] propose an approach which in general allows to overcome this problem. The idea of this method is to use simulation and further estimation of the stationary probability of the corresponding configurations, where the number of stable users gradually increases. In brief, in terms of our model and network parameters, this approach can be described as follows (the corresponding example for 3 users is given in [8]). First for a fixed (for example identical) permutation of all users we assume that all N users are 'saturated' (that is, infinitely loaded) and find stability condition for user 1 based on the well-known negative drift condition (and some argumentation discussed earlier in this work)

$$p_1 < P_c^{(1)}(\mathbf{N}). \tag{18}$$

Then, assuming that user 1 is now stable under (18) (and $N \geq 2$), we assume that users $2, ..., N$ are saturated and then find the stability condition of user 2, again using negative drift condition and the stationary distribution of the 1st user (idle or busy) which is explicitly available from condition (18). Starting with user 3, the stationary distribution of the buffers becomes analytically unavailable and we find the sufficient stability condition (as a negative drift) for user $i + 1$ using the estimates of the joint stationary distributions of the buffers of users $1, ..., i$, found on the previous steps. This algorithm is then repeated until the stability conditions for all users are found. In general this procedure can be applied to each permutation of the set of users and the maximal stability region is obtained as the union of the corresponding stability regions.

It is worth mentioning that in order to apply the described approach, it is essential that the service allocation satisfies a monotonicity property (in our setting, the departure rates are decreasing in the number of active users). However in general our model does not satisfies this property because of a high impact of the users locations. By this reason, in a future research, we plan to modify this approach in such a way to overcome this non-monotonicity to reduce the difference between stability zones demonstrated by Fig. 3.

6   Conclusion

In this work we observe a cellular network model with uplink traffic and retransmissions where the key feature is the non-homogeneous users with state-dependent departure rates. We perform stability analysis of the system, namely, we formulate and prove the necessary stability condition for each user. We also perform a simulation study in order to verify the sufficient conditions obtained in [8] and observe a good coincidence between the theoretical and empirical results for highly loaded network (cell). Moreover, we show via the simulation that the sufficient condition indeed works as a stability criteria in a heavily loaded system, which gives a good potential for applications since the modern dense networks are set to be heavily loaded. Finally, we outline a further work (again based on [8]) which concerns the refinement of the sufficient stability conditions in order to delimit more precisely the stability zone when the network is not highly loaded.

References

1. AL-jaff, D., Bayat, O., Sallomi, A.H.: Analysis of interference and signal quality in cellular wireless networks (2020)
2. Afify, L.H., ElSawy, H., Al-Naffouri, T.Y., Alouini, M.: Unified stochastic geometry model for MIMO cellular networks with retransmissions (2016)
3. Andrews, J.G.: A primer on cellular network analysis using stochastic geometry (2016). ArXiv: 1604.03183
4. Andrews, J.G., Baccelli, F.: A tractable approach to coverage and rate in cellular networks. IEEE Trans. Commun. 59(11), 3122–3134 (2011)
5. Asmussen, S.: Applied Probability and Queues, 2nd edn. Springer, New York (2003). https://doi.org/10.1007/b97236
6. Banerjee, S., Sankararaman, A.: Ergodicity and steady state analysis for Interference Queueing Networks, p. 12 (2020). arXiv:2005.13051
7. Bonald, T., Borst, S., Hegde, N., Jonckheere, M., Proutiere, A.: Flow-level performance and capacity of wireless networks with user mobility. Queue. Syst. 63, 131–164 (2009). https://doi.org/10.1007/s11134-009-9144-7
8. Borst, S.C., Jonckheere, M., Leskelä, L.: Stability of parallel queueing systems with coupled service rates. Disc. Event Dyn. Syst. 18(4), 447–472 (2010)

9. Dhillon, H.S., Novlan, T.D., Andrews, J.G.: Coverage probability of uplink cellular networks. In: IEEE Global Communications Conference (GLOBECOM). Anaheim, CA, USA, pp. 2179–2184 (2012). https://doi.org/10.1109/GLOCOM.2012.6503438
10. Kaj, I., Morozova, T.: Retransmission performance in a stochastic geometric wireless communication model (2023)
11. Min, S., Yu, S.K.: Downlink capacity and base station density in cellular networks. In: Proceedings of the 11th International Symposium on Modeling and Optimization in Mobile, Ad Hoc, and Wireless Networks (WiOpt) (2013)
12. Liu, Q., Zhang, Z.: The analysis of coverage probability, ASE and EE in heterogeneous ultra-dense networks with power control. Dig. Commun. Netw. **6**(4), 524–533 (2020). ISSN 2352–8648. https://doi.org/10.1016/j.dcan.2020.02.002
13. Morozova, T., Kaj, I.: Analysis of a typical cell in the uplink cellular network model using stochastic simulation. In: Proceedings of CITDS (2022)
14. Morozov, E., Steyaert, B.: Stability Analysis of Regenerative Queueing Models. Springer, New York (2021). https://doi.org/10.1007/978-3-030-82438-9
15. Nigam, G., Minero, P., Haenggi, M.: Cooperative retransmission in heterogeneous cellular networks. In: IEEE Global Communications Conference, pp. 1528–1533 (2014)
16. Rajhi, M., Madkhali, H.: Comparison between automatic repeat request (ARQ) protocols and solving the buffer problem. In: International Conference on Computational Science and Computational Intelligence (CSCI) 2020, pp. 195–201 (2020). https://doi.org/10.1109/CSCI51800.2020.00040
17. Sankararaman, A., Baccelli, F., Foss, S.: interference queueing networks on grids (2020)
18. Go, Y., Kune, D.F., Woo, S., Park, K.S., Kim, Y.: Impact of malicious TCP retransmission on cellular traffic accounting (2013)
19. Zhang, X., Andrews, J.G.: Downlink cellular network analysis with multi-slope path loss models. IEEE Trans. Commun. **63**(5), 1881–1894 (2015)

Performance Evaluation of Global Trust Management in MANETs Routing Protocols

Hassan Jari[1,2]([✉]) and Nigel Thomas[1]

[1] Newcastle University, Newcastle upon Tyne, UK
h.a.m.jari2@ncl.ac.uk
[2] Jazan University, Jizan, Saudi Arabia

Abstract. This paper presents a comprehensive trust management strategy for MANETs, concentrating on developing, assessing, and comparing global trust management mechanisms for the Ad-hoc On-demand Distance Vector (AODV) routing protocol. The research introduces a novel trust-based routing protocol called GTAODV, which builds upon the standard AODV protocol. GTAODV incorporates direct, indirect, and global trust mechanisms to evaluate the reliability of nodes when forwarding packets.

This study aims to contribute to ongoing efforts in creating more secure, robust, and efficient ad-hoc networking solutions by focusing on the performance of MANET routing protocols when subjected to black hole attacks. Using Network Simulator 3 (NS-3), the performance of AODV and GTAODV is evaluated and compared. The experimental results demonstrate the effectiveness of the GTAODV protocol in countering black hole attacks, highlighting its potential to enhance the security and reliability of MANETs.

Keywords: MANETs · AODV · Routing protocol · NS-3 · Direct trust · Indirect trust · Global trust

1 Introduction

Mobile Ad-hoc Networks (MANETs), also known as wireless ad-hoc networks, represent an advanced technology that operates without fixed infrastructure or centralised management for communication purposes. A MANET is typically characterised as a mobile network composed of numerous independent nodes, often consisting of mobile devices or other mobile nodes that can self-organise in various configurations and function without strict administrative guidelines. These nodes can join or leave the network at any moment, leading to a constantly changing network topology [8]. In a MANET, each node serves as both a router and a host, necessitating cooperation among nodes to transmit data packets from the source node to the destination node [5]. As a result, these nodes must collaborate to facilitate data transmission across the network, further highlighting the importance of establishing trust and addressing security concerns within MANETs. MANETs employ a variety of protocols, each with its advantages and disadvantages, including the Ad-hoc On-Demand Vector Routing protocol (AODV) and Optimised Link State Routing Protocol (OLSR).

MANETs prove valuable in situations where setting up a conventional wireless network is challenging or unfeasible, such as during disaster relief efforts, military operations, or remote locations. They are also advantageous for Internet of Things (IoT) applications, where devices may be dispersed and need to communicate with each other without relying on a central server. Research into MANETs has recently gained momentum due to their numerous potential applications in everyday life.

1.1   Classification of MANET Routing Protocols

Routing principles categorise protocols based on the features of the methods employed for updating routing data and executing routing schemes. Based on these criteria, three main types of routing protocols can be identified: reactive (on-demand), proactive (table-driven), and hybrid protocols [9]. The following offers an in-depth overview of these three categories of routing protocols.

Proactive Routing Protocols. Proactive routing protocols, also referred to as table-driven routing protocols, consistently maintain updated routing information for all nodes in the network by regularly broadcasting this data throughout the network. Each node stores this information in routing tables, enabling fast and effective routing decisions. Proactive routing protocols require every node to maintain one or more tables containing routing information for all other nodes in the network. When changes in network topology occur, these updates must be propagated across the network. Notable proactive routing protocols include the Wireless Routing Protocol (WRP) and Destination-Sequenced Distance Vector (DSDV) routing protocol.

Reactive Routing Protocols. Reactive routing protocols, also known as on-demand routing protocols, comprise a class of routing protocols that create routes exclusively when needed. Routes are established when a source node intends to send data to a destination node, indicating that these protocols are initiated by a source on an as-needed basis. The primary objective of reactive routing protocols is to minimise the routing overhead and resource utilisation associated with maintaining and updating routing tables for the entire network. Nevertheless, these protocols often entail lengthier routes for transmitting data packets from a source to a destination, resulting in network latency. Notable reactive routing protocols encompass Ad-hoc On-demand Vector Routing (AODV) and Dynamic Source Routing (DSR) protocols.

Hybrid Routing Protocols. Hybrid routing protocols in MANETs merge the advantages of both proactive and reactive routing protocols to facilitate efficient and scalable routing in dynamic network environments. The fundamental concept underpinning hybrid routing protocols is to maintain current routing information for neighbouring nodes using proactive techniques while discovering routes for remote nodes on demand using reactive techniques. This method reduces the routing overhead linked to preserving global routing information while still enabling rapid route discovery for local communication. However, their control mechanisms are more complex, and determining the optimal routing protocol for a specific situation may not be straightforward. The Zone Routing Protocol (ZRP) is among the most used hybrid routing protocols.

1.2   Applications and Uses of Mobile Ad-hoc Networks (MANETs)

MANETs are often employed in disaster recovery and relief situations. For example, communication infrastructure may be severely impaired during a natural disaster, hindering rescue teams' communication and coordination efforts. This flexible, decentralised structure contributes to the dependability of a communication system in challenging environments with limited or compromised infrastructure [9].

Additionally, MANETs can be used in smart cities. In urban settings, MANETs can enable communication between IoT devices, creating interconnected systems that optimise city services [5]. Moreover, a specific application of MANETs is the vehicular ad-hoc networks (VANETs). VANETs employ the same principle as MANETs [8]. They enable vehicles to exchange information with each other and with roadside infrastructure. This can lead to improved traffic management through real-time updates on road conditions, accidents, and congestion [8].

1.3   Ad-hoc On-Demand Distance Vector (AODV) Protocol

The AODV routing protocol functions as a reactive mechanism, enabling self-initiating, dynamic, and multi-hop routing among participating nodes. AODV assists nodes in quickly discovering routes to new destinations, as it eliminates the need to maintain information about inactive routes. Using an on-demand approach for route discovery, AODV employs a destination sequence number to determine the most recent path, thereby guaranteeing the route's freshness [1]. Moreover, it ensures loop-free routes and addresses disruptions in routing connections [1]. The AODV protocol includes various message types-route requests (RREQs), route replies (RREPs), route errors (RERRs)-and a HELLO packet for navigation within the ad hoc network. AODV can elucidate network topology by broadcasting a HELLO message to neighbouring mobile nodes. Additionally, a HELLO message can detect an invalid link by broadcasting it to mobile nodes in the network.

2   Trust Management Mechanisms and Related Work

Probabilistic trust and reputation techniques employ probability calculus and advanced statistical methods to build trust models [2]. These models represent trust and reputation as probability distributions rather than fixed values, accounting for uncertainty and randomness within the system. By leveraging probability theory, these models can be simplified for better understanding and analysis while maintaining the capacity to use various derivation methods such as Bayesian networks and Markov Chain and Hidden Markov models for extracting valuable information and making predictions [10].

Beta distributions and Bayesian inference are commonly used in trust and reputation modelling [3]. These approaches rely on binary ratings that only accept positive or negative inputs. These methods are known for their simplicity and adaptability, making them popular in various computing domains. The beta distribution is useful for modelling trust and reputation as it accommodates system uncertainty, while Bayesian inference enables trust score updates based on new evidence. By employing these techniques, researchers can simulate agent interactions and decision-making processes to create resilient and efficient systems.

Sen et al. [7] proposed a trust-based mechanism to detect malicious nodes that drop packets in a network. This approach leverages the reputation of neighbouring nodes and accounts for the erosion of trust over the network's lifespan. The suggested mechanism employs public and private key pairs to safeguard user identities from attacks. However, this solution might not be suitable for larger networks due to potential scalability issues.

In a study by Marti et al. [6], a reputation-based trust management approach was presented to improve the security of routing in MANETs. This scheme consists of two components: a watchdog that monitors node behaviour gathers reputation data, and takes appropriate actions, such as isolating misbehaving nodes. The aim of this approach is to combine direct observations to generate trust values for secure routing by extending the Dynamic Source Routing (DSR) protocol. However, this approach has limitations, as it relies solely on direct observation and is only used with the DSR protocol.

3   Vulnerabilities and Security Issues in MANETs

One of the primary issues in securing MANETs is the absence of a centralised authority. In conventional wired networks, a central authority can enforce security policies and monitor the network for any suspicious activities. However, there is no central authority in a MANET, and every node must depend on its own security mechanisms for protection. Owing to their distinct features, MANETs are more vulnerable to security breaches than traditional fixed networks. There are several vulnerabilities and security issues that must be addressed to ensure the proper functioning and reliability of the network.

- Limited Resources: Nodes in a MANET typically have limited resources, such as battery power, processing power, and memory. These constraints make it challenging to implement complex security mechanisms that require significant computational power or storage.
- Node Cooperation: In MANETs, nodes rely on each other to forward packets and maintain network connectivity. However, nodes may act selfishly or maliciously by intentionally refusing to forward packets, altering routing information, or dropping packets.

Numerous security attacks and breaches can impact MANETs. The followings are some of the attacks that impact MANETs negatively. This research used a black hole attack to investigate and examine the impact of malicious nodes on the AODV protocol.

- Black Hole Attack: In a black hole attack, a malicious node claims to have the shortest path to the destination, causing other nodes to route their packets through it [4]. Once the malicious node has attracted the traffic, it drops all the packets, leading to data loss and disrupting network communication. This attack can be executed by compromising a node's routing table or sending false route reply messages. Countermeasures against black hole attacks include using trust-based routing protocols and monitoring the behaviour of nodes to detect malicious activity.

4   Proposed Global Trust Management Mechanisms for AODV Protocol

4.1   Global Trust and Overview of Proposed Protocol

GTAODV is an improved version of the AODV routing protocol that integrates global trust principles for making routing decisions. This global trust mechanism depends on a designated central or leader node responsible for calculating trust values for each node participating in the network. The leader node periodically collects reports from participating nodes and processes them to determine each node's trust value within the network. These trust values are then distributed periodically to all participating nodes. Each node uses the received global trust values to make routing decisions, aiming to boost the routing protocol's performance in terms of security, reliability, and efficiency.

In this context, trust is represented as a continuous value that ranges from 0 to 1. A value of 0 indicates complete distrust, 1 signifies absolute trustworthiness, and 0.5 represents uncertainty regarding a node's behaviour. Employing a continuous value to portray trust allows for a more accurate representation of the inherent uncertainty related to trust compared to a binary variable, which can only express trust or distrust without accounting for varying degrees of trustworthiness.

4.2   Proposed Global Trust Mechanism in the AODV Protocol

In the centralised global trust management method, a designated central node gathers observational data from participating nodes in the network. This data comprises direct and indirect trust values about other nodes evaluated by the participating nodes. The indirect trust values are determined by combining recommendations from neighbouring nodes. The leader node processes the received data using specific algorithms to calculate the global trust value for each node in the network. Subsequently, the leader node periodically distributes the computed global trust values to all participating nodes. These nodes then use the global trust values to identify the most reliable route for transmitting packets from a source node to a destination node.

The following terminology and conventions are used throughout this paper: The reliability of a node (r) indicates the probability that the node consistently offers dependable service during the packet routing process. In contrast, node unreliability (n) denotes the likelihood that the packet routing service provided by the node is unreliable. Furthermore, node uncertainty (u) refers to the probability that the node's reliability for packet routing cannot be conclusively determined. The acronym (rnu) collectively denotes these three values, representing reliability, unreliability, and uncertainty. Direct observations are symbolised by dt_rnu, while indirect observations are represented by it_rnu, and global observations are indicated by gt_rnu.

Each node calculates direct trust observations using various parameters, and indirect trust measures are determined by combining direct trust reports received from multiple neighbouring nodes. After calculating the indirect (rnu) values using direct observations, each participating node forwards the results to the designated leader node. The leader node computes global trust values and distributes them to all participating nodes. In this manner, the leader node assists in making informed routing decisions based on the trustworthiness and reliability of the nodes involved in the network. This approach aims to enhance the security, efficiency, and overall performance of the packet routing process within the network.

4.3   Calculating a Node's rnu using Direct Observations (dt_rnu_i)

To compute direct trust, each node in the network observes its neighbouring nodes for specific events that signify their packet forwarding reliability. These events are recorded as either positive (α) or negative (β) observations concerning the neighbouring node. Bayesian Inference is then employed to calculate values of reliability and trustworthiness for each neighbouring node. Bayesian Inference is a statistical method that leverages Bayes' theorem to update the probability of a hypothesis as more evidence or information becomes available [10]. Table 1 shows the parameters used for node monitoring.

Given that only two parameters in Table 1 are considered, the Beta distribution function is selected for modelling node behaviour. Let x and y represent two neighbouring nodes in the network, and node x has generated a total of n reports

Table 1. Trust Observation Parameters

Sr.	Observation Parameter	Frequency of Recording the Observation	Positive Observation (α)	Negative Observation (β)
1	Packet forwarding ability	For each observed data packet	α++ for each data packet forward	β++ for each data packet drop
2	Node Battery	At beginning of a new data transmission session	α++ if node's Battery Power > MBT	β++ if node's Battery Power <= MBT
3	Node's participation in network routing activities	For each observed RREP packet	α++ for the node which initiated control packet	β++ for the node which dropped a control packet. Also, β++ for a node caused a route error
4	Node's packet forwarding queue capacity	At beginning of a new data transmission session	α++ if more than MEQ of queue capacity is empty	β++ if available queue capacity is less than equal to MEQ

about node y. Assume that T_{new} denotes the probability of node y exhibiting positive behaviour at time $n+1$. The posterior distribution of successful cooperation between nodes x and y can be represented using a Beta distribution function, with the density function specified in Eq. 1:

$$Beta(T_{old}|\alpha, \beta) = \frac{\tau(\alpha + \beta + 2)}{\tau(\alpha + 1)\tau(\beta + 1)} T_{old}^{\alpha}(1 - T_{old})^{\beta} \qquad (1)$$

In this Equation, T_{old} is the old value of trust of node x on y. Then an updated value of trust T_{new} is calculated as follows:

$$T_{new} = E(Beta(T_{old}|\alpha, \beta)) = \frac{\alpha}{\alpha + \beta} \qquad (2)$$

Using Eq. 2 and taking into consideration the level of uncertainty u, direct node reliability (dt_r) is calculated using Eq. 3:

$$dt\_r = \frac{\alpha}{\alpha + \beta}(1 - u) \qquad (3)$$

Using Eq. 2 and taking into consideration the level of uncertainty, direct node unreliability expectation (dt_n) is calculated using Eq. 4:

$$dt\_n = \frac{\beta}{\alpha + \beta}(1 - u) \qquad (4)$$

Here the direct node uncertainty (dt_u) is calculated using Eq. 5:

$$dt\_u = \frac{12\alpha\beta}{(\alpha + \beta)^2(\alpha + \beta + 1)} \qquad (5)$$

The value of dt_rnu_i represents the combined reliability, unreliability, and uncertainty of node i calculated using direct monitoring.

4.4   Calculating a Node's rnu using Indirect Observations (it_rnu_i)

Nodes within the network continuously exchange their direct trust reports concerning the reliability of neighbouring nodes. Indirect trust is determined by combining these direct trust reports shared by the neighbouring nodes. Each node regularly disseminates its calculated (dt_rnu_i) values to all neighbouring nodes in this process. Upon receiving these reports, each node processes and synthesises them to compute its indirect trust (it_rnu_i) values, representing the node's reliability, unreliability, and uncertainty. These values are obtained from direct and indirect reports and calculated using a weighted average method. Equation 6 demonstrates the computation of node reliability using indirect reports (it_r_i), while Eq. 7 presents the calculation of node unreliability using indirect reports (it_n_i). Finally, Eq. 8 depicts the calculation of node uncertainty using indirect reports (it_u_i).

$$it\_r_i = \frac{\sum_{0<j<=N} r_{ji} * w_j}{W} \tag{6}$$

$$it\_n_i = \frac{\sum_{0<j<=N} n_{ji} * w_j}{W} \tag{7}$$

$$it\_u_i = 1 - (it\_r_i + it\_n_i) \tag{8}$$

where N is number of nodes in the network, r_ji is the indirect reliability of node i which is reported by node j, n_{ji} is the indirect unreliability of node i which is reported by node j. w_j is the weight assigned to node j depending on past interactions and W is the cumulative weight. The value of w_j is calculated using the following Eqs. 9-11:

$$\alpha\_sum_j = \sum_{i=1}^{N} \alpha_{i,j} \tag{9}$$

$$\beta\_sum_j = \sum_{i=1}^{N} \beta_{i,j} \tag{10}$$

$$w_j = \sum_{j=1}^{N} \alpha\_sum_j + \beta\_sum_j \tag{11}$$

where $\alpha_{i,j}$ is the value of α reported by node i about node j and $\beta_{i,j}$ is the value of β reported by node i about node j.

4.5   Calculating a Node's *rnu* Using Global Trust Observations (gt_rnu_i)

The leader node gathers $it_r nu_i$ reports from all participating nodes in the network. These observations are then synthesised, and $gt_r nu_i$ is calculated using Dempster-Shafer Theory (DST). The leader node employs DST to combine the evidence received from various nodes within the network.

Let

$r_1(i)=$ basic probability value indicating node 1's reliability value for a target node i;

$r_2(i) =$ basic probability value indicating node 2's value of reliability for the target node i;

$n_1(i)=$ basic probability value indicating node 1's unreliability value for a target node i;

$n_2(i) =$ basic probability value indicating node 2's unreliability value for the target node i;

$u_1(i)=$ basic probability value indicating node 1's uncertainty value for a target node i;

$u_2(i) =$ basic probability value indicating node 2's uncertainty value for the target node i.

The updated values of reliability, unreliability and uncertainty for target node i, which are calculated using the above terms and Dempster-Shafer theory (DST) are as follows:

$$gt\_r_i = r_1(i) \oplus r_2(i) = \frac{1}{C}\{r_1(i) * r_2(i) + r_1(i) * r_2(u) + r_1(u) * r_2(i)\} \quad (12)$$

$$gt\_n_i = n_1(i) \oplus n_2(i) = \frac{1}{C}\{n_1(i) * n_2(i) + n_1(i) * n_2(u) + n_1(u) * n_2(i)\} \quad (13)$$

$$gt\_u_i = u_1(i) \oplus u_2(i) = \frac{1}{C}\{u_1(i) * u_2(i)\} \quad (14)$$

where $C = r_1(i) * r_2(i) + r_1(i) * u_2(i) + u_1(i) * r_2(i) + n_1(i) * n_2(i) + n_1(i) * u_2(i) + u_1(i) * n_2(i) + u_1(i) * u_2(i)$

4.6   Integration of Global Trust in the AODV Protocol

The proposed method integrates direct, indirect, and global trust mechanisms into the Ad-hoc On-Demand Distance Vector (AODV) protocol. The modified AODV protocol is referred to as Global Trust AODV (GTAODV).

$$Trust\_value_i = \frac{\rho}{\text{No. of Hops to Destination}} + (1 - \rho) * NRV(gt\_r_i) \quad (15)$$

Where NRV is the *Node Reliability Value*

Different values of ρ were employed to assign varying weights: first, to the hop count from a source node to the destination node, and second, to node reliability. The objective of this process is to optimise the performance of the trust mechanism and identify the best performance scenario. The experimental results concluded that the optimal weightings for modifying the standard AODV protocol are 70% for node reliability and 30% for hop count. As a result, in this paper, ρ is set to 0.30, which indicates that a weight of 70% is assigned to node reliability, and a weight of 30% is assigned to hop count. This mechanism was then integrated into the AODV protocol, resulting in the Global Trust AODV (GTAODV) protocol. The GTAODV protocol employs a technique wherein a packet is forwarded from a source node to a destination node through a neighbouring node with the highest trust value. This mechanism ensures that the packet is routed through the most reliable node, enhancing the network's security and efficiency. By using trust values to determine the node to which the packet is forwarded, the GTAODV protocol can defend against attacks such as black hole attacks and other security threats.

5   Performance Evaluation of AODV and GTAODV in the Presence of a Black Hole Attack

A black hole attack is a type of network security breach where a malicious node disrupts the normal operation of a routing protocol by falsely claiming to have the shortest path to the destination [4]. This attack is particularly prevalent in MANETs using the AODV routing protocol due to its inherent trust in nodes' trustworthiness, making it more susceptible to such attacks [4]. Understanding the impact of black hole attacks on AODV can help identify security vulnerabilities, enabling the development of more effective countermeasures to mitigate these attacks. Furthermore, GTAODV is an enhanced version of the AODV protocol that integrates centralised trust-based mechanisms to improve the security of the routing process. Evaluating the impact of black hole attacks on GTAODV aids in assessing the effectiveness of these trust-based mechanisms in detecting and isolating malicious nodes and, consequently, preventing an attack from compromising the network.

5.1   Experimental Set-Up and Performance Metrics

The Network Simulator 3 (NS-3) was used to simulate the AODV and GTAODV protocols. The parameters listed in Table 2 were chosen in the simulation of a realistic and complex wireless ad-hoc network environment. The simulations were run multiple times with different numbers of malicious nodes so as to examine the impact of the attack on overall network performance. The results can be used to determine which routing protocol is more resilient against black hole attacks and may help future protocol improvements or network configurations aiming to mitigate the effects of such attacks more effectively.

The use of 95% Confidence Intervals enhances the results of the analysis, providing a clear understanding of the differences between the two protocols under the conditions described.

Packet Delivery Ratio (PDR) is a key performance metric in networking, which measures the ratio of successfully received data packets at the destination node to the total number of data packets sent by the source node. Throughput is another crucial performance metric in networking that represents the rate at which data packets are successfully transmitted and received across the network.

The end-to-end delay represents the total time it takes for a data packet to travel from its source node to its destination node. This delay comprises several components, including transmission, propagation, queuing, and processing delays. Routing overheads refer to the additional network resources and bandwidth consumed by the control messages and routing protocol operations required to establish and maintain routes between nodes in the network.

Table 2. Simulation Parameters

Routing protocols	*AODV, GTAODV*
Type of Threat	*Black hole Attack*
Packet Size	512 *Bytes*
Simulation Time	360 *Seconds*
Simulation Area	1000 * 1000 m^2
Number of Nodes	100
Number of Malicious Nodes	5,10,15,20,25
Node Movement Speed	10 m/s
Node Movement	Random Way Point
MAC Protocol	IEEE 802.11b
Transmission Range	250 *Meter*
Number of Simulation Runs	10
Confidence Interval	95%
Traffic Type	UDP

5.2 Evaluation of Packet Delivery Ratio and Throughput When Varying the Number of Malicious Nodes

The performance results for the AODV and GTAODV routing protocols in the presence of black hole attacks are displayed in Figs. 1 and 2, offering valuable insights into the protocols' robustness and efficiency when confronted with a growing number of malicious nodes in the network.

Figure 1 illustrates PDR as a function of the number of malicious nodes in the network and compares the performance of the AODV and GTAODV routing protocols. PDR is presented with 95% Confidence Intervals for both protocols.

The AODV protocol experiences a significant decrease in PDR as the number of malicious nodes rises. This performance degradation is due to the absence of an effective mechanism to detect and avoid malicious nodes, resulting in more dropped or lost packets [6]. Conversely, GTAODV maintains a relatively high PDR despite the increasing number of malicious nodes. This can be attributed to the trust-based mechanism integrated into GTAODV, which enables the protocol to evaluate node trustworthiness based on monitoring the parameters mentioned earlier. By identifying and avoiding malicious or less reliable nodes, GTAODV can sustain a higher PDR, ensuring more efficient and reliable data transmission.

Figure 2 demonstrates the results for throughput in KBps as a function of the number of malicious nodes in the network, comparing the performance of the AODV and GTAODV routing protocols. Throughput measures the rate of successful data packet delivery across the network. Higher throughput implies improved network performance and utilisation of available resources. Throughput is depicted with 95% confidence intervals for both protocols. The throughput of the AODV protocol declines significantly as the number of malicious nodes grows, highlighting the protocol's susceptibility to black hole attacks. The reduction in throughput can be attributed to AODV's inability to differentiate between malicious and non-malicious nodes, leading to inefficient routing and decreased data transmission rates. In contrast, the throughput of the GTAODV protocol remains relatively stable and considerably higher than that of the AODV protocol, even as the number of malicious nodes increases. The trust-based mechanism in GTAODV allows the protocol to effectively identify and avoid malicious nodes, resulting in more efficient routing and better utilisation of network resources. This leads to a higher throughput, which is essential for ensuring optimal network performance.

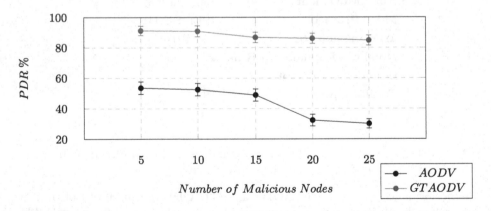

Fig. 1. PDR vs. Number of Malicious Nodes with 95% Confidence Intervals

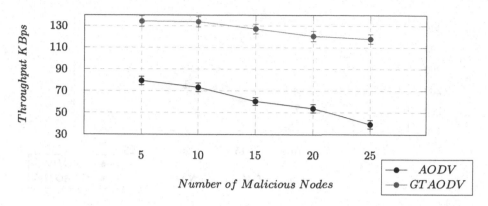

Fig. 2. Throughput vs. Number of Malicious Nodes with 95% Confidence Intervals

5.3   Evaluation of End-to-End Delay and Routing Overheads When Varying the Number of Malicious Nodes

Figures 3 and 4 display the results for end-to-end delay and routing overheads, respectively, for AODV and GTAODV with different numbers of malicious nodes, including their 95% confidence intervals.

Figure 3 shows that the end-to-end delay for AODV with 5 malicious nodes is 186.80 ms. For GTAODV, the delay is 261.846 ms. When the number of malicious nodes increases to 25, the delay for AODV goes up to 257.74 ms (± 9.09 ms), while that for GTAODV also rises to 319.972 ms (± 11.94 ms). These values demonstrate that although GTAODV surpasses AODV in PDR and throughput, it experiences greater end-to-end delays with any number of malicious nodes. This can be attributed to the additional trust mechanism in GTAODV, which increases the processing time at each node during route discovery and packet forwarding, leading to extended delays. Nevertheless, the trade-off for higher delays is enhanced security and performance in terms of PDR and throughput.

Figure 4 illustrates that the routing overheads for AODV with 5 malicious nodes amount to 1843.64 bytes. Conversely, the routing overheads for GTAODV reach 2403.876 bytes. When the number of malicious nodes increases to 25, the routing overheads for AODV climb to 2342.436 bytes (± 62.32 bytes), while those for GTAODV also go up to 3074.435 bytes (± 103.217 bytes). These values suggest that compared to AODV, GTAODV incurs higher routing overheads in the presence of any number of malicious nodes. This can also be attributed to the added trust mechanism in GTAODV, which necessitates extra control packets and processing during route discovery and maintenance. Although this results in increased overheads, the enhanced security and performance observed in PDR and throughput make it a worthwhile trade-off in situations where security is a top priority.

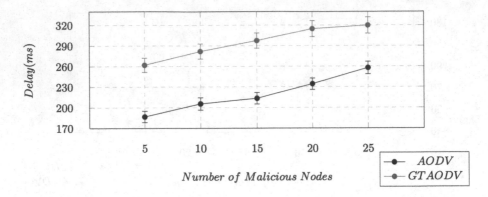

Fig. 3. End-to-End Delay vs. Number of Nodes with 95% Confidence Intervals

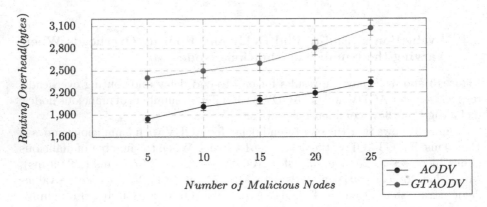

Fig. 4. Routing Overhead vs. Number of Malicious Nodes with 95% Confidence Intervals

6   Conclusion

One of the primary vulnerabilities in MANETs stems from their open peer-to-peer architecture. MANET routing protocols assume all nodes cooperate without maliciously disrupting the protocol's operation. Consequently, MANETs lack a distinct line of defence from a security design standpoint. Due to the wireless nature of these networks, they inherit security threats faced by both wireless and wired networks, and even more attacks arise from their unique characteristics. Black hole attacks in MANET routing can significantly disrupt communication between nodes.

This research proposes a trust-based approach to combat black hole attacks in MANET routing. The suggested algorithm employs trust values, derived from node reliability, to make routing decisions. Reliability values are estimated based on nodes' past behaviour during communication processes. The proposed algorithm uses direct, indirect, and global trust model.

The comparison of AODV and GTAODV protocols based on PDR and throughput reveals that the trust-based mechanism in GTAODV provides notable benefits in network performance and resilience against black hole attacks. By incorporating trust evaluation into the routing process, GTAODV can effectively identify and avoid malicious nodes, ensuring higher PDR and throughput, and ultimately leading to more robust and efficient mobile ad-hoc networks. While GTAODV experiences higher end-to-end delays and routing overheads compared to AODV, it exhibits superior resilience against black hole attacks, maintaining higher PDRs and throughput with various numbers of malicious nodes. As a result, the choice between AODV and GTAODV should be determined based on the network's specific needs, focusing on balancing security, performance, and resource usage.

References

1. Jari, H., Alzahrani, A., Thomas, N.: A novel indirect trust mechanism for addressing black hole attacks in MANET. In: Proceedings of the 11th ACM Symposium on Design and Analysis of Intelligent Vehicular Networks and Applications, pp. 27–34. DIVANet (2021). https://doi.org/10.1145/3479243.3487296
2. Kannan, G., Prasant, M.: Trust computations and trust dynamics in mobile ad-hoc networks: a survey. IEEE Commun. Surv. Tut. 2(14), 279–298 (2012)
3. Menaka, R., Ranganathan, V., Sowmya, B.: Improving performance through reputation based routing protocol for manet. Wirel. Pers. Commun. 94(4), 2275–2290 (2016). https://doi.org/10.1007/s11277-016-3396-4
4. Nitin, K., Monika, S.: A comprehensive taxonomy of schemes to detect and mitigate blackhole attack and its variants in MANETs. Comput. Sci. Rev. 32, 24–44 (2019). https://doi.org/10.1016/j.cosrev.2019.03.001
5. Ramphull, D., Mungur, A., Armoogum, S., Pudaruth, S.: A review of mobile Ad hoc NETwork (MANET) Protocols and their applications. In: 5th International Conference on Intelligent Computing and Control Systems (ICICCS), pp. 204–211. IEEE, Madurai, India (2021). https://doi.org/10.1109/ICICCS51141.2021.9432258
6. Sergio, T., Giuli, K., Mary B.: Mitigating routing misbehavior in mobile ad hoc networks. In: Proceedings of the 6th Annual International Conference on Mobile Computing and Networking (MobiCom '00), pp. 255–265. Association for Computing Machinery, New York, NY, USA (2000). https://doi.org/10.1145/345910.345955
7. Sen, J., Chowdhury, R., Sengupta, I.: A distributed trust mechanism for mobile ad hoc networks. In: 2006 International Symposium on Ad Hoc and Ubiquitous Computing, pp. 62–67. IEEE, Mangalore India (2006)
8. Subir, S., Basavaraju, T., Puttamadappa, C.: Ad hoc mobile wireless networks principles, protocols, and applications, 2nd edn. CRC Press, New York (2016)
9. Vu, Q., Vi, N., Dao, L., Le, N.: Routing algorithms for MANET IoT networks. a comprehensive survey. Wirel. Pers. Commun. 125, 3501–3525 (2022). https://doi.org/10.1007/s11277-022-09722-x
10. Yadollahzadeh-Tabari, M.: Wirel. Pers. Commun. 2(118), 1087–1109 (2021)

Author Index

M. Iacono et al. (Eds.): EPEW/ASMTA 2023, LNCS 14231, pp. 399–400, 2023.
https://doi.org/10.1007/978-3-031-43185-2

Printed in the United States
by Baker & Taylor Publisher Services